Circulating Non-coding RNAs as Diagnostic and Prognostic Markers of Human Diseases

Circulating Non-coding RNAs as Diagnostic and Prognostic Markers of Human Diseases

Guest Editors

Kyriacos N. Felekkis
Christos Papaneophytou

MDPI

Basel • Beijing • Wuhan • Barcelona • Belgrade • Novi Sad • Cluj • Manchester

Guest Editors

Kyriacos N. Felekkis
Department of Life Sciences
University of Nicosia
Nicosia
Cyprus

Christos Papaneophytou
Department of Life Sciences
University of Nicosia
Nicosia
Cyprus

Editorial Office
MDPI AG
Grosspeteranlage 5
4052 Basel, Switzerland

This is a reprint of the Special Issue, published open access by the journal *International Journal of Molecular Sciences* (ISSN 1422-0067), freely accessible at: https://www.mdpi.com/journal/ijms/special_issues/2KHN66N6TU.

For citation purposes, cite each article independently as indicated on the article page online and as indicated below:

Lastname, A.A.; Lastname, B.B. Article Title. *Journal Name* **Year**, *Volume Number*, Page Range.

ISBN 978-3-7258-5667-1 (Hbk)
ISBN 978-3-7258-5668-8 (PDF)
https://doi.org/10.3390/books978-3-7258-5668-8

Contents

About the Editors

Kyriacos N. Felekkis

Kyriacos N. Felekkis is a professor of Pathology and Laboratory Medicine, in the Department of Life Sciences at the School of Life and Health Sciences of the University of Nicosia. He is currently the Dean of the School and Director of the Non-Coding RNA Research Laboratory (ncr-RELSA). Prof. Felekkis obtained his PhD degree from Boston University School of Medicine in 2004 in Pathology and Laboratory Medicine. Since 2011, he has been a TRF in the Department of Life and Health Sciences. He served as the Head of Department and chairperson of the Departmental Postgraduate Programme Committee at the University of Nicosia. He is a member of the Senate and an elected member of the University Council. He is also the Chair of the Student Disciplinary Committee (SDC). In 2015, Prof. Felekkis was appointed a member of the Cyprus National Bioethics Committee by the Council of Ministers. His research interest centers around the study of microRNAs and their role in the pathogenesis of human disease as well as their contribution in gene expression regulation. His is also interested in AI applications in personalized medicine and diagnosis. Prof. Felekkis has close collaboration with several research institutions in Cyprus, Greece, Europe and USA. He has coordinated numerous local and European research projects, and he is an expert reviewer for Horizon and COST Association proposals. His work has been published in various International Journals and books.

Christos Papaneophytou

Christos Papaneophytou is a professor of Biochemistry in the Department of Life Sciences at the School of Life and Health Sciences, University of Nicosia, where he also serves as the Associate Dean. Prof. Papaneophytou's research primarily focuses on drug discovery for the treatment of rheumatoid arthritis, cancer, and respiratory tract infections caused by rhinoviruses, adenoviruses, and SARS-CoV-2. He has a strong interest in identifying inhibitors for the less-studied proteases of human rhinoviruses and adenoviruses, which are crucial for the replication and virulence of these viruses. His work also involves investigating phytochemicals that may inhibit enzymes involved in the pathogenesis of diseases such as cancer and respiratory infections, as well as targeting quorum sensing mechanisms in bacteria as a novel approach to antibiotic therapy. Additionally, Prof. Papaneophytou focuses on identifying prognostic, diagnostic, and therapeutic biomarkers, including non-coding RNAs and proteins, for a variety of diseases, with a particular interest in cancer, cardiovascular diseases, and osteoarthritis. He is also interested in the application of artificial intelligence for personalized medicine and diagnostic purposes. He is a co-founder of the Non-Coding RNA Research Laboratory (ncr-RELSA) at the University of Nicosia and serves as its Associate Director. Prof. Papaneophytou has authored numerous peer-reviewed publications and book chapters and has been a frequent presenter at international scientific conferences. He is involved in several research projects funded by both international and national bodies. Furthermore, he serves as a reviewer for multiple scientific journals and as a guest editor for several high-impact journals.

Preface

This Reprint, titled Circulating Non-coding RNAs as Diagnostic and Prognostic Markers of Human Diseases, presents a curated collection of eight original research articles and five reviews that examine the diagnostic, prognostic, and mechanistic roles of circulating non-coding RNAs (c-ncRNAs) in human health and disease. The scope spans oncology, cardiovascular, metabolic, and inflammatory disorders, highlighting the clinical relevance of microRNAs (miRNAs), long non-coding RNAs (lncRNAs), and circular RNAs (circRNAs) as stable, accessible biomarkers detectable in biofluids such as blood and urine. Collectively, the contributions demonstrate how c-ncRNAs can support non-invasive liquid biopsy applications across early detection, risk stratification, and therapeutic monitoring. The aim of this Reprint is to provide an integrated view of current advances that bridge discovery and translation. Beyond biomarker identification and validation, several studies illuminate the mechanisms by which c-ncRNAs regulate key cellular processes and contribute to disease pathophysiology, thereby strengthening their potential for clinical impact. Methodological work featured here addresses pre-analytical and analytical variability, standardization, and study design—factors essential for reproducibility and for moving c-ncRNA assays toward clinical implementation.

Our motivation in assembling this collection is twofold: first, to showcase high-quality research that advances the precision, reliability, and interpretability of c-ncRNA measurements; and second, to catalyze further studies that connect molecular insight with actionable clinical endpoints. In the era of precision medicine, c-ncRNAs offer dynamic readouts of disease activity and treatment response, positioning them as promising tools for personalized patient care. This Reprint is intended for researchers and clinicians working in molecular medicine, cardiology, oncology, immunology, and respiratory medicine; for laboratory scientists and bioinformaticians developing RNA-based diagnostics; for industry professionals engaged in assay and device development; and for graduate students seeking an authoritative entry point into the rapidly evolving field of circulating RNA biomarkers. We hope the articles collected here will serve both as a reference and as an impetus for future innovation and clinical translation.

Kyriacos N. Felekkis and Christos Papaneophytou
Guest Editors

International Journal of
Molecular Sciences

MDPI

Editorial

Special Issue "Circulating Non-Coding RNAs as Diagnostic and Prognostic Markers of Human Diseases"

Kyriacos Felekkis [1,2] and Christos Papaneophytou [1,2,*]

[1] Department of Life Sciences, School of Life and Health Sciences, University of Nicosia, Nicosia 2417, Cyprus; felekkis.k@unic.ac.cy

[2] Non-Coding RNA Research Laboratory, School of Life and Health Sciences, University of Nicosia, Nicosia 2417, Cyprus

[*] Correspondence: papaneophytou.c@unic.ac.cy; Tel.: +357-22841941

Biomarkers are measurable biological molecules that reflect normal physiological processes, pathological states, or responses to environmental exposures and therapeutic interventions [1]. Their clinical significance resides in their capacity to facilitate early disease detection, guide diagnostic decision making, predict patient prognosis, and monitor therapeutic responses. Over the past decade, research into non-coding RNAs (ncRNAs), including microRNAs (miRNAs), long non-coding RNAs (lncRNAs), circular RNAs (circRNAs), and PIWI-interacting RNAs (piRNAs), has expanded exponentially, revealing robust associations with diverse pathological conditions spanning cancer, cardiovascular disorders, and autoimmune diseases [2]. This growing field has witnessed substantial progress in leveraging ncRNAs as diagnostic and prognostic biomarkers, thereby establishing new paradigms for precision patient management from early disease detection through monitoring of recurrence and therapeutic response [3].

Furthermore, circulating ncRNAs (c-ncRNAs) have emerged as particularly compelling biomarker candidates due to their remarkable stability in biological fluids, ready accessibility in samples such as blood and urine, and distinctive disease- and tissue-specific expression profiles [4]. Beyond their favorable detection characteristics, c-ncRNAs function as critical regulators of fundamental cellular processes, including proliferation, apoptosis, and cell cycle progression. Consequently, they participate in disease pathogenesis through multiple mechanisms, serving variously as causative agents, modulators of pathological processes, or downstream effectors of disease states. The translational potential of c-ncRNAs as non-invasive, liquid-biopsy-based biomarkers is further enhanced by their capacity to provide mechanistic insights into disease pathophysiology and to dynamically reflect therapeutic responses, including those to emerging targeted therapies and immunotherapeutic approaches. While early investigations encountered challenges related to intra- and inter-individual variability in ncRNA expression, recent methodological advances have substantially mitigated these limitations, enabling numerous discoveries to progress toward clinical implementation [5].

In the current era of personalized medicine, elucidating the multifaceted roles of ncRNAs in health and disease represents a research imperative of paramount importance. This Special Issue, titled "Circulating Non-Coding RNAs as Biomarkers of Human Diseases," features eight original research articles and five reviews that collectively advance our understanding of the diagnostic, prognostic, and mechanistic significance of c-ncRNAs in clinical practice. Below is an overview of the contributions included in this Special Issue.

The article by Escuin et al. (contribution 1) provides a proof-of-principle study comparing miRNA expression profiles across matched plasma, tumor tissue, and sentinel lymph node (SLN) samples from 30 estrogen-receptor-positive early breast cancer patients. The authors identified a striking inverse relationship between circulating and tissue miRNAs, with most plasma miRNAs showing downregulation. Importantly, miR-642a-3p and miR-223 were found to be upregulated in patients with metastatic SLNs. Gene ontology and pathway enrichment analyses linked these miRNAs to processes central to metastasis, including epithelial–mesenchymal transition, epithelial cell proliferation, and regulation of receptor tyrosine kinase. These findings highlight the complexity of interpreting circulating miRNA (c-miRNAs) signatures and underscore the potential of specific plasma miRNAs as surrogate markers for lymph node involvement in early breast cancer.

In the second article, Córdoba-Lanús et al. (contribution 2) present a longitudinal study assessing c-miRNAs as early biomarkers of lung cancer (LC) in patients with chronic obstructive pulmonary disease (COPD). Over six years of follow-up, four miRNAs (miR-1246, miR-206, miR-224-5p, and miR-194-5p) were found to be dysregulated in COPD patients who later developed LC. Validation in an independent cohort confirmed altered expression of both miR-1246 and miR-206 up to three years before LC diagnosis, implicating their role in cancer-related pathways. Notably, miR-206 dysregulation was evident well before clinical onset, highlighting its promise as a predictive biomarker for early identification of high-risk COPD patients and enabling targeted screening and intervention.

The article by Gil-Martínez et al. (contribution 3) examines two inflammation-related miRNAs—hsa-miR-26a-1-3p and hsa-miR-376a-3p—in asthma phenotypes and obesity. Elevated levels of miR-26a-1-3p and reduced levels of miR-376a-3p were observed in obese asthma patients, with both linked to key pathogenic pathways including p53 signaling and extracellular matrix interactions. Validation of target genes in lung tissues further supported their functional relevance. These findings suggest that miR-26a-1-3p and miR-376a-3p hold promise as biomarkers for asthma phenotyping, particularly in distinguishing eosinophilic inflammation and obesity-associated disease, with potential implications for personalized therapy.

The fourth article by Ai et al. (contribution 4) investigates the role of circTTN, a circular intronic RNA, in skeletal muscle development, with particular focus on myoblast proliferation and differentiation. In C2C12 myoblasts, circTTN was shown to act as a negative regulator of myogenesis, as its overexpression suppressed both proliferation and differentiation. Mechanistically, circTTN was shown to recruit the Pur-beta (PURB) protein to the promoter region of its host gene, Titin (TTN), thereby repressing TTN transcription. As PURB is itself an inhibitor of myoblast growth, the circTTN–PURB interaction establishes a regulatory axis that amplifies this inhibitory effect. These findings reveal a previously unrecognized mechanism by which circular RNAs can regulate host gene expression and muscle development, providing new insights into circRNA-mediated transcriptional control with potential implications for muscle disorders and regenerative medicine.

The work by Ng et al. (contribution 5) evaluates serum miRNAs as potential biomarkers for non-alcoholic fatty liver disease (NAFLD) in patients with colorectal polyps. By analyzing expression ratios of eight candidate miRNAs, the authors developed a diagnostic panel that improved markedly in accuracy when patients with other metabolic disorders were excluded (AUC = 0.8337). These findings highlight the promise of c-miRNA ratios for early NAFLD detection in high-risk populations and their potential role in guiding screening strategies.

The sixth article by, Iwańczyk et al. (contribution 6), examines the diagnostic potential of c-miRNAs in coronary artery aneurysmal disease (CAAD). In a cohort of 105 patients stratified into CAAD, coronary artery disease (CAD), and normal coronary artery (NCA) groups, the authors validated candidate miRNAs identified in a previous work by their group [6]. Among them, miR-451a and miR-328-3p demonstrated the most substantial diagnostic value, with miR-451a elevated in CAAD versus CAD and miR-328-3p increased in CAAD compared with NCA. Notably, integrating these biomarkers into conventional risk models significantly improved diagnostic accuracy. These findings highlight the promise of miR-451a and miR-328-3p as non-invasive tools for distinguishing CAAD from related coronary conditions.

Nopp et al.'s work (contribution 7) is a multi-phase study evaluating c-miRNAs as prognostic biomarkers for major adverse cardiovascular events (MACEs) in atrial fibrillation (AF). Within a prospective registry of 347 patients, small RNA sequencing and staged validation identified miR-411-5p as consistently associated with cardiovascular mortality and adverse outcomes. These findings position c-miR-411-5p as a promising non-invasive biomarker for risk stratification in AF, with potential to support more personalized approaches to cardiovascular care.

The article by Avgeros et al. (contribution 8) examines plasma miR-146a and miR-155 as non-invasive biomarkers for Mycosis Fungoides (MFs), the most common cutaneous T-cell lymphoma, and studied the influence of genetic variants on their expression. Both miRNAs were elevated in MF patients, including early-stage cases, and showed higher levels in advanced disease, with miR-155 notably increased in patients with skin tumors or erythroderma. SNP analysis revealed associations between specific genotypes and altered miRNA expression, linking genetic variation to MF susceptibility. These findings highlight miR-146a and miR-155 as promising biomarkers for MF diagnosis and staging, while underscoring the role of host genetics in modulating disease risk.

The article by Beňačka et al. (contribution 9) is a review that provides a broad overview of ncRNAs—including miRNAs, piRNAs, lncRNAs, and others—as emerging diagnostic, prognostic, and therapeutic targets. Beyond their fundamental roles in regulating gene expression and cellular processes, ncRNAs are implicated in cancer, cardiovascular, neurodegenerative, and autoimmune diseases. The authors emphasize that ncRNAs, owing to their stability and disease-specific signatures, represent attractive liquid biopsy markers, while therapeutic approaches such as antisense oligonucleotides, RNA mimics, and CRISPR-based tools further underscore their potential in precision medicine.

The tenth article in this Special Issue, a review by Felekkis et al. (contribution 10), examines the potential of c-miRNAs as biomarkers in osteoarthritis (OA), with emphasis on knee disease. Key candidates such as miR-146a, miR-21, and miR-140 are involved in inflammation, cartilage degradation, and chondrogenesis, and their stability in extracellular vesicles makes them attractive for non-invasive testing. While promising for early diagnosis, disease monitoring, and personalized therapy, the authors highlight persistent challenges—including technical variability and the need for large-scale validation—that must be addressed to translate c-miRNAs into clinical practice.

Dzamar et al.'s study (contribution 11) is a systematic review on the role of miRNAs and exosomal miRNAs in cholesteatoma, a destructive cystic lesion of the temporal bone characterized by abnormal keratinocyte proliferation and bone resorption. Screening four major databases, the authors identified 18 studies reporting consistent dysregulation of multiple miRNAs, including downregulation of exosomal miR-17, miR-10a-5p, miR-125b, and miR-34a, and upregulation of exosomal miR-106b-5p, miR-21-3p, and miR-199a, among others. These miRNAs regulate key processes such as proliferation, apoptosis, differentiation, and tissue remodeling, underscoring their dual potential as biomarkers and therapeutic targets. The review also highlights the unique role of exosomes as carriers of pathogenic miRNAs and as promising vehicles for future treatment strategies.

The review article by Kierbiedź-Guzik and Sozańska (contribution 12) discusses recent advances in understanding the role of miRNAs as biomarkers for asthma therapy, emphasizing their potential to enable personalized clinical management. The authors discuss how specific miRNAs can predict treatment response, detect early exacerbations, and monitor patient compliance across therapies ranging from glucocorticosteroids and β-mimetics to biologics. Given their regulatory role in cytokine production, immune modulation, airway remodeling, and inflammation, miRNA expression profiles are closely tied to disease severity and therapeutic outcomes. By mapping miRNA-driven molecular pathways implicated in asthma, the review underscores opportunities for precise therapeutic targeting and individualized treatment strategies. While promising, the authors stress that further validation is needed to fully integrate miRNA-based tools into routine asthma care.

The article by Pieri et al. (contribution 13) is a review that explores how cardio-specific miRNAs become deregulated after SARS-CoV-2 infection and examines their potential role in the development of cardiovascular diseases. Altered expression levels of c-miRNAs such as miR-146a, miR-27a-5p, miR-21, and miR-133a have been linked to cardiac remodeling, arrhythmias, and atherosclerosis in post-COVID-19 patients. The authors propose that viral-driven inflammation and immune dysregulation may trigger miRNA-mediated pathways leading to cardiovascular damage, underscoring the need for further studies to validate these findings and explore their biomarker and therapeutic potential.

In summary, the articles in this Special Issue highlight the versatility and clinical promise of c-ncRNAs as biomarkers across cancer, cardiovascular, metabolic, and inflammatory diseases. From predicting lung cancer years before diagnosis to stratifying cardiovascular risk and refining disease phenotyping, these studies illustrate the power of ncRNAs to inform diagnosis, prognosis, and therapeutic monitoring. While challenges in standardization and clinical translation remain, the convergence of technological advances, mechanistic insights, and growing clinical evidence positions c-ncRNAs at the forefront of precision medicine, with the potential to transform patient care through improved diagnostic accuracy, prognostic precision, and personalized therapy.

Author Contributions: Writing—review and editing, C.P. and K.F. All authors have read and agreed to the published version of the manuscript.

Funding: This research received no external funding.

Institutional Review Board Statement: Not applicable.

Informed Consent Statement: Not applicable.

Data Availability Statement: No new data was created in this study.

Acknowledgments: We would like to thank all contributing authors and reviewers for their scholarly excellence and commitment to this Special Issue.

Conflicts of Interest: The authors declare no conflicts of interest.

List of Contributions:

1. Escuin, D.; López-Vilaró, L.; Bell, O.; Mora, J.; García-Valdecasas, B.; Moral, A.; Clos, M.; Boronat, L.; Arqueros, C.; Barnadas, A. Circulating miRNA Expression Is Inversely Correlated with Tumor Tissue or Sentinel Lymph Nodes in Estrogen Receptor-Positive Early Breast Cancer Patients. *Int. J. Mol. Sci.* **2023**, *24*, 13293. https://doi.org/10.3390/ijms241713293.
2. Córdoba-Lanús, E.; Domínguez de-Barros, A.; Oliva, A.; Mayato, D.; Gonzalvo, F.; Remírez-Sanz, A.; Zulueta, J.J.; Celli, B.; Casanova, C. Circulating miR-206 and miR-1246 as Markers in the Early Diagnosis of Lung Cancer in Patients with Chronic Obstructive Pulmonary Disease. *Int. J. Mol. Sci.* **2023**, *24*, 12437. https://doi.org/10.3390/ijms241512437.
3. Gil-Martínez, M.; Lorente-Sorolla, C.; Rodrigo-Muñoz, J.M.; Naharro, S.; García-de Castro, Z.; Sastre, J.; Valverde-Monge, M.; Quirce, S.; Caballero, M.L.; Olaguibel, J.M.; et al. Obese Asthma Phenotype Is Associated with hsa-miR-26a-1-3p and hsa-miR-376a-3p Modulating the IGF Axis. *Int. J. Mol. Sci.* **2023**, *24*, 11620. https://doi.org/10.3390/ijms241411620.

4. Ai, N.; Yu, Z.; Xu, X.; Liufu, S.; Wang, K.; Huang, S.; Li, X.; Liu, X.; Chen, B.; Ma, H.; et al. Circular Intronic RNA circTTN Inhibits Host Gene Transcription and Myogenesis by Recruiting PURB Proteins to form Heterotypic Complexes. *Int. J. Mol. Sci.* **2023**, *24*, 9859. https://doi.org/10.3390/ijms24129859.
5. Ng, L.; Sin, R.W.-Y.; Cheung, D.H.; Leung, W.-K.; Man, A.T.-K.; Lo, O.S.-H.; Law, W.-L.; Foo, D.C.-C. Serum microRNA Levels as a Biomarker for Diagnosing Non-Alcoholic Fatty Liver Disease in Chinese Colorectal Polyp Patients. *Int. J. Mol. Sci.* **2023**, *24*, 9084. https://doi.org/10.3390/ijms24109084.
6. Iwańczyk, S.; Lehmann, T.; Cieślewicz, A.; Malesza, K.; Woźniak, P.; Hertel, A.; Krupka, G.; Jagodziński, P.P.; Grygier, M.; Lesiak, M.; et al. Circulating miRNA-451a and miRNA-328-3p as Potential Markers of Coronary Artery Aneurysmal Disease. *Int. J. Mol. Sci.* **2023**, *24*, 5817. https://doi.org/10.3390/ijms24065817.
7. Nopp, S.; van der Bent, M.L.; Kraemmer, D.; Königsbrügge, O.; Wojta, J.; Pabinger, I.; Ay, C.; Nossent, A.Y. Circulatory miR-411-5p as a Novel Prognostic Biomarker for Major Adverse Cardiovascular Events in Patients with Atrial Fibrillation. *Int. J. Mol. Sci.* **2023**, *24*, 3861. https://doi.org/10.3390/ijms24043861.
8. Avgeros, C.; Patsatsi, A.; Dimitriadis, D.; Malousi, A.; Koletsa, T.; Papathemeli, D.; Syrnioti, A.; Avgerou, P.; Lazaridou, E.; Tzimagiorgis, G.; et al. Dysregulation of Plasma miR-146a and miR-155 Expression Profile in Mycosis Fungoides Is Associated with rs2910164 and rs767649 Polymorphisms. *Int. J. Mol. Sci.* **2023**, *24*, 271. https://doi.org/10.3390/ijms24010271.
9. Beňačka, R.; Szabóová, D.; Guľašová, Z.; Hertelyová, Z.; Radoňak, J. Non-Coding RNAs in Human Cancer and Other Diseases: Overview of the Diagnostic Potential. *Int. J. Mol. Sci.* **2023**, *24*, 16213. https://doi.org/10.3390/ijms242216213.
10. Felekkis, K.; Pieri, M.; Papaneophytou, C. Exploring the Feasibility of Circulating miRNAs as Diagnostic and Prognostic Biomarkers in Osteoarthritis: Challenges and Opportunities. *Int. J. Mol. Sci.* **2023**, *24*, 13144. https://doi.org/10.3390/ijms241713144.
11. Dżaman, K.; Czerwaty, K.; Reichert, T.E.; Szczepański, M.J.; Ludwig, N. Expression and Regulatory Mechanisms of MicroRNA in Cholesteatoma: A Systematic Review. *Int. J. Mol. Sci.* **2023**, *24*, 12277. https://doi.org/10.3390/ijms241512277.
12. Kierbiedź-Guzik, N.; Sozańska, B. miRNAs as Modern Biomarkers in Asthma Therapy. *Int. J. Mol. Sci.* **2023**, *24*, 11499. https://doi.org/10.3390/ijms241411499.
13. Pieri, M.; Vayianos, P.; Nicolaidou, V.; Felekkis, K.; Papaneophytou, C. Alterations in Circulating miRNA Levels after Infection with SARS-CoV-2 Could Contribute to the Development of Cardiovascular Diseases: What We Know So Far. *Int. J. Mol. Sci.* **2023**, *24*, 2380. https://doi.org/10.3390/ijms24032380.

References

1. Califf, R.M. Biomarker definitions and their applications. *Exp. Biol. Med.* **2018**, *243*, 213–221. [CrossRef] [PubMed]
2. Poliseno, L.; Lanza, M.; Pandolfi, P.P. Coding, or non-coding, that is the question. *Cell Res.* **2024**, *34*, 609–629. [CrossRef] [PubMed]
3. Eldakhakhny, B.; Sutaih, A.M.; Siddiqui, M.A.; Aqeeli, Y.M.; Awan, A.Z.; Alsayegh, M.Y.; Elsamanoudy, S.A.; Elsamanoudy, A. Exploring the role of noncoding RNAs in cancer diagnosis, prognosis, and precision medicine. *Noncoding RNA Res.* **2024**, *9*, 1315–1323. [CrossRef] [PubMed]
4. Ghani, M.U.; Du, L.; Moqbel, A.Q.; Zhao, E.; Cui, H.; Yang, L.; Ke, X. Exosomal ncRNAs in liquid biopsy: A new paradigm for early cancer diagnosis and monitoring. *Front. Oncol.* **2025**, *15*, 1615433. [CrossRef] [PubMed]
5. Toden, S.; Goel, A. Non-coding RNAs as liquid biopsy biomarkers in cancer. *Br. J. Cancer* **2022**, *126*, 351–360. [CrossRef] [PubMed]
6. Iwańczyk, S.; Lehmann, T.; Cieślewicz, A.; Radziemski, A.; Malesza, K.; Wrotyński, M.; Jagodziński, P.P.; Grygier, M.; Lesiak, M.; Araszkiewicz, A. Circulating microRNAs in patients with aneurysmal dilatation of coronary arteries. *Exp. Ther. Med.* **2022**, *23*, 404. [CrossRef] [PubMed]

International Journal of
Molecular Sciences

MDPI

Article

Circulating miRNA Expression Is Inversely Correlated with Tumor Tissue or Sentinel Lymph Nodes in Estrogen Receptor-Positive Early Breast Cancer Patients

Daniel Escuin [1,*,†], Laura López-Vilaró [1,2,†], Olga Bell [1], Josefina Mora [3], Bárbara García-Valdecasas [4], Antonio Moral [4,5], Montserrat Clos [4], Laia Boronat [6], Cristina Arqueros [6] and Agustí Barnadas [1,5,6,7,*]

[1] Institut d'Investigació Biomèdica Sant Pau (IIB-Sant Pau), 08041 Barcelona, Spain; llopezv@santpau.cat (L.L.-V.); obell@santpau.cat (O.B.)
[2] Department of Pathology, Hospital de la Santa Creu i Sant Pau, 08041 Barcelona, Spain
[3] Department of Biochemistry, Hospital de la Santa Creu i Sant Pau, 08041 Barcelona, Spain; jmora@santpau.cat
[4] Department of Surgery, Hospital de la Santa Creu i Sant Pau, 08041 Barcelona, Spain; bgvaldecasas@santpau.cat (B.G.-V.); amoral@santpau.cat (A.M.); mclos@santpau.cat (M.C.)
[5] Faculty of Medicine, Universitat Autònoma de Barcelona (UAB), 08193 Bellaterra, Spain
[6] Department of Medical Oncology, Hospital de la Santa Creu i Sant Pau, 08041 Barcelona, Spain; lboronat@santpau.cat (L.B.); carqueros@santpau.cat (C.A.)
[7] Centro de Investigación Biomédica en Red Cáncer (CIBERONC), 28029 Madrid, Spain
* Correspondence: descuin@santpau.cat (D.E.); abarnadasm@santpau.cat (A.B.)
† These authors contributed equally to this work.

Abstract: The deregulation of microRNAs (miRNAs) is associated with the various steps of the metastatic process. In addition, circulating miRNAs are remarkably stable in peripheral blood, making them ideal noninvasive biomarkers for disease diagnosis. Here, we performed a proof-of-principle study to determine whether tumor-tissue-derived miRNAs are traceable to plasma in ER-positive early breast cancer patients. We performed RNA-sequencing on 30 patients for whom plasma, sentinel lymph nodes (SLNs) and tumor tissue were available. We carried out differential expression, gene ontology and enrichment analyses. Our results show that circulating miRNAs are inversely expressed compared with tumor tissue or SLNs obtained from the same patients. Our differential expression analysis shows the overall downregulation of circulating miRNAs. However, the expression of miR-643a-3p and miR-223 was up-regulated in patients with positive SLNs. Furthermore, gene ontology analysis showed the significant enrichment of biological processes associated with the regulation of epithelial cell proliferation and transcriptional regulation commonly involved in the promotion of metastases. Our results suggest the potential role of several circulating miRNAs as surrogate markers of lymph node metastases in early breast cancer patients. Further preclinical and clinical studies are required to understand the biological significance of the most significant miRNAs and to validate our results in a larger cohort of patients.

Keywords: miRNAs; early breast cancer; sentinel lymph node; metastasis; biomarkers

1. Introduction

Cancer metastases are responsible for the majority of breast cancer deaths. Despite intensive research in this field, our understanding of the molecular events that drive metastatic progression remains largely incomplete. In recent years, microRNAs (miRNAs) have emerged as important regulators of many cellular processes, including the various steps of the metastatic process. miRNAs are small (19–25 nt), noncoding RNA (ncRNA) molecules that regulate the expression of target genes by binding to complementary regions of mRNAs to repress their translation or regulate their degradation. Many cellular pathways are affected by the regulatory function of miRNAs, and several human pathologies, including cancers, are associated with the deregulation of miRNAs [1] and their metastases [2].

The development of miRNA signatures in liquid biopsies for the diagnosis of a variety of different cancer types is an active field of research [3]. This is in part due to the fact that circulating miRNAs and other small RNA molecules are stable in peripheral blood [4,5], making them ideal noninvasive biomarkers for disease diagnosis. Our recent research has focused on the involvement of miRNAs in the development of locoregional metastases in patients with early breast cancer [6]. We showed a good correlation in the miRNA expression profile between sentinel lymph nodes (SLNs) and matched primary tumor samples, in agreement with a previous study [7]. The down-regulation of various miRNAs was significantly differentially expressed in patients with locoregional metastasis. In addition, we recently profiled the expression of miRNAs in a small number of plasma samples collected before surgery from patients with early-stage breast cancer who were not previously treated [8]. Our results showed an overall downregulation of miRNAs that, in many cases, are direct targets of proteins that promote metastasis. One important caveat in our previous studies was a lack of correlation between miRNA expression in plasma, tumor tissues and SLNs obtained from the same patients. This is of utmost importance in order to develop potential circulating-based biomarkers. First, to elucidate whether circulating miRNAs are derived from tumor-tissue miRNAs and, second, to establish whether plasma miRNAs accurately reflect the heterogeneity of a primary tumor and its subclones [3].

Here, we performed RNA-sequencing to profile the expression of miRNAs in 30 early breast cancer patients for whom plasma, tumor tissue and SLN samples were available. Our results show an inverse relationship between circulating miRNAs and tumor/SLN samples. Nonetheless, several circulating miRNAs were associated with metastatic SLN status via a significant enrichment of the biological processes associated with the regulation of epithelial–mesenchymal transition, cell proliferation and transcriptional regulation. Our data highlight the potential use of circulating miRNAs as surrogate markers of locoregional metastases in breast cancer.

2. Results

2.1. Patients

A total of 30 female patients were included in this study. We analyzed 30 plasma samples, 29 SLNs (2 SLNs were collected from the same patients) and nine tumor tissues. We paired plasma and SLN samples for 27 patients; paired plasma and tumor tissue for 9 patients; and all three samples available for 7 patients (Table S1). The main clinicopathological characteristics of the patients are described in Table 1. A total of 12 (40%) patients had SLN-positive tumors; seven SLNs were diagnosed as micrometastasis (23%); and five were diagnosed as macrometastasis (17%).

2.2. RNA-Sequencing

All samples passed pre- and post-sequencing quality checks, which confirmed average read quality and base quality Q-scores of >30 (99.9% correct) [9] and the expected read length distribution for miRNAs (Figure S1). The mean read numbers for plasma and tissues were 18.5 $\pm$ 0.9 million and 13.8 $\pm$ 0.6 million, respectively. Following sequencing and trimming, all reads containing identical insert sequences and UMI sequences (insert-UMI pair) were collapsed into a single read and passed into the analysis pipeline. This allowed for the true quantification of the miRNAs by eliminating library amplification bias and a better representation of the RNA molecules in the sample. We obtained an average of 1.97 million and 2.24 million collapsed reads for plasma and tissues, respectively, which resulted in good miRNA mapping reads with a very dominant miRNA peak in all samples, indicating good sample/data quality (Figure S2A). Overall, we obtained average genome mapping rates of 65% and 95% for plasma and tissues, respectively (Figure S2B). After mapping and counting relevant entries in mirbase_22, the expression of known miRNAs was calculated using the TPM method (Table S2). We did not identify any sequences identical to those of known miRNAs in miRBase_22 for other organisms.

Table 1. Basic patient and tumor characteristics.

Variable		N (%)
Patients		30 (100)
Age, years	Mean + SD	63.9 ± 7.6
	Median (range)	62.4 (52.6–83.8)
Tumor stage	IA	16 (53.3)
	IIA	9 (30)
	IIB	5 (16.7)
Tumor grade	I	10 (33.3)
	II	20 (66.7)
Node status	Negative	18 (60)
	Positive	12 (40)
	Micrometastasis	7 (23.3)
	Macrometastasis	5 (16.7)
Tumor focality	Unifocal	18 (60)
	Multifocal	10 (33.3)
	Multicentric	2 (6.7)
Estrogen receptor	ER+ PR+	27 (90)
	ER+ PR−	3 (10)
Molecular subtype	Luminal A	20 (66.7)
	Luminal B	10 (33.3)
Ki67	<14%	21 (70)
	≥14%	9 (30)
Surgery	Mastectomy	8 (26.7)
	Lumpectomy	22 (73.3)
Lymphovascular invasion	Negative	29 (96.7)
	Positive	1 (3.3)

2.3. Unsupervised Clustering Analysis

We investigated whether patients were assigned to biological groups based on their miRNA expression. We performed a supervised two-way hierarchical clustering of miRNAs and samples using the 50 miRNAs with the largest coefficients of variation based on TMM counts. Interestingly, the expression of circulating miRNAs showed an inverse correlation with either tumor tissues or SLNs obtained from the same patients. That is, for all patients studied, we observed that any miRNAs overexpressed in tissues were downregulated in plasma and vice versa (Figures 1A and S3). We obtained similar results using a principal component analysis (PCA), which clearly clustered plasma samples into a different group (Figure 1B). In addition, we performed Spearman's correlation analysis (r_s) between samples obtained from each patient. Our data show that both tumor and SLN samples from the same patient have a high correlation index, with an average value of $r_s = 0.78$ among all patients. Interestingly, we obtained two SLNs from patients 1 and 20. The latter patient had both SLNs diagnosed as macrometastases, whereas the former patient had one macrometastasis and one micrometastasis. In both cases, we showed the highest correlations with $r_s = 0.875$ and $r_s = 0.864$, respectively (Figure 1C). In contrast, both tumors and SLNs showed little correlation when compared with plasma samples from the same patient, with averages of $r_s = 0.61$ and $r_s = 0.59$, respectively (Figures 1C and S4).

2.4. Differentially Expressed Circulating miRNAs

We focused on circulating miRNAs to investigate whether differentially expressed plasma miRNAs were associated with the locoregional metastasis status of the patients. First, we analyzed plasma samples according to positive ($n = 12$) or negative ($n = 18$) SLN metastasis status. We found 34 miRNAs with a significant differential expression ($p < 0.05$). However, only the upregulation of hsa-miR-642a-3p remained significant after correcting for multiple testing (q = 0.034) (Figure 2A). Overall, we found four miRNAs upregulated and seven miRNAs downregulated with an absolute log2 fold change of ≥1 (Table S3). Similar results were found when patients were sub-classified into positive macrometastasis (n = 5) or micrometastasis (n = 8) SLNs and compared with patients with negative SLNs (n = 18). We found that hsa-miR-642a-3p remained upregulated in patients with micrometastasis SLNs ($p < 0.05$) but not in patients with macrometastasis SLNs. In contrast, we found that hsa-miR-223 was upregulated (q = 0.03) in patients with positive macrometastasis (Figure 2B). Since our series included patients with luminal A and luminal B tumors (Table 1), we analyzed the differential expression between these two subgroups. Our results show 12 miRNAs with an absolute log2 fold change of ≥1 when comparing luminal B against luminal A tumors. Ten miRNAs were downregulated and two miRNAs were upregulated in luminal B tumors. However, none of them remained significant after correcting the p-values for multiple testing (q > 0.05) (Figure 2D, Table S3).

2.5. *Biological Significance and Enriched Analysis of Circulating miRNAs*

We performed a biological significance analysis using a subset of miRNAs ($p < 0.05$ and absolute log fold change >0.3). Table 2 shows a summary of the top targeted genes for each miRNA investigated. A complete list of validated target genes and all evidence types are provided in Table S4. In addition, we carried out an enrichment analysis of the GO and Reactome Pathways databases using the identified set of *Homo sapiens* target genes to understand how our data are related to biological functions. The results are summarized in Figure 3 and Table 3. Our data show that differentially expressed miRNAs were associated with biological processes (BP) markedly focused on gland development, gene expression regulation via receptor tyrosine kinases, epithelial cell motility and proliferation processes ($p < 0.001$). A complete list of the analyses for each comparison is included in Table S5. Differentially expressed miRNAs were enriched in cellular component (CC) terms located in membrane fractions, WNT signalosome and transcription factor complexes. Moreover, differentially expressed miRNAs were enriched in molecular function (MF) terms associated with RNA polymerase II, chromatin and ubiquitin-binding ($p < 0.001$) (Table 3).

2.6. *Clinicopathological Correlation with Circulating miRNAs*

Our series included 30 patients with ER+ early breast cancer, and we reported recurrence in two (7%) patients. The median follow-up time was 5.2 years (range 2.3–7.3 years). At the last follow-up, all were reported to be alive. We investigated whether the differentially expressed miRNAs correlated with the patients' clinicopathological parameters. The expressions of miR-6724-5p, miR-27a-3p, miR-423-3p and miR-140-5p were significantly higher in patients with high Ki67. The expression of miR-16-2 was significantly higher in older patients (>60 years). The expressions of miR-6724-5p, miR-320d and miR-423-3p were significantly higher in luminal B tumors. The increased expression of miR-182-5p was associated with higher-grade tumors, and low miR-3157-5p expression was associated with the presence of tumor multifocality. Given the low number of events, we were unable to perform any survival analysis on our cohort of patients.

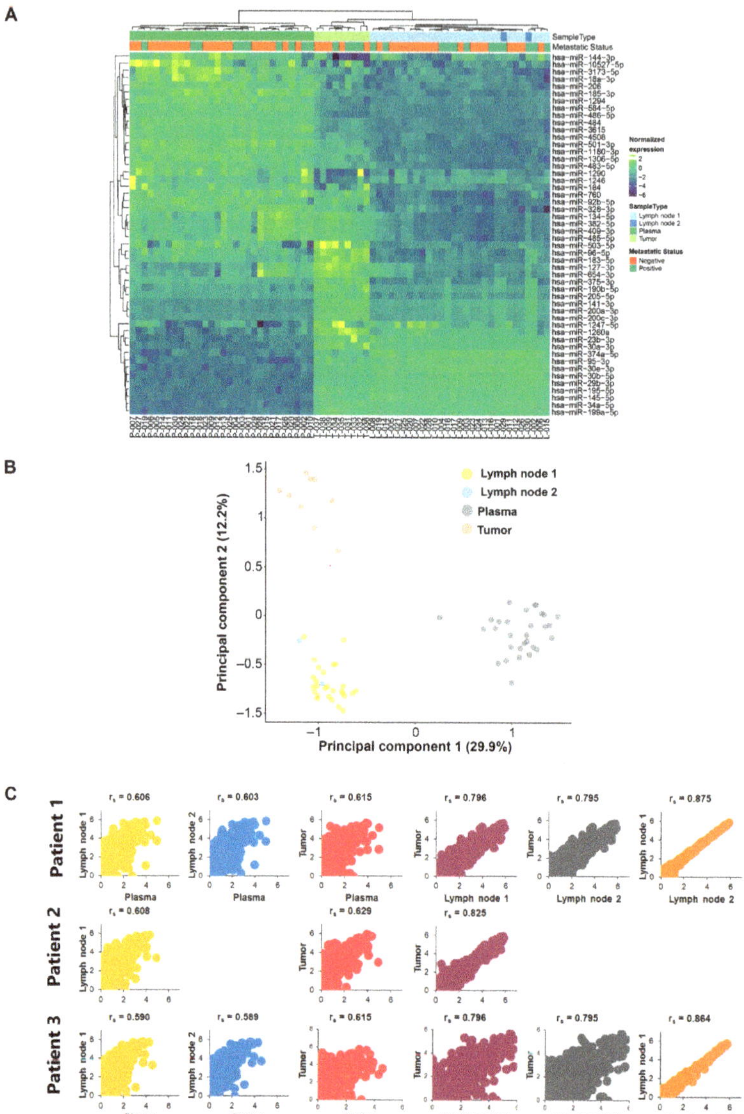

Figure 1. Class discovery associated with SLN metastatic status. The analysis was performed using the 50 miRNAs with the largest coefficients of variation based on the trimmed mean of M-values (TMM counts). (**A**) Heat map and unsupervised hierarchical clustering. Each row represents one miRNA, and each column represents one sample. The color represents the relative expression level of a miRNA across all samples. The color scale shows the expression level above (purple) or below (green) the mean. (**B**) Principal component analysis (PCA) shows sample clusters arising naturally based on the miRNA expression profile. Tumors are shown in light brown, plasma samples in grey and SLNs in yellow. Patients with more than 1 SLN analyzed are depicted as blue. (**C**) Representative scatterplots of selected patients showing a sample-to-sample comparison between different tissues from the same patient as indicated. The scatterplots show the log expression of miRNA expression and the Spearman correlation coefficient (r_S) for each comparison.

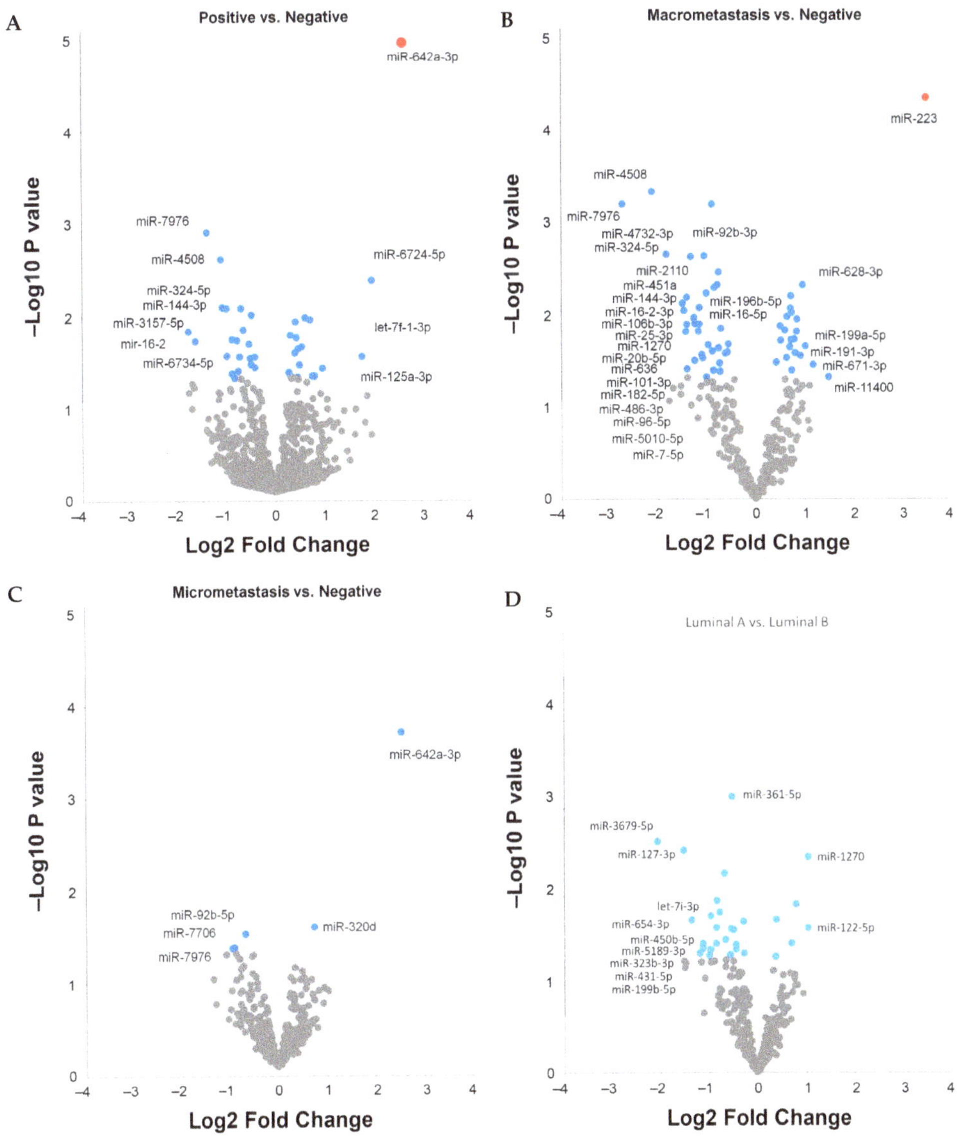

Figure 2. Differentially expressed circulating miRNAs. The volcano plots show differentially expressed miRNAs in plasma samples according to the patients' locoregional metastatic status as indicated (**A–C**) or the molecular subtype (**D**). The data show the logarithmic relationship between non-adjusted p-values (y-axis) and the fold change expression (x-axis). Corrected q-values of <0.05 are shown as red dots; non-adjusted p-values of <0.05 are shown as blue dots; and non-significant p-values of >0.05 are shown as gray dots. Only miRNAs with an absolute log2 fold change of $\geq$1 are labeled.

Table 2. Biological significance analysis. We analyzed a subset of differentially expressed miRNAs, and we show the top 5 validated targets for each miRNA.

Mature miRNA	Target Genes	Targets	Log FC	*p*-Value
hsa-miR-192-5p	*DHFR, WNK1, RB1, ACVR2B, AKT1*	41	−1.439	0.001
hsa-miR-376c-3p	*BCL2, BMI1, RUNX2, NR5A2, GRB2*	11	−2.432	0.001
hsa-miR-192-5p	*DHFR, WNK1, RB1, ACVR2B, AKT1*	41	−1.439	0.001
hsa-miR-193b-5p	*STMN1*	1	−1.572	0.003
hsa-miR-99a-5p	*FGFR3, AKT1, CAPNS1, FKBP5, MTOR*	15	−0.967	0.005
hsa-miR-885-5p	*CASP3, CDK2, CTNNB1, IGF1R, MCM5*	5	−1.910	0.008
hsa-miR-9-5p	*NTRK3, PRDM1, REST, FOXO1, NFKB1*	82	1.456	0.009
hsa-miR-15b-3p	*CCND3, IGF1R, RECK, TRIM31, SIRT4*	5	−0.647	0.010
hsa-miR-92b-3p	*CDKN1C, ITGA6, ITGAV, SMAD3, SMAD7*	13	−0.624	0.010
hsa-miR-425-5p	*CCND1, BTK, FGFR3, SMAD2, PTEN*	9	−0.484	0.010
hsa-let-7a-5p	*NF2, TRIM71, HMGA2, CASP3, DICER1*	55	−0.437	0.011
hsa-miR-320d	*GNAI1, RBFOX2*	2	0.616	0.013
hsa-miR-194-5p	*ACVR2B, BMI1, BMP1, CDH2, CDKN1B*	23	−0.826	0.013
hsa-miR-486-5p	*ARHGAP5, CDK4, DOCK3, FBN1, FOXO1*	14	−0.491	0.015
hsa-miR-629-5p	*HNF4A, TRIM33*	2	0.471	0.015
hsa-miR-125b-5p	*ERBB2, ERBB3, BAK1, NTRK3, SMO*	122	0.599	0.018
hsa-miR-92a-3p	*ITGA5, BMPR2, CD69, CDH1, DNMT1*	38	−0.498	0.018
hsa-miR-27a-3p	*SP1, SP3, SP4, RUNX1, FOXO1*	67	0.362	0.018
hsa-miR-223-3p	*FOXO3, FOXO1, FBXW7, IGFR1, MEF2C*	67	0.812	0.018
hsa-miR-625-3p	*MAP2K6*	1	−0.623	0.019
hsa-miR-96-5p	*FOXO1, IRS1, AQP5, CELSR2, ODF2*	31	−0.848	0.021
hsa-miR-122-5p	*CCNG1, BCL2L2, ADAM17, NDRG3, ADAM10*	67	−1.286	0.021
hsa-miR-378a-3p	*RUNX1, CDK6, CYP19A1, GLI3, GRB2*	19	−0.514	0.022
hsa-miR-25-3p	*FBXW7, ATP2A2, CDH1, CDKN1C, CYP2B6*	29	−0.648	0.023
hsa-miR-483-3p	*CDK4, CCN2, IGF1, SMAD4, RASGRF1*	10	−1.207	0.023
hsa-miR-642a-3p	*GRK2, CACNA1B, CDC25B, CDKN1A, FOXN3*	82	0.796	0.023
hsa-miR-34a-5p	*E2F3, VEGFA, BCL2, MYCN, SIRT1*	127	−0.980	0.024
hsa-miR-7-5p	*IRS2, IRS1, EGFR, RAF1, PAK1*	46	−0.737	0.031
hsa-let-7b-5p	*HMGA2, CDK6, CDC25A, ACTG1, ACVR1*	40	−0.402	0.031
hsa-miR-30c-5p	*UBE2I, BCL9, CAMK2D, CASP3, RUNX2*	39	−0.475	0.033
hsa-miR-885-3p	*BMPR1A, CD274*	2	−1.663	0.034
hsa-miR-24-3p	*CDKN2A, ACVR1B, DHFR, MYC, E2F2*	84	0.301	0.034
hsa-miR-365a-3p	*HDAC4, ACVR1, BAX, CCND1, BCL2*	10	−1.078	0.035
hsa-miR-20b-5p	*ESR1, BRCA1, CDKN1A, EFNB2, EPHB4*	20	−0.860	0.036
hsa-miR-126-3p	*SPPL2A, ZSWIM1, PIGS, ZNF257, ZNF543*	44	0.316	0.038
hsa-miR-101-5p	*PRKDC, ATM, EZH2, FOS, STMN1*	12	−1.359	0.040
hsa-miR-23b-3p	*PLAU, MET, CA2, RUNX2, CCNG1*	31	0.338	0.041
hsa-miR-942-5p	*CDKN1A, GSK3B, IFI27, NFKBIA, SFRP4*	6	−0.348	0.043
hsa-miR-451a	*MIF, ADAM10, AKT1, BCL2, CDKN2D*	22	−0.707	0.043
hsa-miR-339-3p	*FOXO1, MCL1, NFKB1*	3	0.483	0.044
hsa-miR-140-3p	*CD38, COL4A1, FN1, GPC1, ITGA6*	9	−0.389	0.044
hsa-miR-485-3p	*NTRK3, MAT1A, NFYB, PPARGC1A, SLC40A1*	6	−0.993	0.045
hsa-miR-324-5p	*GLI1, SMO, ETS1, SP1, MTFR1*	6	−0.860	0.047
hsa-let-7c-5p	*MYC, HMGA2, TGFBR1, BCL2L1, CASP3*	28	0.506	0.048
hsa-miR-155-5p	*AGTR1, RHOA, ETS1, MEIS1, FOXO3*	112	0.432	0.049

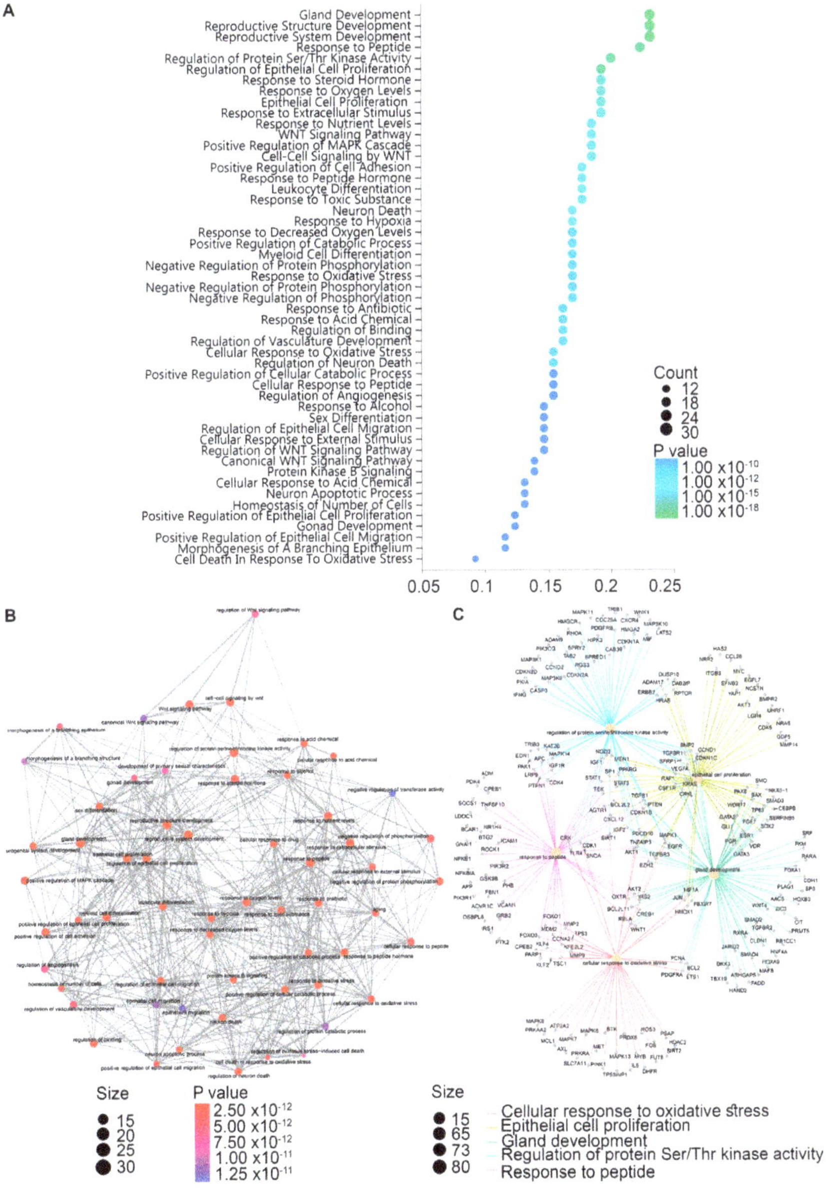

Figure 3. Gene ontology (GO) enrichment analysis for significant biological processes associated with positive SLNs. (**A**) Dot plot graph shows the 50 most significant biological-process GO terms (*y*-axis) and the ratio between the number of expressed miRNAs associated with the GO term and the number of significantly differentially expressed genes associated with the GO term (*x*-axis). The colors of the nodes indicate the *p*-value and the size of the nodes denotes the number of miRNAs associated with a specific GO term. (**B**) Enrichment map of the top 60 miRNAs. Pathways are grouped by similarity. The size of each node indicates the number of miRNAs found in a pathway, whereas the color of the node is based on the significance of the *p*-value (**C**) Neural network shows the GO terms for the biological processes associated with patients with positive SLNs.

Table 3. Gene ontology analysis. Gene set enrichment analysis using terms within GO categories (biological process, cellular component, molecular function) was applied to extract biological meaning from the identified differentially expressed transcripts and predicted mRNA targets. The top 10 GO-significant categories associated with differentially expressed circulating miRNAs in patients with positive SLNs are shown. "Counts" refers to the ratio between the number of enriched differentially expressed miRNAs and the total number of miRNAs assigned to these terms.

GO ID	GO Term	Counts	q-Value
	BIOLOGICAL PROCESS		
0048732	Gland development	30/130	1.7×10^{-18}
0048608	Reproductive structure development	30/130	1.7×10^{-18}
0061458	Reproductive system development	30/130	1.7×10^{-18}
1901652	Response to peptide	29/130	6.62×10^{-16}
0050678	Regulation of epithelial cell proliferation	25/130	6.73×10^{-15}
0048545	Response to steroid hormone	25/130	8.22×10^{-15}
0070482	Response to oxygen levels	25/130	1.67×10^{-14}
0050673	Epithelial cell proliferation	25/130	1.17×10^{-13}
0071900	Regulation of protein serine/threonine kinase activity	26/130	1.72×10^{-13}
0097305	Response to alcohol	19/130	1.1×10^{-12}
0045785	Positive regulation of cell adhesion	23/130	1.16×10^{-12}
	CELLULAR COMPONENT		
0005667	Transcription factor complex	19/131	1.14×10^{-11}
0044798	Nuclear transcription factor complex	14/131	1.25×10^{-10}
0090575	RNA polymerase II transcription factor complex	13/131	3.02×10^{-10}
0000785	Chromatin	18/131	7.87×10^{-07}
0045121	Membrane raft	14/131	1.51×10^{-06}
0098857	Membrane microdomain	14/131	1.51×10^{-06}
0098589	Membrane region	14/131	2×10^{-06}
0000790	Nuclear chromatin	13/131	8.06×10^{-06}
0044454	Nuclear chromosome part	15/131	2.58×10^{-05}
1990909	WNT signalosome	4/131	2.58×10^{-05}
	MOLECULAR FUNCTION		
0000978	RNA polymerase II proximal promoter sequence-specific DNA binding	22/130	4.96×10^{-10}
0000987	Proximal promoter sequence-specific DNA binding	22/130	5.91×10^{-10}
0019902	Phosphatase binding	15/130	$5.91 x^{-10}$
0001228	DNA-binding transcription activator activity, RNA polymerase II-specific	19/130	1×10^{-08}
0031625	Ubiquitin protein ligase binding	16/130	1.49×10^{-08}
0019903	Protein phosphatase binding	12/130	1.87×10^{-08}
0044389	Ubiquitin-like protein ligase binding	16/130	2.48×10^{-08}
0003682	Chromatin binding	19/130	3.39×10^{-08}
0001085	RNA polymerase II transcription factor binding	10/130	4.99×10^{-07}
0033613	Activating transcription factor binding	8/130	8.48×10^{-07}

3. Discussion

In recent years, miRNAs have emerged as important regulators of the various steps of the metastatic process [10]. Currently, lymph node affection remains the most important prognosis factor in breast cancer [11], and the presence of metastasis in SLNs is still the recommended procedure for the axillary staging of early breast cancer. Our recent research focused on the involvement of miRNAs in the development of locoregional metastases in patients with early-stage breast cancer. Our initial study focused on profiling the expression of miRNAs via RNA-seq in paired tumor tissue and lymph nodes from the same patients [6], whereas the second was a small observational study of plasma samples in breast cancer patients [8]. Both studies reported an association between several miRNAs and the locoregional metastatic outcomes of patients. However, we could not establish whether the expression of circulating miRNAs resembled that of primary tumors or SLNs since both sample cohorts were obtained from different patients.

Therefore, in this study, we sought to address this issue by analyzing plasma, SLNs and tumor tissues derived from the same patients and correlating the data with the metastatic status of these patients. In total, we RNA-sequenced 30 plasmas, 30 SLNs and nine tumor tissues ($n = 68$ samples) from 30 patients with ER+ early breast cancer. Overall, our data show that the miRNA expression in tumors and SLNs is similar, with minor changes that are likely due to histological differences between the SLN and the tumor. In contrast, the expression of circulating miRNAs showed an inverse correlation with tumor and SLN samples. For instance, any specific miRNAs that were upregulated in tumor tissues or SLNs were downregulated in plasma samples and vice versa. In fact, when we performed a sample-to-sample comparison on the same patients, the plasma samples showed a low Spearman correlation when compared with both tumor tissues or SLNs. Our data agree with previous publications, suggesting little correlation between the tumor and serum expression of miRNAs in breast cancer using qPCR [12]. However, we cannot rule out the possibility that blood sample collection may influence the expression of circulating miRNAs. In this regard, there is evidence showing that, in metastatic colorectal cancer, circulating miRNA expression changes before and after passing through the metastatic site (liver), as if somehow some miRNAs are retained specifically in the liver, while others are not [13]. In addition, transcriptome analysis revealed non-identical miRNA profiles between arterial and venous plasma [14]. Similarly, comparisons between arterial and venous miRNAs in hypertension-versus-control conditions also revealed a prominent disease association between circulating miRNAs and their target genes in arteries but not in veins [15]. The overall picture that emerges from these studies is that the blood collection point affects the miRNA expression profile. Increasing evidence also shows that tumor-draining veins constitute a better source of biomarkers than peripheral veins in colon cancer patients [16]. For instance, a recent small exploratory study indicated that tumor-proximal venous samples are highly enriched in miRNAs, ctDNA mutations and circulating tumor cells among oncological biomarkers and may allow for more robust molecular analysis than peripheral vein samples [17,18]. Whether something similar occurs in breast cancer patients remains to be answered. In our study, all plasma samples were obtained from a peripheral vein, and this could, at least in part, explain the observed discrepancies between circulating miRNAs and paired tumors or SLNs.

Despite all the aforementioned data, we identified several circulating miRNAs associated with the locoregional metastatic statuses of patients. Overall, we found that most differentially expressed miRNAs associated with positive SLNs were downregulated ($p < 0.05$), in agreement with our previous study, including eight miRNAs (339-5p, miR-29b-3p, 139-5p, 144-3p, 423-3p,584-5p, 99b-3p and 101-3p) that we previously described [8]. However, their differential expression did not remain significant after adjusting for a false discovery rate. Interestingly, our results show the upregulation of miR-642a-3p and miR-223 ($q < 0.05$) in patients with positive locoregional metastases. The upregulation of miR-642a-3p has been reported to mediate the HDAC inhibitor-mediated downregulation of HER2 and apoptosis in breast cancer cell lines [19]. Others have associated the upregulation of miR-642a-3p as an invasion/metastasis associated miRNA via the mTOR and PI3K-AKT signaling pathways in gallbladder cancer [20], whereas, in breast cancer, miR-642a-3p expression increases in the breast milk extracellular vesicles of mothers with obesity and is associated with cancer signaling pathways [21]. In the case of miR-223, its role in human cancer is controversial, and it has been found recently to act as both a tumor-suppressor gene and an oncomiR [22]. This dual role may mainly depend on its target in different locations of tissues and cells, and it has been described in breast [23], pancreatic [24] and prostate [25] human cancers. Our data show an upregulation of miR-223 in early-breast cancer patients with positive macrometasis in the SLNs and, thus, support its role as an oncomiR. Interestingly, our enrichment analysis showed that miR-233 top target genes (i.e., FOXO3, MEF2C and FBXW7) have already been described as mediating miR-233 oncomiR characteristics. For instance, FoxO3a, a well-defined tumor suppressor gene in the forkhead transcription factor O subfamily (FoxO), has been shown to be downregulated by the increased expression of miR-233 in pancreatic cancer cell lines [24]. Finally, our GO analysis shows that both miR-642a-3p and miR-223 are involved in GO biological processes related to gland and reproductive development, the regulation of receptor tyrosine kinases and epithelial cell proliferation among the top significant terms, adding further evidence of their role in promoting tumor metastasis.

To conclude, in this study, we provided evidence that circulating miRNAs have an inverse correlation with the expression observed in tumor and SLN samples derived from the same patients. Therefore, we could not fully establish whether the expression of circulating miRNAs resembles that of the primary tumor or the SLNs. This is of utmost importance in developing potential circulating-based biomarkers because (1) the origin of tissue-derived miRNAs traceable in plasma and (2) to what extent circulating miRNAs accurately reflect the heterogeneity of a tumor and its subclones (or all tumor lesions in metastatic patients) need to be established [3]. Nonetheless, we described several miRNAs associated with the locoregional metastatic status of patients with breast cancer, which have been reported to be direct targets of proteins that promote metastasis. Further studies are required with a larger number of patients to demonstrate whether the sensitivity and specificity of plasma miRNAs may be improved through selective venous sampling in breast cancer patients and to unveil the molecular mechanisms of the miRNAs described in this study.

4. Materials and Methods

4.1. Patients

We studied 30 patients with early breast cancer treated with surgery. Male patients were excluded from this study. None of the patients had prior treatment with surgery, chemotherapy or radiation. All patients had confirmed diagnoses based on the histopathology of tumor biopsies and intraoperative SLNs evaluated using the OSNA assay [26]. All tumors were invasive ductal carcinomas (IDCs) with or without in situ components. All patients were estrogen receptor-positive and HER2-negative. We collected the following clinical and pathological parameters: age, menopausal status, personal and familiar disease precedents, clinical follow-up, tumor stage determined according to the UICC system [27], histological grade determined using the Elston–Ellis grading system [28], tumor histology, presence of associated carcinoma in situ, presence of vascular and lymphatic invasion, tumor-infiltrating lymphocytes, tumor focality, tumor necrosis, proliferation of non-tumoral tissue.

4.2. Sample Processing

All samples were collected from early breast cancer patients who had not received any previous treatment. Peripheral blood was withdrawn before surgery. Approximately 10–15 mL of peripheral blood was collected for plasma processing in EDTA tubes. Plasma tubes were processed within 2 h of collection, spun at $1200\times g$ for 10 min and inspected for the absence of hemolysis, as previously described [8,29]. Tumors and SLNs were processed as previously described [6]. All samples were aliquoted in 1.5 mL tubes and stored at −80 °C until further processing.

4.3. RNA Isolation

RNA was isolated from 300 µL of plasma samples using the miRNeasy serum/plasma advanced kit (Qiagen, Germantown, MD, USA) according to the manufacturer's instructions. RNA was eluted in a volume of 14 µL. RNA was isolated from tumor tissues and SLNs using the miRNeasy kit (Qiagen, Germantown, MD, USA) according to the manufacturer's instructions and eluted in a volume of 30 µL. The RNA integrity (RIN) level was measured for each RNA sample using the Agilent TapeStation (Santa Clara, CA, USA).

All samples used in this study had a RIN value > 7. A range of spike-ins was added to all samples prior to RNA isolation. A pre-sequencing quality check using q-PCR was performed on all samples to control for the quality of the RNA and the inhibition of enzymatic reactions in downstream reactions, as previously described [8].

4.4. RNA-Sequencing

Next-generation sequencing (NGS) and genome annotation were performed as previously described [6,8]. Genome annotation was performed using the QIAGEN CLC Genomics Server v20.0.4 (Qiagen, Germantown, MD, USA). Following sequencing, Cutadapt (1.9.1) [30] was used to trim adaptor sequences. A quality check (QC) was performed to ensure Q-scores > 30 (>99.9% correct) for our data [9]. Reads with the correct length were analyzed for the presence of UMIs using Cutadapt v1.9.1 (Dortmund, Germany) and then collapsed by UMIs into FASTQ files. This approach eliminated library amplification bias and allowed for the true identification of miRNAs. The Bowtie2 software v2.2.6 (San Diego, CA, USA) was used to map the reads. The mapping criterion for aligning reads to spike-ins, abundant sequences and miRBase_22 was for reads to have perfect matches with the reference sequences. To map the genome, one mismatch was allowed in the first 32 bases of the read. Small insertions and deletions (INDELs) were not allowed. The resulting sequences were annotated using human assembly GRCh38 (Ensembl) and the miRBase_22 database. Transcripts per million (TPM) was used as a normalization procedure to correct for differences in sequencing depth and to quantify each RNA species.

4.5. Differential Expression Analysis

Differential expression analyses were performed using the TMM normalization method [31] and the EdgeR v3.17 statistical software package (Bioconductor; Victoria, Australia). We used TMM-normalized quantification R scripts [32] to perform principal component analysis (PCA) and unsupervised hierarchical clustering analysis. We selected the top 50 miRNAs with the largest coefficient of variation across all samples to obtain a cluster of samples. The data were normalized to TMM and converted to log2 scale.

4.6. Biological Significance Analysis, Gene Ontology (GO) and Enrichment Analysis

A search of validated gene targets was performed with multiMiR Bioconductor's package (http://multimir.ucdenver.edu last accessed on 7 July 2023) [33], which enables the retrieval of miRNA–target interactions from different databases like mirecords, mirtarbase and tarbase. Only miRNA–target interactions with strong experimental evidence (i.e., luciferase assay or Western blot) were considered. Gene ontology (GO) [34] and Reactome Pathway [35] analyses were performed with R/Bioconductor's Cluster Profiler package, v3.12.0 [36], using the experimentally verified targets of significantly differentially expressed miRNAs as inputs, as previously described [6,8].

4.7. Statistics

Differentially expressed miRNAs from RNA-sequencing data were detected using an exact test based on conditional maximum likelihood (CML), included in the R/Bioconductor package EdgeR (Victoria, Australia) [37]. p-values from RNA-sequencing were corrected (q-values) for multiple testing using the Benjamini–Hochberg procedure [38]. A false discovery rate (FDR) of $q < 0.05$ was considered significant. In all group comparisons, missing expression values were treated as zero. Differences in total numbers of miRNAs between groups were analyzed using two-sided parametric t-tests. A clinicopathological analysis was performed using Student's t-test for the comparison of quantitative variables and the X2 or Fisher exact tests for the comparison of qualitative variables. The Cox regression model was used to perform the multivariate analysis. A two-sided p-value ≤ 0.05 was considered significant.

Supplementary Materials: The following supporting information can be downloaded at https://www.mdpi.com/article/10.3390/ijms241713293/s1.

Author Contributions: Conceptualization, D.E., L.L.-V. and A.B.; resources, investigation and methodology, D.E., L.L.-V., O.B., J.M., B.G.-V., A.M., M.C., L.B., C.A. and A.B.; formal analysis, D.E., L.L.-V. and A.B.; writing—original draft preparation, D.E., L.L.-V. and A.B.; manuscript review, D.E., L.L.-V., O.B., J.M., B.G.-V., A.M., M.C., L.B., C.A. and A.B.; editing, D.E. and L.L.-V.; project administration, D.E. and A.B.; funding acquisition, D.E. and A.B. All authors have read and agreed to the published version of the manuscript.

Funding: This research was funded by grants PI13/00110 to D.E. and PI19/00362 to D.E. and A.B. from the Instituto de Salud Carlos III (ISCIII, Madrid, Spain) and co-funded by the European Regional Development Fund (Brussel, Belgium)grant CB16/12/00471 to A.B. from the Thematic Area for Cancer (CIBERONC, Madrid, Spain); grant 2021_SGR-00827 to A.B., D.E., C.A. and L.B. from the Catalonian Support to Research Groups (Barcelona, Catalunya, Spain).

Institutional Review Board Statement: This study was conducted in accordance with the Declaration of Helsinki and approved by the Clinical Research Ethics Committee of the "Institut d'Investigació Biomèdica Sant Pau".

Informed Consent Statement: Informed consent was obtained from all subjects involved in the study.

Data Availability Statement: The original contributions presented in this study are publicly available at the Sequence Research Archive under ID PRJNA748920 and are available for download here: http://www.ncbi.nlm.nih.gov/bioproject/748920 (accessed on 22 July 2021).

Acknowledgments: We thank the patients and their caregivers for their participation.

Conflicts of Interest: The authors declare no conflict of interest.

References

1. Dai, R.; Ahmed, S.A. MicroRNA, a new paradigm for understanding immunoregulation, inflammation, and autoimmune diseases. *Transl. Res.* **2011**, *157*, 163–179. [CrossRef] [PubMed]
2. Nicoloso, M.S.; Spizzo, R.; Shimizu, M.; Rossi, S.; Calin, G.A. MicroRNAs--the micro steering wheel of tumour metastases. *Nat. Rev. Cancer* **2009**, *9*, 293–302. [CrossRef]
3. Heitzer, E.; Haque, I.S.; Roberts, C.E.S.; Speicher, M.R. Current and future perspectives of liquid biopsies in genomics-driven oncology. *Nat. Rev. Genet.* **2018**, *20*, 71–88. [CrossRef] [PubMed]
4. Mitchell, P.S.; Parkin, R.K.; Kroh, E.M.; Fritz, B.R.; Wyman, S.K.; Pogosova-Agadjanyan, E.L.; Peterson, A.; Noteboom, J.; O'Briant, K.C.; Allen, A.; et al. Circulating microRNAs as stable blood-based markers for cancer detection. *Proc. Natl. Acad. Sci. USA* **2008**, *105*, 10513–10518. [CrossRef] [PubMed]
5. Pozniak, T.; Shcharbin, D.; Bryszewska, M. Circulating microRNAs in Medicine. *Int. J. Mol. Sci.* **2022**, *23*, 3996. [CrossRef] [PubMed]
6. Escuin, D.; Lopez-Vilaro, L.; Bell, O.; Mora, J.; Moral, A.; Perez, J.I.; Arqueros, C.; Ramon, Y.C.T.; Lerma, E.; Barnadas, A. MicroRNA-1291 Is Associated With Locoregional Metastases in Patients With Early-Stage Breast Cancer. *Front. Genet.* **2020**, *11*, 562114. [CrossRef] [PubMed]
7. Gravgaard, K.H.; Lyng, M.B.; Laenkholm, A.V.; Sokilde, R.; Nielsen, B.S.; Litman, T.; Ditzel, H.J. The miRNA-200 family and miRNA-9 exhibit differential expression in primary versus corresponding metastatic tissue in breast cancer. *Breast Cancer Res. Treat.* **2012**, *134*, 207–217. [CrossRef]
8. Escuin, D.; Lopez-Vilaro, L.; Mora, J.; Bell, O.; Moral, A.; Perez, I.; Arqueros, C.; Garcia-Valdecasas, B.; Ramon, Y.C.T.; Lerma, E.; et al. Circulating microRNAs in Early Breast Cancer Patients and Its Association With Lymph Node Metastases. *Front. Oncol.* **2021**, *11*, 627811. [CrossRef]
9. Cock, P.J.; Fields, C.J.; Goto, N.; Heuer, M.L.; Rice, P.M. The Sanger FASTQ file format for sequences with quality scores, and the Solexa/Illumina FASTQ variants. *Nucleic Acids Res.* **2010**, *38*, 1767–1771. [CrossRef]
10. Valastyan, S. Roles of microRNAs and other non-coding RNAs in breast cancer metastasis. *J. Mammary Gland. Biol. Neoplasia* **2012**, *17*, 23–32. [CrossRef]
11. Shek, L.L.; Godolphin, W. Model for breast cancer survival: Relative prognostic roles of axillary nodal status, TNM stage, estrogen receptor concentration, and tumor necrosis. *Cancer Res.* **1988**, *48*, 5565–5569. [PubMed]
12. Chan, M.; Liaw, C.S.; Ji, S.M.; Tan, H.H.; Wong, C.Y.; Thike, A.A.; Tan, P.H.; Ho, G.H.; Lee, A.S. Identification of circulating microRNA signatures for breast cancer detection. *Clin. Cancer Res.* **2013**, *19*, 4477–4487. [CrossRef] [PubMed]
13. Monzo, M.; Santasusagna, S.; Moreno, I.; Martinez, F.; Hernandez, R.; Munoz, C.; Castellano, J.J.; Moreno, J.; Navarro, A. Exosomal microRNAs isolated from plasma of mesenteric veins linked to liver metastases in resected patients with colon cancer. *Oncotarget* **2017**, *8*, 30859–30869. [CrossRef]

14. Xu, W.; Zhou, Y.; Xu, G.; Geng, B.; Cui, Q. Transcriptome analysis reveals non-identical microRNA profiles between arterial and venous plasma. *Oncotarget* **2017**, *8*, 28471–28480. [CrossRef] [PubMed]
15. Jin, L.; Li, M.; Wang, H.; Yin, Z.; Chen, L.; Zhou, Y.; Han, Y.; Cui, Q.; Zhou, Y.; Xue, L. Transcriptome analysis of arterial and venous circulating miRNAs during hypertension. *Sci. Rep.* **2021**, *11*, 3469. [CrossRef]
16. Santasusagna, S.; Moreno, I.; Navarro, A.; Martinez Rodenas, F.; Hernandez, R.; Castellano, J.J.; Munoz, C.; Monzo, M. Prognostic Impact of miR-200 Family Members in Plasma and Exosomes from Tumor-Draining versus Peripheral Veins of Colon Cancer Patients. *Oncology* **2018**, *95*, 309–318. [CrossRef]
17. Tamrazi, A.; Sundaresan, S.; Gulati, A.; Tan, F.J.; Wadhwa, V.; Bartlett, B.R.; Diaz, L.A.J. Endovascular image-guided sampling of tumor-draining veins provides an enriched source of oncological biomarkers. *Front. Oncol.* **2023**, *13*, 916196. [CrossRef]
18. Damascelli, B.; Ticha, V.; Repetti, E.; Dorji, T. Beyond Standard Practice in Liquid Biopsy: Selective Venous Sampling. *J. Vasc. Interv. Radiol.* **2021**, *32*, 668–671. [CrossRef]
19. Shi, Y.; Jia, Y.; Zhao, W.; Zhou, L.; Xie, X.; Tong, Z. Histone deacetylase inhibitors alter the expression of molecular markers in breast cancer cells via microRNAs. *Int. J. Mol. Med.* **2018**, *42*, 435–442. [CrossRef]
20. Cao, J.; Shao, H.; Hu, J.; Jin, R.; Feng, A.; Zhang, B.; Li, S.; Chen, T.; Jeungpanich, S.; Topatana, W.; et al. Identification of invasion-metastasis associated MiRNAs in gallbladder cancer by bioinformatics and experimental validation. *J. Transl. Med.* **2022**, *20*, 188. [CrossRef]
21. Cho, Y.E.; Vorn, R.; Chimenti, M.; Crouch, K.; Shaoshuai, C.; Narayanaswamy, J.; Harken, A.; Schmidt, R.; Gill, J.; Lee, H. Extracellular vesicle miRNAs in breast milk of obese mothers. *Front. Nutr.* **2022**, *9*, 976886. [CrossRef]
22. Gao, Y.; Lin, L.; Li, T.; Yang, J.; Wei, Y. The role of miRNA-223 in cancer: Function, diagnosis and therapy. *Gene* **2017**, *616*, 1–7. [CrossRef]
23. Yang, M.; Chen, J.; Su, F.; Yu, B.; Su, F.; Lin, L.; Liu, Y.; Huang, J.D.; Song, E. Microvesicles secreted by macrophages shuttle invasion-potentiating microRNAs into breast cancer cells. *Mol. Cancer* **2011**, *10*, 117. [CrossRef]
24. Huang, R.; Song, X.; Wang, C.M. MiR-223 regulates CDDP resistance in pancreatic cancer via targeting FoxO3a. *Eur. Rev. Med. Pharmacol. Sci.* **2019**, *23*, 7892–7898. [CrossRef]
25. Wei, Y.; Yang, J.; Yi, L.; Wang, Y.; Dong, Z.; Liu, Z.; Ou-yang, S.; Wu, H.; Zhong, Z.; Yin, Z.; et al. MiR-223-3p targeting SEPT6 promotes the biological behavior of prostate cancer. *Sci. Rep.* **2014**, *4*, 7546. [CrossRef] [PubMed]
26. Tsujimoto, M.; Nakabayashi, K.; Yoshidome, K.; Kaneko, T.; Iwase, T.; Akiyama, F.; Kato, Y.; Tsuda, H.; Ueda, S.; Sato, K.; et al. One-step nucleic acid amplification for intraoperative detection of lymph node metastasis in breast cancer patients. *Clin. Cancer Res.* **2007**, *13*, 4807–4816. [CrossRef] [PubMed]
27. Webber, C.; Gospodarowicz, M.; Sobin, L.H.; Wittekind, C.; Greene, F.L.; Mason, M.D.; Compton, C.; Brierley, J.; Groome, P.A. Improving the TNM classification: Findings from a 10-year continuous literature review. *Int. J. Cancer* **2014**, *135*, 371–378. [CrossRef] [PubMed]
28. Elston, C.W.; Ellis, I.O. Pathological prognostic factors in breast cancer. I. The value of histological grade in breast cancer: Experience from a large study with long-term follow-up. *Histopathology* **1991**, *19*, 403–410. [CrossRef] [PubMed]
29. Appierto, V.; Callari, M.; Cavadini, E.; Morelli, D.; Daidone, M.G.; Tiberio, P. A lipemia-independent NanoDrop((R))-based score to identify hemolysis in plasma and serum samples. *Bioanalysis* **2014**, *6*, 1215–1226. [CrossRef]
30. Martin, M. Cutadapt removes adapter sequences from high-throughput sequencing reads. *EMBnet. J.* **2011**, *17*, 3. [CrossRef]
31. Robinson, M.D.; Oshlack, A. A scaling normalization method for differential expression analysis of RNA-seq data. *Genome Biol.* **2010**, *11*, R25. [CrossRef]
32. R_Development_Core_Team. *R: A Language and Environment for Statistical Computing*; The R Foundation for Statistical Computing: Vienna, Austria, 2011; ISBN 3-900051-07-0. Available online: http://www.R-project.org/ (accessed on 6 August 2021).
33. Ru, Y.; Kechris, K.J.; Tabakoff, B.; Hoffman, P.; Radcliffe, R.A.; Bowler, R.; Mahaffey, S.; Rossi, S.; Calin, G.A.; Bemis, L.; et al. The multiMiR R package and database: Integration of microRNA-target interactions along with their disease and drug associations. *Nucleic Acids Res.* **2014**, *42*, e133. [CrossRef]
34. Gene Ontology, C. Gene Ontology Consortium: Going forward. *Nucleic Acids Res.* **2015**, *43*, D1049–D1056. [CrossRef] [PubMed]
35. Fabregat, A.; Jupe, S.; Matthews, L.; Sidiropoulos, K.; Gillespie, M.; Garapati, P.; Haw, R.; Jassal, B.; Korninger, F.; May, B.; et al. The Reactome Pathway Knowledgebase. *Nucleic Acids Res.* **2018**, *46*, D649–D655. [CrossRef] [PubMed]
36. Yu, G.; Wang, L.G.; Han, Y.; He, Q.Y. ClusterProfiler: An R package for comparing biological themes among gene clusters. *OMICS* **2012**, *16*, 284–287. [CrossRef]
37. Robinson, M.D.; McCarthy, D.J.; Smyth, G.K. edgeR: A Bioconductor package for differential expression analysis of digital gene expression data. *Bioinformatics* **2010**, *26*, 139–140. [CrossRef] [PubMed]
38. Benjamini, Y.; Hochberg, Y. Controlling the False Discovery Rate: A Practical and Powerful Approach to Multiple Testing. *J. R. Stat. Society. Ser. B* **1995**, *57*, 289–300. [CrossRef]

International Journal of *Molecular Sciences*

MDPI

Article

Circulating miR-206 and miR-1246 as Markers in the Early Diagnosis of Lung Cancer in Patients with Chronic Obstructive Pulmonary Disease

Elizabeth Córdoba-Lanús [1,2,3,*], Angélica Domínguez de-Barros [2], Alexis Oliva [2,4], Delia Mayato [5], Francisca Gonzalvo [5], Ana Remírez-Sanz [6], Javier J. Zulueta [7,8,9], Bartolomé Celli [10] and Ciro Casanova [1,5,11]

1 Department of Internal Medicine, Dermatology and Psychiatry, Universidad de La Laguna, 38071 San Cristóbal de La Laguna, Spain; casanovaciro@gmail.com
2 Instituto Universitario de Enfermedades Tropicales y Salud Pública de Canarias, Universidad de La Laguna, 38296 San Cristóbal de La Laguna, Spain; angelica4arealejos@gmail.com (A.D.d.-B.); amoliva@ull.edu.es (A.O.)
3 Centro de Investigación Biomédica en Red de Enfermedades Infecciosas (CIBERINFEC), Instituto de Salud Carlos III, 28029 Madrid, Spain
4 Department of Pharmaceutical Technology, Universidad de La Laguna, 38206 Santa Cruz de Tenerife, Spain
5 Pulmonary Department-Research Unit, Hospital Universitario Nuestra Señora de Candelaria, 38010 Santa Cruz de Tenerife, Spain; deliamayato@gmail.com (D.M.); franciscagonzalvo@hotmail.com (F.G.)
6 CIMA, Centro de Investigación Médica Aplicada, Universidad de Navarra, 31008 Pamplona, Spain; aremirez@unav.es
7 Navarra Institute for Health Research (IdISNA), 31008 Pamplona, Spain; jzulueta@unav.es
8 Pulmonary, Critical Care and Sleep Medicine Division, Mount Sinai Morningside Hospital, New York, NY 10029, USA
9 Centro de Investigación Biomédica en Red de Cáncer (CIBERONC), 28029 Madrid, Spain
10 Pulmonary Critical Care Medicine Division, Brigham and Women's Hospital, Harvard Medical School, Boston, MA 02115, USA; bcelli@copdnet.org
11 Centro de Investigación Biomédica en Red de Enfermedades Respiratorias (CIBERES), Instituto de Salud Carlos III, 28029 Madrid, Spain
* Correspondence: acordoba@ull.edu.es; Tel.: +34-922316502 (ext. 6357)

Abstract: Lung cancer (LC) is the most common cause of cancer death, with 75% of cases being diagnosed in late stages. This study aimed to determine potential miRNAs as biomarkers for the early detection of LC in chronic obstructive pulmonary disease (COPD) cases. Ninety-nine patients were included, with registered clinical and lung function parameters followed for 6 years. miRNAs were determined in 16 serum samples from COPD patients (four with LC and four controls) by next generation sequencing (NGS) at LC diagnosis and 3 years before. The validation by qPCR was performed in 33 COPD-LC patients and 66 controls at the two time points. Over 170 miRNAs ($\geq$10 TPM) were identified; among these, miR-224-5p, miR-206, miR-194-5p, and miR-1246 were significantly dysregulated ($p < 0.001$) in COPD-LC 3 years before LC diagnosis when compared to the controls. The validation showed that miR-1246 and miR-206 were differentially expressed in COPD patients who developed LC three years before ($p = 0.035$ and $p = 0.028$, respectively). The in silico enrichment analysis showed miR-1246 and miR-206 to be linked to gene mediators in various signaling pathways related to cancer. Our study demonstrated that miR-1246 and miR-206 have potential value as non-invasive biomarkers of early LC detection in COPD patients who could benefit from screening programs.

Keywords: lung cancer; COPD; circulating miRNAs; biomarkers

1. Introduction

Lung cancer is the second most prevalent cancer and the most common cause of cancer death [1], with more than 2.2 million new cases of lung cancer in 2020 and 1.80 million deaths globally.

Non-small-cell lung cancer (NSCLC), which accounts for 85% of lung cancers [2], has a poor prognosis (5-year overall survival rate < 15%), and the main contributing factor to this is that more than 75% of lung cancer cases are diagnosed in late stages because of the lack of screening programs for persons at risk [2]. Low-dose computed tomography (LDCT) is the lung cancer screening (LCS) tool most useful in detecting LC at its early stages [3]. Although it is effective in reducing mortality, it has some drawbacks such as the high rate of false positive results that drives the performance of invasive procedures and the overall high cost for the health system.

Early detection offers clear advantages in treating and monitoring the disease by improving cure rates, so alternative approaches to improve the identification of those cases are needed. Patients who suffer from chronic obstructive pulmonary disease (COPD), secondary to tobacco use, are at increased risk of developing LC, with an incidence that is 2–5 times that of smokers without COPD [4,5]. Genetics and epigenetics seem to play important roles as contributors to LC development [5–7]. Therefore, the finding of non-invasive, reliable biomarkers that can help increase the probability of earlier LC detection in individuals at increased risk, as COPD patients are, seems like a fertile area to explore.

At present, cancer-related biomarkers are not specific to tissues and have poor early diagnosis and prognostic value; therefore, they cannot be used as targeted therapy for cancers. MicroRNAs (miRNAs) have emerged as good potential biomarkers of early diagnosis, the monitoring of progression, and the treatment of many cancers [8]. miRNAs are short non-coding RNAs, 19 to 25 nucleotides in length, involved in virtually all cellular processes, including development, apoptosis, and cell cycle regulation [9,10]. Moreover, miRNAs have shown great stability in different specimens and tissues [11]. After their biogenesis, mature miRNAs interact with certain messenger RNAs (mRNAs) in the cytoplasm during post-transcriptional regulation in an induced RNA-silencing multiprotein complex (RISC) [11]. The dysregulation of miRNAs caused by either genetic or epigenetic mechanisms has been described in various diseases ranging from cancer [12] to autoimmune diseases [13], lung diseases [14], and viral infections [15]. In addition, it has been proposed that gene therapy using miRNAs might be used to block the progression of cancers [16] by inhibiting the overexpression of oncogenic miRNAs or replacing those that are effective on tumor-suppressive genes. Thus, miRNAs can serve as either biomarkers or therapeutic targets in cancer, such as NSCLC.

Several studies have shown that a miRNA profile can be used to discriminate subjects with NSCLC from controls [17–19]. However, there is scarce information about miRNAs' usefulness as early markers of LC in patients at risk, and, as far as we know, not one report in COPD patients exists. This study was performed in a well-phenotyped COPD cohort with a follow-up of more than 10 years, with the main objective of characterizing a miRNA profile capable of early detection of those patients developing LC during the years of observation.

2. Results

2.1. miRNA Screening Study

A total of eight individuals (COPD patients with LC, and COPD patients who did not develop LC during follow-up as controls) were included in the screening study with biological samples in two timelines of analysis (at the time of LC diagnosis and 3 years before). The cases and controls were male, age-matched (mean age of 64 years) and presented similar smoking habits. The main clinical characteristics of these individuals are shown in Tables S1 and S2. Most of the COPD patients (90%) had moderate airway obstruction. Smokers with COPD and LC had higher FEV_1 and lower K_{CO} values than COPD controls, but a similar proportion of emphysema was visualized by CT scan. LC patients' histological subtypes were non-small-cell lung cancer (NSCLC) (75% adenocarcinoma, 25% squamous carcinoma).

Over one hundred and seventy miRNAs ($\geq$10 TPM) were identified by next-generation sequencing (NGS) in the 16 samples analyzed (Figure 1). Between these, hsa-miR-224-5p was found significantly upregulated and hsa-miR-194-5p was found significantly downregulated in COPD patients with LC three years before LC diagnosis when compared to COPD cases that did not develop LC during follow-up (log2FC = 2.81 and log2FC = −4.24, respectively; $p < 0.0001$, FDR < 0.001) (Figure 2). hsa-miR-206 was found significantly dysregulated in COPD individuals 3 years before these patients had a LC diagnosis (log2FC = −5.25, $p < 0.001$), although the correction for multiple testing was FDR > 0.01 (Table 1).

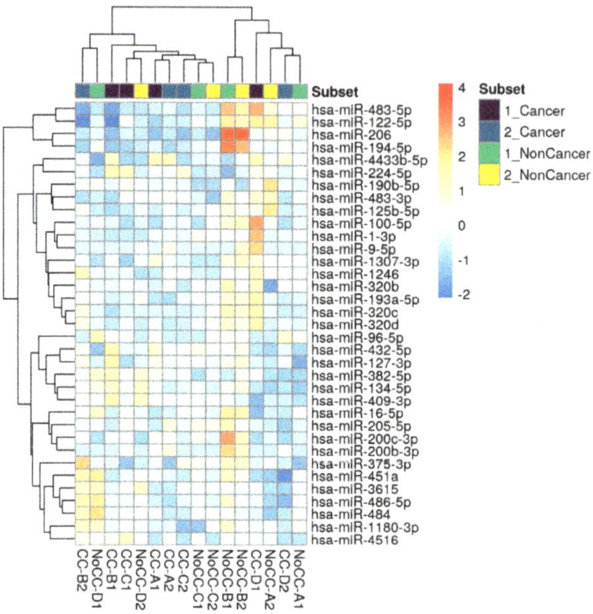

Figure 1. Top genes with the highest variance across samples were selected for hierarchical clustering. Each row represents one gene, and each column represents one sample. The color represents the difference of the count value to the row mean (−2 to 4).

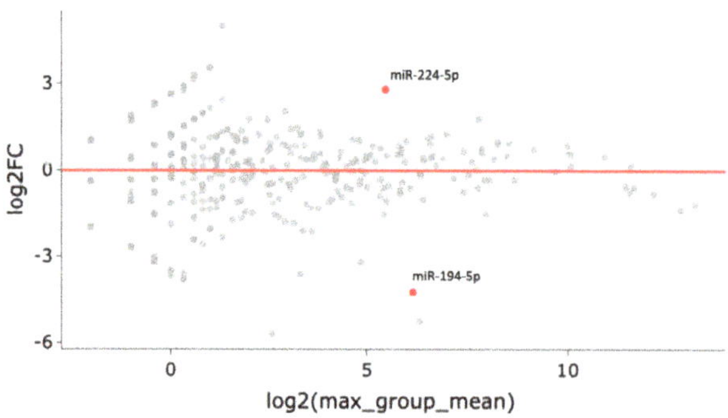

Figure 2. Each gene's fold change (FC) is plotted against its mean expression among all samples three years before LC diagnosis. All significantly differentially expressed genes are marked in red. Significant changes are defined as p-value < 0.001, FDR < 0.01, and Log2FC > 2.

At the time of LC diagnosis, patients with COPD who developed LC presented dysregulated hsa-miR-194-5p and hsa-miR-206 expression levels when compared to controls ($p = 0.0036$ and $p < 0.0001$, respectively), while hsa-miR-1246 maintained increased levels of expression ($p = 0.0032$, but FDR > 0.05) (Table 1).

Table 1. Main dysregulated circulating miRNAs from patients with COPD and LC vs. patients with COPD who did not develop LC during follow-up at two timelines (the time of LC diagnosis and 3 years before).

Name	Log2FC	*p*-Value	FDR
3 years before LC diagnosis			
hsa-miR-224-5p	2.81	**0.000003**	**0.0024**
hsa-miR-194-5p	−4.24	**0.000024**	**0.0082**
hsa-miR-206	−5.25	**0.000105**	**0.0241**
hsa-miR-26a-5p	−1.91	**0.001207**	0.2683
At LC diagnosis			
hsa-miR-206	−4.36	**0.000799**	**0.0553**
hsa-miR-194-5p	−2.89	**0.003618**	0.9977
hsa-miR-1246	2.25	**0.003178**	0.9977

Significant changes are defined as *p*-value < 0.001, FDR < 0.01, and Log2FC > 1.5. Abbreviations: FC, fold change; FDR, false discovery rate.

2.2. Validation Study

The validation of the miRNA expression analysis was performed on all 99 participants included in the study for the two timeline points: at the time of LC diagnosis (33 LC cases, 66 controls), and three years before (21 LC cases and 30 controls). The main clinical characteristics of cases and controls are shown in Tables 2 and 3. The patients were primarily men (85%), with a mean age of 63 years, and heavy smokers. The patients with COPD and LC had higher FEV_1 and lower K_{CO} values than the COPD controls, but a similar proportion of emphysema. The tumor histological subtypes present in the LC patients were mostly non-small-cell lung cancer (NSCLC) (61% adenocarcinoma, 27% squamous carcinoma, 3% microcytic carcinoma, and two cases of undifferentiated carcinoma).

Table 2. Baseline demographical and clinical characteristics of COPD patients who developed lung cancer (LC) at time of diagnosis and those who did not.

Variable	COPD with LC N = 33	COPD without LC N = 66	*p*-Value
Age *	63 ± 9	63 ± 9	-
Sex (male%)	85	85	-
BMI *	28 ± 5	28 ± 5	0.719
Smoking habit (pack-year) ‡	64 ± 24	61 ± 27	0.631
Active smoker (%)	50	70	0.218
FEV_1 (L) *	2.24 ± 0.74	1.67 ± 0.72	**0.001**
FEV_1 (% pred) *	79 ± 21	62 ± 26	**0.001**
FVC (% pred) *	105 ± 20	91 ± 24	**0.005**
FEV1/FVC (% pred) *	58 ± 11	52 ± 14	**0.018**
PaO_2 *§	72 ± 6	71 ± 12	0.872
K_{CO} *§	68 ± 25	84 ± 27	0.084
IC/TLC (%) *§	34 ± 8	34 ± 9	0.881
6MWD (mts) *§	486 ± 115	497 ± 101	0.747
Dyspnea mMRC **	0 (0–1)	1 (0–2)	0.242
BODE index **§	1 (0–2)	0 (0–2)	0.191
Charlson index **§	1 (1–1)	0 (0–1)	0.740
Emphysema (%) †§	67	57	0.424
Lung cancer stage (%)			
I	62	-	
II	9	-	
III	19	-	
IV	10	-	

* Data are presented as mean ± SD. ** Data are presented as median (25th–75thpc). ‡ Number of packs of cigarettes smoked per day × number of years of smoking. BMI: body mass index; FEV_1: forced expiratory volume in one second; FVC: forced vital capacity; % pred: percent predicted; PaO_2: partial oxygen tension; K_{CO}: transfer factor coefficient of the lung for carbon monoxide, which is DL_{CO}/VA; IC/TLC: inspiratory capacity to total lung capacity ratio; 6MWD: six-minute walking distance test. BODE index: body mass index, airflow obstruction, dyspnea, and exercise capacity. † Emphysema diagnosed by CT scan. § This measure was available for 11 individuals with COPD and LC. *p*-value < 0.05 (in bold).

Table 3. Demographical and clinical characteristics of smoker COPD patients who did and did not develop lung cancer (LC) during follow-up, three years before diagnosis.

Variable	COPD with LC N = 21	COPD without LC N = 30	p-Value
Age *	60 ± 9	60 ± 9	-
Sex (male%)	85	85	-
Smoking habit (pack-year) ‡	65 ± 20	67 ± 27	0.554
BMI	28 ± 5	28 ± 4	0.662
FEV_1 (L) *	2.20 ± 0.78	1.51 ± 0.56	**0.004**
FEV_1 (% pred) *	76 ± 22	54 ± 19	**0.007**
FVC (% pred) *	102 ± 21	88 ± 22	0.035
FEV1/FVC (% pred) *	58 ± 12	48 ± 13	**0.032**
PaO_2 *§	74 ± 5	71 ± 10	0.457
K_{CO} *§	72 ± 16	84 ± 26	**0.045**
IC/TLC (%) *§	36 ± 8	36 ± 9	0.961
6MWD (mts) *§	532 ± 59	508 ± 88	0.467
Dyspnea mMRC **	0 (0–12)	1 (0–2)	0.242
BODE index **§	1 (0–1)	1(0–2)	0.645
Charlson index **§	1 (1–2)	0(0–1)	0.058

* Data are presented as mean ± SD. ** Data are presented as median (25th–75thpc). ‡ Number of packs of cigarettes smoked per day × number of years smoking. BMI: body mass index; FEV_1: forced expiratory volume in one second; FVC: forced vital capacity; % pred: percent predicted; PaO_2: partial oxygen tension; K_{CO}: transfer factor coefficient of the lung for carbon monoxide, which is DL_{CO}/VA; IC/TLC: inspiratory capacity to total lung capacity ratio; 6MWD: six-minute walking distance test; BODE index: body mass index, airflow obstruction, dyspnea, and exercise capacity; § 11 COPD individuals with LC were analyzed for these variables.

In the validation experiments, miR-224 was expressed in less than 85% of the studied population, so it was excluded for further analysis. miR-1246 was overexpressed in patients with COPD who developed LC 3 years before LC diagnosis when compared to COPD controls who did not develop LC (log2FC = 2.63, $p = 0.032$) (Figure 3). A significant difference in miR-1246 expression was also found at the time of LC diagnosis (log2FC = −4.79, $p = 0.022$) (Figure S1). miR-206 was also found dysregulated between cases and controls 3 years before (log2FC = −2.205, $p = 0.027$) (Figure 3), and maintained at the time of LC diagnosis (log2FC = −1.624, $p = 0.023$) (Figure S1).

The expression of miR-194-5p did not differ significantly between the study groups in either of the two timelines.

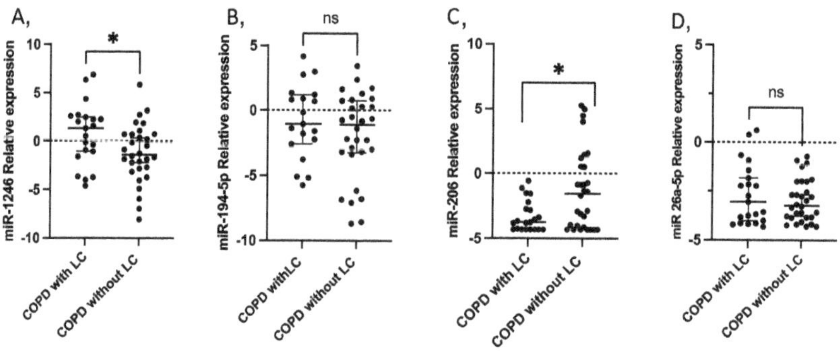

Figure 3. Circulating miRNA expression levels in serum samples from 21 COPD patients with LC three years before diagnosis compared with 30 COPD without LC as controls. (**A**) miR-1246 ($p = 0.035$); (**B**) miR-194-5p; (**C**) miR-206 ($p = 0.028$); (**D**) miR-26-5p. Lines represent the median with an interquartile range. * $p < 0.05$; ns: non-significant.

ROC curve analysis showed that miR-1246 and miR-206 expression in serum could discriminate between COPD controls and patients with COPD who will develop LC within the next three years (AUC: 0.68 (95% confidence interval (CI) = 0.62~0.85), $p = 0.035$; and AUC: 0.68 (95% CI = 0.63~0.83), $p = 0.030$, respectively) (Figure 4).

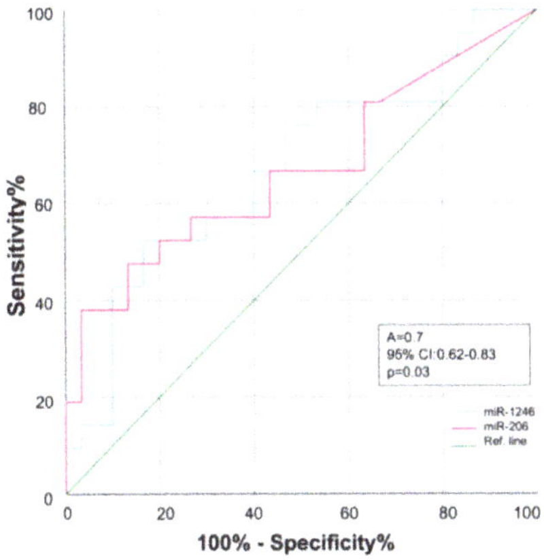

Figure 4. Receiver operating characteristics (ROC) curve analysis. Diagnostic performance of miR-1246 (blue line) and miR-206 (red line) for early discrimination of COPD patients who will develop lung cancer within the next three years. AUC: area under the curve; 95% CI: 95% confidence interval.

2.3. *Clinical Relations*

The levels of expression of miR-26a-5p and miR-194-5p in COPD patients who would go on to develop LC within the next 3 years correlated significantly with the levels of arterial oxygen PaO_2 ($r = 0.74$, $p = 0.035$, and $r = 0.97$, $p < 0.001$, respectively), although this measure was only available in a small sample of individuals (Figure S2). No relationship was found between any other miRNA expression in COPD individuals with LC and other clinical variables.

2.4. *Functional Annotation Analysis and Gene Target Prediction*

Our functional analysis, through the KEGG pathways, revealed miR-1246 to be significantly associated with target genes enriched in several functions and signaling pathways. The most significant ones are reported in Table 4A: the viral carcinogenesis, the apoptosis pathway, the TP53 signaling pathway, and the central carbon metabolism in cancer. On the other hand, miR-206 was shown to be significantly associated with the enriched target genes detailed in Table 4B: glycosphingolipid biosynthesis, miRNAs in cancer, proteoglycans in cancer, and transcriptional misregulation in cancer.

Table 4. Principal enriched KEGG pathways of predicted target genes of miR-1246 (**A**) and miR-206 (**B**).

(A)			
#	**KEGG Pathway**	**Genes**	***p*-Value ***
1	Viral carcinogenesis	PIK3CB, BAX, CDK6, TP53, KAT2b, CREB5, CCNE1	1.7095e-05
2	Oocyte meiosis	SLK, SMC1	0.0052
3	Apoptosis	CASP7, PIK3CB, BAX, BCL2, TP53, PPP3CA, CFLAR	0.0069
4	Thyroid hormone signaling pathway	MED14, PIK3CB, MED13, NOTCH2, TP53, MED1, KAT2B	0.0206
5	Central carbon metabolism in cancer	PIK2CB, TPB53, PKD1	0.0206

Table 4. *Cont.*

		(A)	
#	**KEGG Pathway**	**Genes**	***p*-Value ***
6	p53 signaling pathway	BAX, CDK6, TP53, TP53I3, CCNE1, CCNG2	0.0437
7	Adrenergic signaling in cardiomyocytes	PIK3CB, PPP2CA, BCL2, CALM2, ATP2B1, PPP2R1B	0.0437
8	Glioma	PIK3CB, CDK6, CALM2, TP53	0.0437
		(B)	
#	**KEGG Pathway**	**Genes**	***p*-Value ***
1	Glycosphingolipid biosynthesis-lacto and neolacto series	FUT3, FUT9	4.395×10^{8}
2	MicroRNAs in cancer	PCD4, MET, ROCK1, DICER1, CDC25C, CCND2, EGFR, NOTCH2, KRAS, CDK6, PRKCE, HDAC4, IKBKB, FOXP1, BCRA1, PIM1, TIMP3, NOTCH3, VMP1, MAPK1, GLS2, PDGFA	0.00101
3	Estrogen signaling pathway	ESR1, ASCY1, ATF2, CREB5, EGFR, KRAS, CALM2, SOS1, KCNJ6, IPTR3, MAPK1, PRKACB	0.0012
4	Gap junction	ADCY1, GRM5, DRD1, EGFR, KRAS, SOS1, GjA1, ITPR3, MAPK1, PRKACB, PDGFA	0.0012
5	Dorso-ventral axis formation	ETS1, CPEB1, EGFR, NOTCH2, KRAS, SOS1, NOTCH3, MAPK1	0.0043
6	Proteoglycans in cancer	ESR1, ACTB, PDC4, MET, ROCK1, CBL, FRS2, EGFR, KRAS, CBLC, TIMP3, SOS1, DDX5, IGF1, FN1, VMP1, ITPR3, MAPK1, PRKACB	0.0065
7	Glioma	CDK4, EGFR, KRAS, CDK6, CALM2, SOS1, IGF1, MAPK1, PDGFA	0.0121
8	Transcriptional misregulation in cancer	MET, PAX7, CCND2, PBX1, WT1, CDK4, H3F3A, DDX5, MAF, MAX, IGF1, PPARG, H3F3B, KMT2A, GOLPH3, MEIS1, PAX3, GRIA3, PDFGFA, COMMD3-BMI1	0.0226

Resulting pathways in relation to Tarbase v.7, TargetScan, and microT-CDS using miRPath v.3. KEGG, Kyoto Encyclopedia for Genes and Genomes. * $p < 0.05$.

The GO analysis of the collection of target genes of miR-1246 together with miR-206 was enriched in several functions, including the cellular nitrogen metabolic process, the biosynthetic process, the protein modification process, gene expression, the cellular component or protein assembly, membrane organization, and the mitotic cell cycle (Table S3).

3. Discussion

The presence of chronic obstructive pulmonary disease (COPD) is an important risk factor for developing lung cancer. Considerable effort is being made to identify biomarkers that, in combination with conventional tests, can help identify individuals more likely to develop LC and candidates for close monitoring and early diagnosis of LC. This study is the first to describe the altered expression of two circulating miRNAs, hsa-miR1246 and hsa-miRNA-206, that can be detected in the serum samples of patients with COPD up to three years before lung cancer diagnosis and could help identify subjects at high risk for LC development who are likely to benefit from screening programs.

3.1. miR-1246 and Predictive Risk of Lung Cancer

In our study, miR-1246 was found overexpressed in individuals with COPD and LC three years before diagnosis, and this differential expression was maintained at the time of diagnosis when compared to COPD controls who did not develop LC. Moreover, we could detect increased levels of this miRNA in patients who developed LC in relation to the presence of emphysema, a well-known risk factor of LC [6]. Our findings are in accordance with Yang et al., who, in a small group of patients, identified miR-1246 to be the most upregulated miRNA in the serum of patients with LC when compared to healthy controls [20]. However, this study was only conducted at the time of diagnosis, thus not informing its value as a toll for early screening for LC. Interestingly, these authors also reported that this miRNA increased the migration and invasiveness of A549 lung cancer cells. Specifically, they observed that E-cadherin expression decreased and vimentin and transforming growth factor-β expression increased, indicating the possible role of miR-1246 in the Wnt/-βcatenin pathway and its relation to cancer progression and metastasis. Similarly, in the study by Kim et al., miR-1246 was found to be overexpressed in sphere-forming cells, and an anti-miR-1246 strategy was effectively shown to suppress the proliferation, sphere-formation, colony-forming ability, and invasiveness of cancerous cells [21]. In addition, other authors have documented a role for miR-1246 in the progression of non-small-cell lung cancer [22].

Our study's functional in silico analysis suggests that miR-1246 may be linked to gene mediators in various signaling pathways related to cancer. When exploring the KEGG pathway analysis in relation to the miR-1246 target genes, it was enriched, for example, in the P53 signaling pathway. Also, miR-1246 has been identified as a novel p53 target miRNA, as shown by the study of Liao et al., who reported p53 to inhibit DYRK1A expression through the induction of miR-1246 in non-small-cell lung cancer cells [23]. Several studies indicate that P53, a typical tumor suppressor gene, is involved in NSCLC development and progression [24]. BCL2L2 is a predicted target of miR-1246. It this context, increases in Bcl-2 protein expression have been shown to contribute to the development of a wide variety of human cancers, including lung cancer, and may act as a resistance factor against several anticancer agents [25]. Another predicted target of miR-1246 is HER4 (ERBB4,) a tyrosine kinase receptor. EGFR is a plasma membrane glycoprotein that belongs to this family of receptors. EGFR dimerization may result in cancer cell proliferation, the inhibition of apoptosis, invasion, metastasis, and tumor-induced neovascularization [26].

3.2. miRNA-206 and Predictive Risk of Lung Cancer

Another important finding was that miR-206 dysregulation can be detected early in patients with COPD who will develop lung cancer three years before diagnosis. The low expression of miR-206 is related to lung cancer invasion and metastasis [27,28]. miR-1 expression, which is related to miR-206, was found to be decreased in lung cancers [29]. It was reported that the treatment with HDAC as an inhibitor of lung cancer cells could induce the expression of repressed miR-1 by the downregulation of oncogenes such as *MET*, *PIM1*, *FOXP1*, and *HDAC4* [29]. *MET* is a common target in three of the pathways detected for miR-206 in the in silico analysis of the present study: miRNAs in cancer, proteoglycans in cancer, and transcriptional misregulation in cancer.

Interestingly, Yang et al. found that forced miR-206 expression restored gefitinib sensitivity in IL6-induced gefitinib-resistant EGFR-mutant lung cancer cells by inhibiting the IL6/JAK1/STAT3 pathway [30]. Furthermore, studies performed on NSCLC in A549 cells showed that the miR-206 antagomir therapy decreased tumor volume and the formation of intra-tumoral capillary tubes, and increased apoptosis by blocking the 14-3-3zeta/STAT3/HIF-1alpha/VEGF signaling pathway [31]. miR-206 has been reported to function as a tumor metastasis suppressor in different type of cancers, including colon and gastric cancers [32,33].

3.3. miRNAs and Their Relation to Pulmonary Function

We found a strong correlation between miR-26a-5p and miR-194-5p expression and partial oxygen arterial tension (PaO_2) in COPD patients who went on to develop LC within the next 3 years. Other studies have reported increased levels of miR-26a in COPD patients with chronic hypoxia [34] and in relation to the apnea process [35]. These findings highlight the importance of the hypoxemic condition in stimulating the expression of certain miRNAs. It is well known that within the mechanisms relating epigenetics and oxidative stress is the transcription factor hypoxia-inducible factor 1 α (HIF-1α) [36]. Hypoxia, or a hypoxic microenvironment, triggers the angiogenesis process, and the cellular response to hypoxia is primarily regulated by HIF-1α. Since miR-26a-5p is released from endothelial cells, its increased expression supports a relation between endothelial dysfunction and the development and progression of cancer within this group of COPD patients at increased LC risk.

The present study has some limitations. First, the present findings refer to circulating biomarkers, not to markers in lung tissue biopsies from patients, as they were not available. However, the main objective was to test the predictive value of these biological tools in non-invasive samples. Circulating miRNAs have been widely used as biomarkers due to their resistance to degradation and ubiquity. Nevertheless, further studies are needed to confirm the co-expression of the proposed miRNAs and their target mRNAs in addition to functional in vitro experiments for miRNA target regulation. Second, the number of subjects with COPD and LC included in the study may be considered small. However, this is a prospective study, where a large and very well-characterized cohort of COPD patients was included, with repeated biological samples and a long follow-up time until the development and diagnosis of LC. This fact helped obtain informative results, laying the foundations to increase the related knowledge of the potential use of miRNA-based approaches in lung cancers. Finally, the AUC values to inform the potential lung cancer risk was 0.7 for both miRNAs. This is not excessively high but could be potentiated by the addition of traditional screening tests such as chest CT or by other potentially early diagnostic markers yet to be defined. However, we do recognize that a validation of the present findings must be performed on another large prospective cohort of LC patients with an analysis of samples at several time points.

In conclusion, our study demonstrated that miR-1246 and miR-206 have potential value as non-invasive biomarkers of early lung cancer detection in COPD patients. A confirmational study on a larger prospective cohort is needed to test this signature, evaluating its use in screening and early lung cancer management.

4. Materials and Methods

4.1. Study Individuals

Ninety-nine individuals with a diagnosis of COPD recruited from the Hospital Universitario N/S de Candelaria, Tenerife, Spain, were included in this study. They are part of a cohort of 263 smokers, recruited and followed annually at this hospital as part of the BODE cohort [37], and a cohort of 3825 individuals from a lung cancer screening program (Pamplona-International early detection program, P-IELCAP) [38]. Of these patients, 13 individuals in Tenerife and 20 in the Pamplona cohort developed LC during follow-up and were enrolled in the study. Of the 33 patients with LC, 21 had two blood samples obtained: one at LC diagnosis and the other 3 years before. From the pool of COPD patients who did not develop LC over follow-up, 66 were age- and gender-matched to be used as controls. All subjects were followed annually for a mean of 72 months.

Inclusion criteria: age > 40 years, smoking history >15 pack-years, post-bronchodilator FEV1/FVC ratio < 0.70, and clinically stable for at least 6 weeks at the time of evaluation. Spirometry, lung volume, diffusion capacity for carbon monoxide, and exercise capacity were measured according to ATS-ERS guidelines [39,40]. Dyspnea was evaluated using the mMRC scale [41], the BODE index was calculated as previously described [42], and co-morbidities were quantified using the Charlson index [43]. A pulmonologist visually scored the baseline for emphysema presence using criteria established by the Fleischner Society [44]. All-cause mortality was recorded. Exclusion criteria included any other respiratory diseases and uncontrolled comorbidities such as malignancy at baseline. None of the patients had received any anti-tumor treatments at the time of sample collection.

The study was approved by the ethical committee board of Hospital Universitario N/S de Candelaria (PI 55/17, CHUNSC-2018-29). Written informed consent was obtained from all participants. The research was conducted in accordance with the Declaration of Helsinki.

4.2. Sample Collection

Serum samples (a total of 162) were collected from 33 COPD patients who developed LC during follow-up and 66 COPD controls without LC at the time of LC diagnosis. Of these, 21 patients had serum samples obtained 3 years before LC diagnosis that were matched with 30 controls with samples at the same timeline. The serum samples were separated from whole blood within 1 h after collection and stored at −80 °C until further use in the genetic study.

4.3. NGS miRNA Screening Assay

Sixteen serum samples corresponding to eight individuals (four COPD patients with LC and four COPD patients who did not develop LC during follow-up as controls) with biological samples in two timelines of analysis (at the time of LC diagnosis and 3 years before) were analyzed for the miRNA screening. RNA was isolated from 200 µL of serum using the miRNeasy Serum/Plasma Advanced Kit (Qiagen). The library preparation was carried out using the QIAseq miRNA Library Kit (QIAGEN). A total of 5µL total RNA was converted into miRNA NGS libraries. miRNAs expression screening was performed by next-generation sequencing (NGS) in an Illumina NextSeq500 platform (Exiqon A/S, Copenhagen, Denmark). The numbers of known miRNAs were calculated by counting data to relevant entries in miRBase v20 software (http://mirbase.org). The miRNA expression was expressed as Tags Per Million (TPM, the number of reads for a particular miRNA). miRNAs that were stably expressed across all samples were identified using NormFinder software [45].

4.4. Validation Assay: Quantitative RT-PCR for miRNA Expression

The top abundant miRNAs that resulted from the NGS screening step were validated by qRT-PCR in all study participants at two moments: at the time of LC diagnosis and 3 years before (a total of 162 serum samples). RNA was isolated from serum samples, and cDNA was synthesized by retro transcription using miRCURY LNA RT kit (Qiagen Inc., Hilden, Germany). To guarantee isolation control, the spike-ins UniSp2, UniSp4, and UniSp5 were used. In cDNA synthesis and amplification, the conjunctions of cel-miR-39-3p and UniSp6 were used as controls, as indicated by the manufacturers (miRCURY LNA RT Kit and RNA Spike-in Kit, Qiagen Inc., Hilden, Germany). The expression of the resulting significant miRNAs from the NGS screening study at the first stage was determined by an RT-PCR analysis using miRCURY LNA Sybr Green PCR Master Mix (Qiagen Inc., Hilden, Germany). A final volume of 10 µL per reaction contained 1X Sybr Green Master Mix, 200 nmol/L specific primer set (miRCURY LNA miRNA PCR assay, Qiagen Inc., Hilden, Germany), and 3 ng cDNA. All samples were performed in triplicate. The experiments were performed on a Step One Plus real-time PCR system (Applied Biosystem, Foster City, CA, USA) under the following conditions: 95 °C 15min, followed by 40 cycles of 94 °C for 15 s, 55 °C 30 s, and 72 °C 30 s. hsa-miR-486-5p and hsa-let7a-5p, which were the best candidates in the NGS screening analysis, were used as reference genes for normalization control. A non-template control was carried out in each experiment. The relative expression analysis of the target miRNAs was performed using the comparative threshold method 2−ΔΔCt [46].

4.5. In Silico Analysis

We integrated the miRNAs has-miR-206 and has-miR-1246 to search for potential targets in three software programs: miRBase v20 (available online: http://www.mirbase.org/), TargetScan v.7.2 (available online: http://www.targetscan.org/vert_72/), and the DIANA miRPath-v3.0 (available online: http://www.microrna.gr/miRPathv3). In the next step, the g:Profiler software (available online: https://biit.cs.ut.ee/gprofiler/gost) was used to contrast the resulting genes proposed as targets of has-miR-206 and has-miR-1246.

Functional and enrichment analyses were performed using the software's Kyoto Encyclopedia of Genes and Genomes (KEGG) (available online: https://www.kegg.jp/) and Gene-ontology (GO) terms [47,48] through the analysis with miRTar Human software (available online: https://mirtarbase.cuhk.edu.cn/~miRTarBase/miRTarBase_2022/php/index.php).

4.6. Statistical Analysis

The sample was characterized using summary statistics: relative frequency of each category, 50th percentiles (5–95th) and non-scale normal, and mean ± SD, as appropriate. The comparisons between cases and controls were carried out using Pearson's chi-square test, Mann–Whitney's U, Wilcoxon, Fisher Exact, and Kruskal–Wallis. The correlations between variables were estimated using the Spearman or Pearson tests. miRNA expression analysis was performed at two time points: at LC diagnosis and 3 years before (median follow-up 36 months). Cases and controls were matched for age and gender.

The R statistical software package [49] and DeSeq2 software [50] were used to perform the miRNA differential expression analysis. TMM normalization of the samples was used, consisting of the trimmed mean of the M-values method based on log-fold and absolute gene-wise changes in the expression levels between samples. These fold change (FC) values were transformed to a Log scale for normalization. A miRNA was considered candidate for validation when Log2FC > 1.5. The Benjamini–Hochberg false discovery rate (FDR) algorithm was used to correct the p-values for multiple testing [50], where FDR is defined as the expected fraction of false rejections among those hypotheses rejected.

In the multivariate logistic regression analysis, the following covariates were included: smoking status (pack-years of smoking), BMI, and the presence of emphysema (visual score by CT scan). The sensitivity and specificity of each miRNA for the early identification of those patients who developed lung cancer were determined by receiver operator characteristic (ROC) curves. The area under the ROC curve (AUC) of each test was calculated for direct comparison.

SPSS 25.0 (IBM Co) and R software were used for all statistical analyses and two-tailed p-values < 0.05 were considered significant.

Supplementary Materials: The following supporting information can be downloaded at: https://www.mdpi.com/article/10.3390/ijms241512437/s1.

Author Contributions: E.C.-L. and C.C. participated in the conception and design, analysis and interpretation, and drafting of the manuscript for important intellectual content. They take responsibility for the integrity of the data and the accuracy of the data analysis. A.D.d.-B., A.O. and J.J.Z. participated in the data analysis and interpretation. A.D.d.-B., D.M., F.G. and A.R.-S. participated in the data acquisition. B.C. participated in the interpretation of the results and helped draft the manuscript for important intellectual content. All authors have read and agreed to the published version of the manuscript.

Funding: This research was funded by the Agustín de Betancourt Program, Cabildo de Tenerife, (Tenerife 2030, FDCAN, MEDI) and CIBERINFEC contract (CB21/13/00100), Instituto de Salud Carlos III, Spain.

Institutional Review Board Statement: The study was conducted in accordance with the Declaration of Helsinki and approved by the Ethics Committee of Hospital Universitario La Candelaria, Tenerife, Spain (PI55/17, CHUNSC-2018-29).

Informed Consent Statement: Informed consent was obtained from all subjects involved in the study.

Data Availability Statement: The datasets supporting the conclusions of this article are included within the article and its additional files.

Acknowledgments: We would like to thank Luis Montuenga from the Department of Pathology, Anatomy and Physiology, School of Medicine and Sciences, and CIMA, University of Navarra, Pamplona, for providing the patient samples. We also would like to thank the IUETSPC–Universidad de La Laguna for the usage of infrastructure equipment and facilities, and particularly Jacob Lorenzo-Morales for the scientific support.

Conflicts of Interest: Elizabeth Córdoba-Lanús, Angélica Domínguez de-Barros, Alexis Oliva, Delia Mayato, Francisca Gonzalvo, Ana Remírez-Saenz, and Javier J. Zulueta declare no financial or personal conflicts of interests. Ciro Casanova Macario has received speaker or consulting fees from AstraZeneca, Bial, Boehringer Ingelheim, Chiesi, GlaxoSmithKline, Menarini, and Novartis, and research grants from GlaxoSmithKline, Menarini, and AstraZeneca in the last 3 years. Bartolomé Celli declares having received compensation for advisory boards and consultation fees from GlaxoSmithKline, Boehringer-Ingelheim, AstraZeneca, Novartis, Pulmonx, Sanofi-Aventis, CHIESI, Menarini, and Gala Therapeutics, and declares having no shares or interest in any company. Neither have received or had any relationship with tobacco money.

References

1. WHO. Leading Causes of Death Globally. 2019. Available online: https://www.who.int/news-room/fact-sheets/detail/the-top-10-causes-of-death (accessed on 18 February 2023).
2. Celli, B. Chronic obstructive pulmonary disease and lung cancer: Common pathogenesis, shared clinical challenges. *Proc. Am. Thorac. Soc.* **2012**, *9*, 74–79. [CrossRef] [PubMed]
3. Criner, G.J.; Agusti, A.; Borghaei, H.; Friedberg, J.; Martinez, F.J.; Miyamoto, C.; Vogelmeier, C.F.; Celli, B.R. Chronic Obstructive Pulmonary Disease and Lung Cancer: A Review for Clinicians. *Chronic. Obstr. Pulm. Dis.* **2022**, *9*, 454–476. [CrossRef]
4. GOLD. Global Strategy for the Diagnosis, Management, and Prevention of Chronic Obstructive Pulmonary Disease—2022 Report. 2022. Available online: https://goldcopd.org/wp-content/uploads/2021/11/GOLD-REPORT-2022-v1.0-12Nov2021_WMV.pdf (accessed on 5 March 2023).
5. Álvarez, F.V.; Trueba, I.M.; Sanchis, J.B.; López-Rodó, L.M.; Suárez, P.M.R.; Escuín, J.S.d.C.; Barreiro, E.; Pintado, M.H.B.; Vicente, C.D.; Aldeyturriaga, J.F.; et al. Executive summary of the SEPAR recommendations for the diagnosis and treatment of non-small cell lung cancer. *Arch. Bronconeumol.* **2016**, *52*, 378–388. [CrossRef]
6. De Torres, J.P.; Marin, J.M.; Casanova, C.; Santiago, C.; Elizabeth, C.L.; Rebeca, B.D.; Javier, Z.; Armando, A.J.; Marina, S.; Manuel, C.; et al. Lung cancer in patients with chronic obstructive pulmonary disease—Incidence and predicting factors. *Am. J. Respir. Crit. Care Med.* **2011**, *184*, 913–919. [CrossRef]
7. Parris, B.A.; O'Farrell, H.E.; Fong, K.M.; Yang, I.A. Chronic obstructive pulmonary disease (COPD) and lung cancer: Common pathways for pathogenesis. *J. Thorac. Dis.* **2019**, *11* (Suppl. 17), S2155–S2172. [CrossRef]
8. Mitchell, P.S.; Parkin, R.K.; Kroh, E.M.; Tewari, M. Circulating microRNAs as stable blood-based markers for cancer detection. *Proc. Natl. Acad. Sci. USA* **2008**, *105*, 10513–10518. [CrossRef] [PubMed]
9. Wu, L.; Fan, J.; Belasco, J.G. MicroRNAs direct rapid deadenylation of mRNA. *Proc. Natl. Acad. Sci. USA* **2006**, *103*, 4034–4039. [CrossRef] [PubMed]
10. Meister, G. miRNAs get an early start on translational silencing. *Cell* **2007**, *131*, 25–28. [CrossRef]
11. Cortez, M.A.; Bueso-Ramos, C.; Ferdin, J.; Lopez-Berestein, G.; Sood, A.K.; Calin, G.A. MicroRNAs in body fluids—The mix of hormones and biomarkers. *Nat. Rev. Clin. Oncol.* **2011**, *8*, 467–477. [CrossRef]
12. Croce, C.M. Causes and consequences of microRNA dysregulation in cancer. *Nat. Rev. Genet.* **2009**, *10*, 704–714. [CrossRef]
13. Dai, R.; Ahmed, S.A. MicroRNA, a new paradigm for understanding immunoregulation, inflammation, and autoimmune diseases. *Transl. Res.* **2011**, *157*, 163–179. [CrossRef] [PubMed]
14. Rupani, H.; Sanchez-Elsner, T.; Howarth, P. MicroRNAs and respiratory diseases. *Eur. Respir. J.* **2013**, *41*, 695–705. [CrossRef] [PubMed]
15. Huang, S.-Y.; Huang, C.-H.; Chen, C.-J.; Chen, T.-W.; Lin, C.-Y.; Lin, Y.-T.; Kuo, S.-M.; Huang, C.-G.; Lee, L.-A.; Chen, Y.-H.; et al. Novel Role for miR-1290 in Host Species Specificity of Influenza A Virus. *Mol. Ther. Nucleic Acids* **2019**, *17*, 10–23. [CrossRef] [PubMed]
16. Esquela-Kerscher, A.; Slack, F.J. Oncomirs—microRNAs with a role in cancer. *Nat. Rev. Cancer* **2006**, *6*, 259–269. [CrossRef]

17. Wang, K.; Chen, M.; Wu, W. Analysis of microRNA (miRNA) expression profiles reveals 11 key biomarkers associated with non-small cell lung cancer. *World J. Surg. Oncol.* **2017**, *15*, 175. [CrossRef]
18. Cai, J.; Wu, J.; Zhang, H.; Fang, L.; Huang, Y.; Yang, Y.; Zhu, X.; Li, R.; Li, F. miR-186 downregulation correlates with poor survival in lung adenocarcinoma, where it interferes with cell-cycle regulation. *Cancer Res.* **2013**, *73*, 756–766. [CrossRef]
19. Huang, P.; Ye, B.; Yang, Y.; Shi, J.; Zhao, H. MicroRNA-181 functions as a tumor suppressor in non-small cell lung cancer (NSCLC) by targeting Bcl-2. *Tumour Biol.* **2015**, *36*, 3381–3387. [CrossRef]
20. Yang, F.; Xiong, H.; Duan, L.; Li, Q.; Li, X.; Zhou, Y. MiR-1246 promotes metastasis and invasion of A549 cells by targeting GSK-3β-mediated Wnt/β-catenin pathway. *Cancer Res. Treat.* **2019**, *51*, 1420–1429. [CrossRef]
21. Kim, G.; An, H.-J.; Lee, M.-J.; Song, J.-Y.; Jeong, J.-Y.; Lee, J.-H.; Jeong, H.-C. Hsa-miR-1246 and hsa-miR-1290 are associated with stemness and invasiveness of non-small cell lung cancer. *Lung Cancer* **2016**, *91*, 15–22. [CrossRef]
22. Zhang, W.C.; Chin, T.M.; Yang, H.; Nga, M.E.; Lunny, D.P.; Lim, E.K.H.; Sun, L.L.; Pang, Y.H.; Leow, Y.N.; Malusay, S.R.Y.; et al. Tumour-initiating cell-specific miR-1246 and miR-1290 expression converge to promote non-small cell lung cancer progression. *Nat. Commun.* **2016**, *7*, 11702. [CrossRef]
23. Liao, J.M.; Zhou, X.; Zhang, Y.; Lu, H. miR-1246, a new link of the p53 family with cancer and Down syndrome. *Cell Cycle* **2012**, *11*, 2624–2630. [CrossRef]
24. Javid, J.; Mir, R.; Julka, P.K.; Ray, P.C.; Saxena, A. Association of p53 and mdm2 in the development and progression of non-small cell lung cancer. *Tumour Biol.* **2015**, *36*, 5425–5432. [CrossRef]
25. Othman, N.; Nagoor, N.H. The role of microRNAs in the regulation of apoptosis in lung cancer and its application in cancer treatment. *BioMed Res. Int.* **2014**, *2014*, 318030. [CrossRef]
26. Ciardiello, F.; Tortora, G. EGFR antagonists in cancer treatment. *N. Engl. J. Med.* **2008**, *358*, 1160–1174; Erratum in *N. Engl. J. Med.* **2009**, *360*, 1579. [CrossRef]
27. Chen, Q.-Y.; Jiao, D.-M.; Yan, L.I.; Wu, Y.; Hu, H.; Song, J.; Yan, J.; Wu, L.; Xu, L.; Shi, J. Comprehensive gene and microRNA expression profiling reveals miR-206 inhibits MET in lung cancer metastasis. *Mol. Biosyst.* **2015**, *11*, 2290–2302. [CrossRef]
28. Wang, X.; Ling, C.; Bai, Y.; Zhao, J. MicroRNA-206 is associated with invasion and metastasis of lung cancer. *Anat. Rec.* **2011**, *294*, 88–92. [CrossRef]
29. Nasser, M.W.; Datta, J.; Nuovo, G.; Kutay, H.; Motiwala, T.; Majumder, S.; Wang, B.; Suster, S.; Jacob, S.T.; Ghoshal, K. Down-regulation of micro-RNA-1 (miR-1) in lung cancer. Suppression of tumorigenic property of lung cancer cells and their sensitization to doxorubicin-induced apoptosis by miR-1. *J. Biol. Chem.* **2008**, *283*, 33394–33405. [CrossRef] [PubMed]
30. Yang, Y.; Wang, W.; Chang, H.; Han, Z.; Yu, X.; Zhang, T. Reciprocal regulation of miR-206 and IL-6/STAT3 pathway mediates IL6-induced gefitinib resistance in EGFR-mutant lung cancer cells. *J. Cell Mol. Med.* **2019**, *23*, 7331–7341. [CrossRef] [PubMed]
31. Xue, D.; Yang, Y.; Liu, Y.; Wang, P.; Dai, Y.; Liu, Q.; Chen, L.; Shen, J.; Ju, H.; Li, Y.; et al. MicroRNA-206 attenuates the growth and angiogenesis in non-small cell lung cancer cells by blocking the 14-3-3zeta/STAT3/HIF-1alpha/VEGF signaling. *Oncotarget* **2016**, *7*, 79805–79813. [CrossRef] [PubMed]
32. Sun, P.; Sun, D.; Wang, X.; Liu, T.; Ma, Z.; Duan, L. MiR-206 is an independent prognostic factor and inhibits tumor invasion and migration in colorectal cancer. *Cancer Biomark.* **2015**, *15*, 391–396. [CrossRef]
33. Ren, J.; Huang, H.J.; Gong, Y.; Yue, S.; Tang, L.-M.; Cheng, S.Y. MicroRNA-206 suppresses gastric cancer cell growth and metastasis. *Cell Biosci.* **2014**, *4*, 26. [CrossRef]
34. Lacedonia, D.; Scioscia, G.; Pia Palladino, G.; Gallo, C.; Carpagnano, G.E.; Sabato, R.; Barbaro, M.P.F. MicroRNA expression profile during different conditions of hypoxia. *Oncotarget* **2018**, *9*, 35114–35122. [CrossRef]
35. Eichhorn, L.; Dolscheid-Pommerich, R.; Erdfelder, F.; Ayub, M.A.; Schmitz, T.; Werner, N.; Jansen, F. Sustained apnea induces endothelial activation. *Clin. Cardiol.* **2017**, *40*, 704–709. [CrossRef] [PubMed]
36. Li, Y.; Seto, E. HDACs and HDAC Inhibitors in Cancer Development and Therapy. *Cold Spring Harb. Perspect. Med.* **2016**, *6*, a026831. [CrossRef] [PubMed]
37. Casanova, C.; de Torres, J.P.; Aguirre-Jaíme, A.; Pinto-Plata, V.; Marin, J.M.; Cordoba, E.; Baz, R.; Cote, C.; Celli, B.R. The progression of chronic obstructive pulmonary disease is heterogeneous: The experience of the BODE cohort. *Am. J. Respir. Crit. Care Med.* **2011**, *184*, 1015–1021. [CrossRef] [PubMed]
38. Mesa-Guzmán, M.; González, J.; Alcaide, A.B.; Bertó, J.; de-Torres, J.P.; Campo, A.; Seijo, L.M.; Ocón, M.M.; Pueyo, J.C.; Bastarrika, G.; et al. Surgical Outcomes in a Lung Cancer-Screening Program Using Low Dose Computed Tomography. *Arch. Bronconeumol. (Engl. Ed.)* **2021**, *57*, 101–106.
39. American Thoracic Society Statement. Lung function testing: Selection of reference values and interpretative strategies. *Am. Rev. Respir. Dis.* **1991**, *144*, 1202–1218. [CrossRef]
40. Macintyre, N.; Crapo, R.O.; Viegi, G.; Johnson, D.C.; van der Grinten, C.P.M.; Brusasco, V.; Burgos, F.; Casaburi, R.; Coates, A.; Enright, P.; et al. Standardisation of the single-breath determination of carbon monoxide uptake in the lung. *Eur. Respir. J.* **2005**, *26*, 720–735. [CrossRef]
41. Mahler, D.A.; Wells, C.K. Evaluation of clinical methods for rating dyspnea. *Chest* **1988**, *93*, 580–586. [CrossRef]
42. Celli, B.R.; Cote, C.G.; Marin, J.M. The body-mass index, airflow obstruction, dyspnea, and exercise capacity index in chronic obstructive pulmonary disease. *N. Engl. J. Med.* **2004**, *350*, 1005–1012. [CrossRef]
43. Charlson, M.; Szatrowski, T.P.; Peterson, J.; Gold, J. Validation of a combined comorbidity index. *J. Clin. Epidemiol.* **1994**, *47*, 1245–1251. [CrossRef]

44. Lynch, D.A.; Austin, J.H.; Hogg, J.C.; Grenier, P.A.; Kauczor, H.U.; Bankier, A.A.; Barr, R.G.; Colby, T.V.; Galvin, J.R.; Gevenois, P.A.; et al. CT-Definable Subtypes of Chronic Obstructive Pulmonary Disease: A Statement of the Fleischner Society. *Radiology* **2015**, *277*, 192–205. [CrossRef] [PubMed]
45. Andersen, C.L.; Jensen, J.L.; Ørntoft, T.F. Normalization of real-time quantitative reverse transcription-PCR data: A model-based variance estimation approach to identify genes suited for normalization, applied to bladder and colon cancer data sets. *Cancer Res.* **2004**, *64*, 5245–5250. [CrossRef]
46. Livak, K.J.; Schmittgen, T.D. Analysis of relative gene expression data using real-time quantitative PCR and the 2(-Delta Delta C(T)) method. *Methods* **2001**, *25*, 402–408. [CrossRef] [PubMed]
47. Ashburner, M.; Ball, C.A.; Blake, J.A.; Botstein, D.; Butler, H.; Cherry, J.M.; Davis, A.P.; Dolinski, K.; Dwight, S.S.; Eppig, J.T.; et al. Gene ontology: Tool for the unification of biology. The gene ontology consortium. *Nat. Genet.* **2000**, *25*, 25–29. [CrossRef]
48. Mi, H.; Muruganujan, A.; Ebert, D.; Huang, X.; Thomas, P.D. PANTHER version 14, more genomes, a new PANTHER GO-slim and improvements in enrichment analysis tools. *Nucleic Acids Res.* **2019**, *47*, D419–D426. [CrossRef] [PubMed]
49. Team R Core and Others. R: A Language and Environment for Statistical Computing. 2013. Available online: https://www.R-project.org/ (accessed on 26 October 2022).
50. Love, M.I.; Huber, W.; Anders, S. Moderated estimation of fold change and dispersion for RNA-seq data with DESeq2. *Genome Biol.* **2014**, *15*, 550. [CrossRef]

International Journal of *Molecular Sciences*

MDPI

Article

Obese Asthma Phenotype Is Associated with hsa-miR-26a-1-3p and hsa-miR-376a-3p Modulating the IGF Axis

Marta Gil-Martínez [1,2], Clara Lorente-Sorolla [1], José M. Rodrigo-Muñoz [1,2], Sara Naharro [1], Zahara García-de Castro [1], Joaquín Sastre [2,3], Marcela Valverde-Monge [2,3], Santiago Quirce [2,4], María L. Caballero [2,4], José M. Olaguibel [2,5] and Victoria del Pozo [1,2,6,*]

1 Immunoallergy Laboratory, Immunology Department, Instituto de Investigación Sanitaria Fundación Jiménez Díaz, Universidad Autónoma de Madrid (IIS-FJD, UAM), 28040 Madrid, Spain; marta.gilm@quironsalud.es (M.G.-M.); clara.lorente@quironsalud.es (C.L.-S.); jose.rodrigom@quironsalud.es (J.M.R.-M.); sara.naharrog@quironsalud.es (S.N.); zahara.garcia@quironsalud.es (Z.G.-d.C.)

2 CIBER de Enfermedades Respiratorias (CIBERES), Instituto de Salud Carlos III (ISCIII), 28029 Madrid, Spain; jsastre@fjd.es (J.S.); marcela.valverde@quironsalud.es (M.V.-M.); squirce@gmail.com (S.Q.); mlcsoto@hotmail.com (M.L.C.); jmolaguibel@telefonica.net (J.M.O.)

3 Allergy Department, Hospital Universitario Fundación Jiménez Díaz, 28040 Madrid, Spain

4 Department of Allergy, Hospital Universitario La Paz, IdiPAZ, 28046 Madrid, Spain

5 Severe Asthma Unit, Department of Allergy, Hospital Universitario de Navarra, NavarraBiomed, 31008 Pamplona, Spain

6 Department of Medicine, Faculty of Medicine, Universidad Autónoma de Madrid (UAM), 28049 Madrid, Spain

* Correspondence: vpozo@fjd.es

Abstract: Clarifying inflammatory processes and categorising asthma into phenotypes and endotypes improves asthma management. Obesity worsens severe asthma and reduces quality of life, although its specific molecular impact remains unclear. We previously demonstrated that hsa-miR-26a-1-3p and hsa-miR-376a-3p, biomarkers related to an inflammatory profile, discriminate eosinophilic from non-eosinophilic asthmatics. We aimed to study hsa-miR-26a-1-3p, hsa-miR-376a-3p, and their target genes in asthmatic subjects with or without obesity to find biomarkers and comprehend obese asthma mechanisms. Lung tissue samples were obtained from asthmatic patients (n = 16) and healthy subjects (n = 20). We measured miRNA expression using RT-qPCR and protein levels (IGF axis) by ELISA in confirmation samples from eosinophilic (n = 38) and non-eosinophilic (n = 39) obese (n = 26) and non-obese (n = 51) asthma patients. Asthmatic lungs showed higher hsa-miR-26a-1-3p and hsa-miR-376a-3p expression than healthy lungs. A study of seven genes regulated by these miRNAs revealed differential expression of *IGFBP3* between asthma patients and healthy individuals. In obese asthma patients, we found higher hsa-miR-26a-1-3p and IGF-1R values and lower values for hsa-miR-376a-3p and IGFBP-3. Hsa-miR-26a-1-3p and IGFBP-3 were directly and inversely correlated with body mass index, respectively. Hsa-miR-26a-1-3p and hsa-miR-376a-3p could be used as biomarkers to phenotype patients with eosinophilic and non-eosinophilic asthma in relation to comorbid obesity.

Keywords: asthma phenotypes/endotypes; eosinophilic asthma; obese asthma; serum microRNAs; biomarkers; IGF axis

1. Introduction

Asthma is a heterogeneous disease of the airways characterised by chronic airway inflammation leading to respiratory symptoms [1]. Asthma comprises an array of phenotypes driven by different mechanistic pathways, known as endotypes [2]. The asthma subtype most commonly associated with the T2 inflammatory profile (T2-high asthma) is severe eosinophilic asthma, in which patients experience recurrent exacerbations despite high doses of inhaled corticosteroids or even occasional oral corticosteroids [3].

Asthma is frequently accompanied by a variety of comorbidities that influence the severity of the disease. These are associated with a clinically significant reduction in quality of life (QoL) and exacerbate the rate of morbidity and the economic burden of the disease [4]. These comorbidities have been categorised based on the involvement of respiratory and/or extra-respiratory domains [5]. Obesity is a major public health concern as well as an extra-respiratory risk factor for both asthma incidence and prevalence. The relationship between obesity and asthma is the result of a complex interplay between biological, physiological, and environmental factors [6]. Nowadays, with the continuous rise in obesity, particularly among children, the obese asthma phenotype is attracting great interest [7]. Both T2 and non-T2 inflammation can be encountered in asthma patients with obesity [8]. Obese asthmatic patients are primarily associated with a non-eosinophilic phenotype and are characterised by a state of low-grade systemic inflammation marked by the activation of M1 macrophages and $CD8^+$ T cells and an increase in inflammatory indicators such as Toll-like receptor 4 (TLR4), interleukin (IL)-1b, IL-6 and IL-17, interferon (IFN)-γ, tumour necrosis factor (TNF)-α, leptin, and resistin. Asthma, which is often severe in patients who are obese, may be somewhat resistant to current treatment options, including corticosteroids, thus limiting possible therapeutic approaches and complicating efforts to control the disease [8–10].

It is unclear which pathophysiological mechanisms underlie the development of obesity as a comorbidity and enhancer of asthma severity and how these mechanisms are related, and further research on the endophenotype of obese patients with severe eosinophilic asthma is needed [10]. Biomarkers are highly useful for phenotyping and endotyping asthma, as they allow for personalised therapies to be offered to patients [11]. Currently, there is increasing interest in using microRNA (miRNA) profiles as biomarkers of disease [12]. miRNAs, which are small, non-coding RNA molecules, are engaged in regulating gene expression by inhibiting protein translation [13]. Previous studies have demonstrated that miRNAs hold great potential as asthma mediators and biomarkers [14]. In previous research, we used serum samples from patients with eosinophilic (n = 40) and non-eosinophilic (n = 36) asthma to evaluate miRNA expression by next-generation sequencing (miRNA-seq), with subsequent validation by RT-qPCR. After these analyses, we demonstrated that hsa-miR-26a-1-3p and hsa-miR-376a-3p were able to discriminate between patients with eosinophilic and non-eosinophilic asthma [11].

The aim of this study was to analyse the expression of these miRNAs and their targets in subjects with eosinophilic vs. non-eosinophilic asthma, with or without comorbid obesity, in order to determine whether these miRNAs are associated with differential inflammatory pathways of asthma.

2. Results

2.1. Characteristics of the Study Population

Descriptive data on the demographic, inflammatory, functional, and clinical characteristics of all individuals studied are summarised in Table S1a (population 1) and Table S1b (population 2).

Contrasting the group of asthmatic patients and control individuals based on lung biopsy samples (Table S1a), age was the only significant difference observed (52.4 vs. 63.7 years, $p < 0.05$).

As evidenced in Table S1b, with respect to functional and clinical parameters, the serum samples of obese and non-obese asthmatic patients showed significant differences in FEV1/FVC (%) (86.5 vs. 78.8, $p < 0.05$) and in severe and moderate disease severity (32% vs. 56.8%, and 44% vs. 20.5%; $p < 0.05$). The functional and clinical parameters of these patients according to the four study groups are shown in Table 1 (population 2, obese and non-obese, further subdivided into eosinophilic and non-eosinophilic), showing significant differences in FeNO levels ($p < 0.0001$) and in the rates of severe asthma ($p < 0.05$). None of the other demographic, inflammatory, functional, or clinical characteristics presented statistically significant differences (Table S1b and Table 1).

Table 1. Demographic, inflammatory, functional, and clinical characteristics of obese and non-obese non-eosinophilic asthmatic patients, and obese and non-obese eosinophilic asthmatic patients (population 2; serum samples analysis).

	Obese Eosinophilic (n = 10)	Non-Obese Eosinophilic (n = 28)	Obese Non-Eosinophilic (n = 16)	Non-Obese Non-Eosinophilic (n = 23)	p-Value
Age (years) ^	50.4 (±13.8)	46.4 (±15.5)	53.9 (±8.4)	47 (±15.7)	N.S.
Female (%)	6 (60%)	20 (71.4%)	13 (81.3%)	15 (65.2%)	N.S.
BMI †	34.5 (32.4–38.2)	23.6 (21.8–26.7)	32.2 (31–33.3)	23.8 (22.6–26)	****
Obesity (%)	10 (100%)	0 (0%)	16 (100%)	0 (0%)	****
Tobacco habit (%)					
Smokers	0 (0%)	3/27 (11.1%)	0/15 (0%)	0/20 (0%)	N.S.
Passive	0 (0%)	4/27 (14.8%)	2/15 (13.3%)	0/20 (0%)	N.S.
Ex-smokers	4 (40%)	8/27 (29.6%)	6/15 (40%)	7/20 (35%)	N.S.
Non-smokers	6 (60%)	12/27 (44.4%)	7/15 (46.7%)	13/20 (65%)	N.S.
Blood eosinophils (cells/μL) †	600 (500–800)	700 (600–845)	150 (100–200)	100 (100–100)	****
Sputum eosinophils (%) †	5% (2.5–20)	0% (0–48)	0% (0–6)	2% (0–5)	N.S.
Atopy (%)	7 (70%)	17 (60.7%)	13 (81.3%)	16 (69.6%)	N.S.
IgE (IU) †	131 (55.6–347)	209 (87.4–753.5)	191 (61–670)	151 (53.3–358.3)	N.S.
FEV_1/FVC (%) †	85% (74.5–104)	78.3% (68–90)	88% (72–94.3)	80.3% (71–84.5)	N.S.
FeNO (ppb) †	29 (21.5–41)	60 (34–82)	20.5 (11.1–30.8)	16 (12–23)	****
Exacerbations (%)	5 (50%)	17 (60.7%)	8 (50%)	9 (39.1%)	N.S.
Severity (%)					
Severe	3 (30%)	18/26 (69.2%)	5/15 (33.3%)	7/18 (38.9%)	*
Moderate	5 (50%)	5/26 (19.2%)	6/15 (40%)	4/18 (22.2%)	N.S.
Mild	2 (20%)	3/26 (11.5%)	4/15 (26.7%)	6/18 (33.3%)	N.S.
Intermittent	0 (0%)	0/26 (25%)	0/15 (0%)	1/18 (5.6%)	N.S.
ACT †	20.5 (15–23)	21 (14.5–23.5)	21 (17–22)	22 (18–25)	N.S.
ICS and LABA (%)	10 (100%)	25 (89.3%)	15 (93.8%)	19 (82.6%)	N.S.
Late asthma onset	8/9 (88.9%)	20/25 (80%)	9/15 (60%)	14/18 (77.8%)	N.S.

Results are expressed as ^ mean (±SD) or † median (IQR); N.S., Non-significant; ****, $p < 0.0001$; *, $p < 0.05$; BMI, body mass index; FEV1, forced expiratory volume measured during the first second; FVC, forced vital capacity; FeNO, fractional exhaled nitric oxide; Ppb, parts per billion; ICS and LABA, inhaled corticosteroids and long-acting β2-agonists; and ACT, asthma control test.

2.2. miRNA and Gene Expression in Lung Tissue Samples

We found statistically significant differences in the expression of hsa-miR-26a-1-3p and hsa-miR-376a-3p in lung biopsies between asthmatic patients and healthy subjects ($p < 0.01$ and $p < 0.001$, respectively), with higher expressions observed in asthmatic patients (Figure 1a). This confirms that serum hsa-miR-26a-1-3p and hsa-miR-376a-3p are linked to asthma, suggesting that these miRNAs could possibly serve as biomarkers of the disease.

We subsequently conducted an in silico analysis using DIANA-miRPath v3.0 to determine the implication of these two differentially expressed miRNAs in biological processes and to identify potential target genes. We found that hsa-miR-26a-1-3p and hsa-miR-376a-3p control several signalling pathways: prion diseases ($p < 0.0001$), p53 signalling pathway and non-small cell lung cancer ($p < 0.05$), regulated by hsa-miR-

26a-1-3p, and extracellular matrix (ECM)-receptor interaction ($p < 0.0001$), influenced by hsa-miR-376a-3p. Of these, the p53 signalling pathway and extracellular matrix (ECM)-receptor interaction are critical pathways related to asthma pathogenesis. Since *CDK6*, *CCND1*, and *IGFBP3*, involved in the p53 signalling pathway, are target genes of hsa-miR-26a-1-3p, and *COL3A1*, *COL6A2*, *COL1A1*, and *COL6A3*, implicated in the extracellular matrix (ECM)-receptor interaction signalling pathway, are targeted by hsa-miR-376a-3p, RT-qPCR was performed on lung biopsy samples from asthmatic patients and healthy individuals. The results of this analysis revealed that only *IGFBP3* was significantly increased in asthmatic individuals in comparison to healthy subjects ($p < 0.01$; Figure 1b).

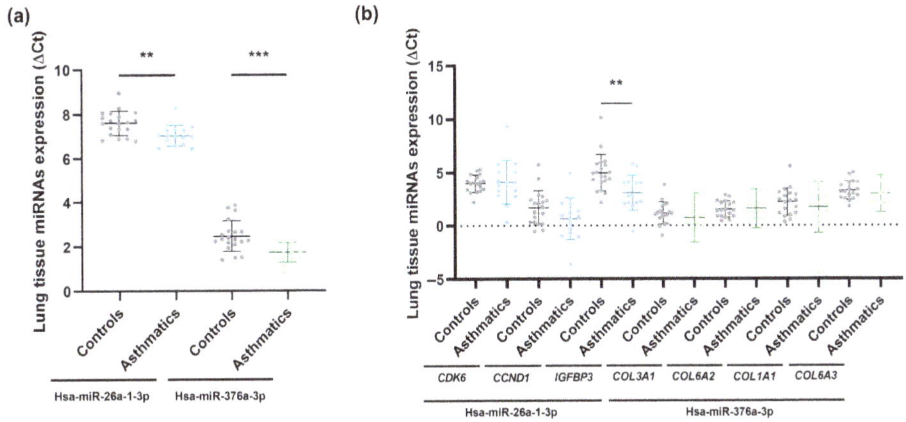

Figure 1. Lung biopsy sample miRNA expression (ΔCt) in asthmatic and non-asthmatic (controls) subjects. Hsa-miR-26a-1-3p and hsa-miR-376a-3p were differentially expressed between the two groups, showing higher expression in asthmatic individuals (**a**); and gene expression (ΔCt) measured in lung tissue samples from asthmatic and non-asthmatic (controls) subjects. Only *IGFBP3* exhibited differential expression between the two groups, being overexpressed in the asthma group (**b**). **, $p < 0.01$; ***, $p < 0.001$.

To establish the association between miRNA molecules and their target gene expression levels (ΔCt), a lung-tissue correlation analysis was carried out. We verified a significant, direct correlation for the total population between hsa-miR-26a-1-3p and its target gene, *IGFBP3* (Pearson r = 0.36, $p < 0.05$), instead of detecting, as usual, an inverse miRNA-target gene correlation. Since miRNAs generally behave as silencers of their target mRNAs, a possible explanation would be that miRNA–miRNA relationships are complex, and the lung tissue samples in which they were studied differ from each other (influence of genetic and/or environmental factors). There may be other miRNAs involved in the modulation of a particular mRNA (IGFBP-3 mRNA) and there could also be an interaction of other target genes of this miRNA (hsa-miR-26a-1-3p) that could be inhibitors of *IGFBP3*, in this case, highlighting the intricacy of post-transcriptional miRNA–mRNA regulation throughout the whole tissue.

2.3. miRNA Expression and Extracellular Protein Concentration in Serum Samples

To further investigate and confirm the relationship between these miRNAs and the *IGFBP3* gene, we analysed the serum expression of two soluble extracellular proteins, IGFBP-3 (encoded by *IGFBP3*, the target gene of hsa-miR-26a-1-3p) and IGF-1R (encoded by *IGF1R*, targeted by hsa-miR-376a-3p), both of which are implicated, in the IGF axis, in patients with eosinophilic and non-eosinophilic asthma. No significant differences in target protein levels were detected between the eosinophilic and non-eosinophilic groups (data not shown), although statistically significant differences were observed in IGFBP-3 ($p < 0.01$) and IGF-1R ($p < 0.05$) levels among obese and non-obese asthmatic patients: IGFBP-3 showed higher levels in non-obese subjects, while IGF-1R levels were greater in obese individuals (Figure 2a,b). Regarding the levels of these proteins when subdividing patients with and without eosinophilic asthma into obese and non-obese subjects, IGFBP-3 was more concentrated in non-obese individuals than in obese subjects ($p < 0.05$) among patients with eosinophilic asthma and those with non-eosinophilic asthma (Figure 2a); on the other hand, IGF-1R was higher in obese individuals than in non-obese subjects ($p = 0.05$ and $p < 0.05$, in that order), and obese patients with eosinophilic asthma had higher amounts of IGF-1R than the rest of the groups (Figure 2b).

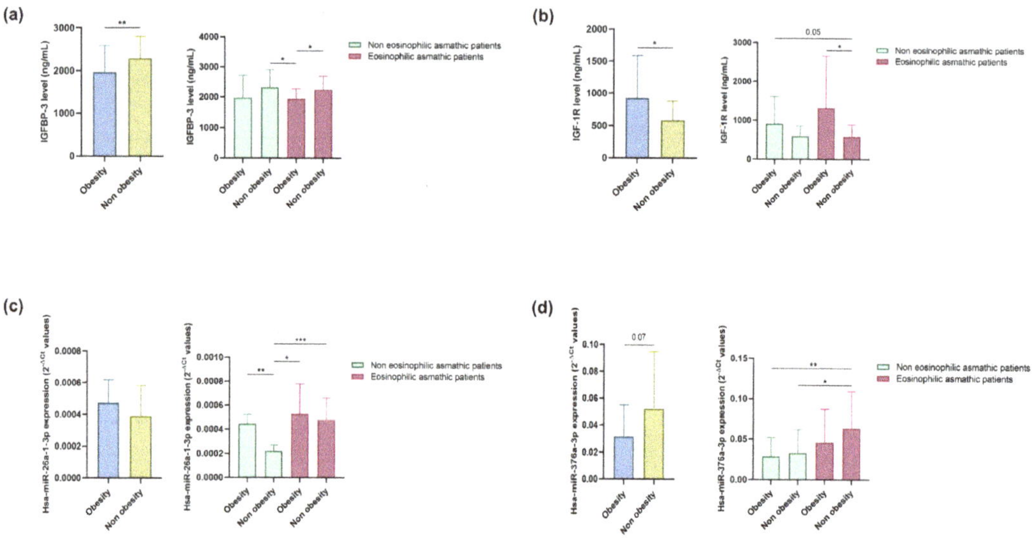

Figure 2. Serum IGFBP-3 (**a**); and IGF-1R (**b**) concentration. Serum hsa-miR-26a-1-3p (**c**); and hsa-miR-376a-3p (**d**) ($2^{-\Delta Ct}$). Differential expression of miRNAs and soluble extracellular proteins was observed between obese and non-obese eosinophilic asthma patients and obese and non-obese non-eosinophilic asthma patients. *, $p < 0.05$; **, $p < 0.01$; ***, $p < 0.001$.

Next, we assessed whether hsa-miR-26a-1-3p and hsa-miR-376a-3p are modulated by obesity status in asthmatic patients with and without eosinophilia. A non-significant increase in hsa-miR-26a-1-3p expression and lower expression of hsa-miR-376a-3p ($p = 0.07$) was found in this group of obese subjects (Figure 2c,d). When we subdivided patients with eosinophilic and non-eosinophilic asthma into obese and non-obese individuals, we observed that both miRNAs were more highly expressed in patients with eosinophilic asthma than in those with non-eosinophilic asthma ($p < 0.05$ and $p < 0.001$, and $p < 0.05$ and $p < 0.01$, respectively; Figure 2c,d). Specifically, hsa-miR-26a-1-3p exhibited higher expression in obese subjects than in non-obese individuals ($p < 0.01$) and in patients with non-eosinophilic asthma (Figure 2c), while hsa-miR-376a-3p levels were highest among non-obese eosinophilic asthmatics.

2.4. Correlation of miRNAs and Soluble Extracellular Proteins with Clinical Parameters

Because the measured soluble extracellular proteins are engaged in the IGF axis, which is metabolism-related, we analysed our data to determine if BMI correlated with serum levels of these proteins and miRNAs (Figure 3). A negative correlation was identified between IGFBP-3 and BMI in the group of patients with eosinophilic asthma (Pearson $r = -0.43$, $p < 0.01$; Figure 3a) and in the total study population (Pearson $r = -0.34$, $p < 0.01$; Figure 3c); while hsa-miR-26a-1-3p correlated directly with BMI in patients with non-eosinophilic disease (Pearson r = 0.86, $p < 0.0001$; Figure 3b) and in the total population (Pearson r = 0.38, $p < 0.05$; Figure 3c).

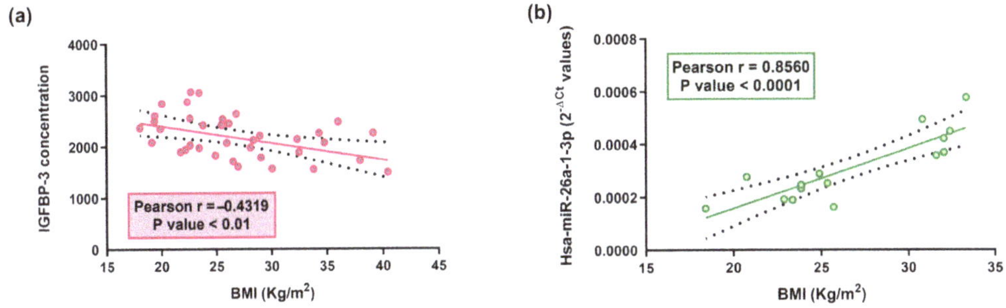

(c)

	IGFBP-3			Hsa-miR-26a-1-3p		
CLINICAL PARAMETER	Eosinophilic asthma	Non-eosinophilic asthma	Both asthma fenotypes	Eosinophilic asthma	Non-eosinophilic asthma	Both asthma fenotypes
BMI (Kg/m²)	**<u>−0.4319</u>**	−0.3030	**<u>−0.3443</u>**	−0.0361	**<u>0.8560</u>**	**<u>0.3767</u>**

Figure 3. Relative IGFBP-3 concentration correlates inversely with BMI in serum samples ($p < 0.01$) in eosinophilic asthmatic patients (**a**); a direct correlation between relative hsa-miR-26a-1-3p expression ($2^{-\Delta Ct}$) and BMI was reported in serum samples ($p < 0.0001$) in non-eosinophilic asthmatic patients (**b**); Dots (red and green) are individual values, black dashed line indicates error bars and red/green solid line shows linear regression (**a**,**b**). Table showing the correlation coefficient (r) between BMI parameter and serum IGBP-3 concentration and hsa-miR-26a-1-3p expression levels ($2^{-\Delta Ct}$) (**c**). The red color indicates negative correlations, whereas green shows positive correlations (**c**). Bold and underlined values are statistically significant ($p < 0.05$).

2.5. ROC Curves to Establish Potential Biomarkers of Inflammatory Endotype in Asthma

ROC curve analyses for the two miRNAs (Table 2) were performed for two-by-two group comparisons (obese and non-obese eosinophilic asthma patients, obese and non-obese non-eosinophilic asthma patients, and the entire population of obese and non-obese asthma patients), and the AUC were estimated.

Table 2. ROC curves for different comparisons between groups of asthmatic patients: obese and non-obese non-eosinophilic asthmatic patients, obese and non-obese eosinophilic asthmatic patients, and obese and non-obese in all asthmatic patients.

	Hsa-miR-26a-1-3p		Hsa-miR-376a-3p	
Comparison Groups	**AUC**	***p*-Value**	**AUC**	***p*-Value**
Obese non-eosinophilic vs. Non-obese non-eosinophilic	**1.00**	**<0.01**	0.50	>0.05
Obese eosinophilic vs. Non-obese eosinophilic	0.56	>0.05	0.63	>0.05
Non-obese non-eosinophilic vs. Non-obese eosinophilic	**0.91**	**<0.001**	**0.71**	**<0.05**
Obese non-eosinophilic vs. Obese eosinophilic	0.50	>0.05	0.64	>0.05
Obese non-eosinophilic vs. Non-obese eosinophilic	0.58	>0.05	**0.75**	**<0.01**
Non-obese non-eosinophilic vs. Obese eosinophilic	**1.00**	**<0.05**	0.61	>0.05
Obese vs. Non-obese	0.66	>0.05	0.60	>0.05

AUC, area under curve. Bold values are statistically significant ($p < 0.05$). Green color intensity indicates a higher positive correlation.

Hsa-miR-26a-1-3p presented an AUC value of 0.91 for the comparison between non-eosinophilic and eosinophilic asthma within the non-obese group ($p < 0.001$). An AUC of 1.00 was found for the comparisons between obese and non-obese, non-eosinophilic asthmatics ($p < 0.01$) and between the non-obese, non-eosinophilic and obese eosinophilic asthma patients ($p < 0.05$). Hsa-miR-376a-3p showed AUC values of 0.71 and 0.75, respectively, for the comparisons between non-eosinophilic and eosinophilic subjects in the non-obese asthma group ($p < 0.05$) and between obese non-eosinophilic and non-obese eosinophilic asthmatic individuals ($p < 0.01$).

3. Discussion

The results of this study reveal that hsa-miR-26a-1-3p and hsa-miR-376a-3p are post-transcriptional modulators of the IGF axis and potential epigenetic biomarkers to differentiate asthma patients by inflammatory condition (eosinophilia and/or obesity).

Severe asthma is an inherently heterogeneous disease that encompasses several clinical phenotypes that may have several shared pathophysiological mechanisms, known as endotypes of asthma [15]. Therefore, inflammatory endotype characterisation should be a crucial component of the algorithm used to assess and manage severe disease [16]. The discovery and use of biomarkers to define phenotypes and endotypes of severe asthma and guide therapeutic strategy is increasingly necessary for clinicians, as an individual approach to disease management and personalised medicine is essential [17]. Eosinophilic phenotypes and T2-associated endotypes with eosinophilic inflammation are currently identified in clinical practice by biomarkers, including peripheral blood and sputum eosinophil count, total serum IgE levels, FeNO value, periostin, and miRNAs [18]. We observed that serum samples from both obese and non-obese patients with eosinophilic asthma showed markedly higher FeNO scores compared to non-eosinophilic asthmatic patients. As previous research has found a relationship between FeNO and the eosinophilic phenotype in asthma, FeNO, in combination with blood eosinophils, can accurately predict eosinophilic airway inflammation [19,20]. Also, as demonstrated by our results (non-obese, non-eosinophilic asthma group), elevated FeNO values and increased blood eosinophil counts are associated with increased severity of asthmatic disease, worse lung function, and an increased risk of asthma exacerbation [21].

It has been established that severe asthma is a complex clinical condition not exclusively related to airway inflammation and response to treatment [22]. Asthma is often associated with various comorbidities which can worsen asthma symptoms and affect asthma severity [23]. For patients with asthma, comorbid conditions lead to an enhanced disease burden and, consequently, a reduced QoL [24]. The involvement of other disorders is likely to lead to polypharmacy, which can negatively impact treatment adherence and asthma control [25,26] Additionally, comorbid conditions complicate the diagnosis and management of asthmatic patients and carry a considerable economic cost [23]. Obesity is a common comorbidity in children and adults with asthma, and both obesity and asthma have risen dramatically in prevalence [27,28]. Obesity is defined as having a BMI of 30 kg/m^2 or greater [29]. Obese asthmatics are generally characterised by more severe asthma, are less responsive to treatment, and have poorer asthma control, according to the asthma control test (ACT) questionnaires [30], although this does not appear to be a determining factor in the risk of asthma exacerbations with regard to disease severity [31]. This absence of a link between obesity status and exacerbations appears to be in line with our results, as both obese asthmatic patient groups, compared to the two non-obese asthmatic patient groups, had similar or higher asthma severity (80% and 73.3%, respectively; 88.5% and 61.1% of patients with severe and moderate asthma), lower ACT scores (20.5 and 21, 21 and 22), and a lower, or roughly similar, presence of exacerbations (50% and 50%, 60.7% and 39.1%). As cited earlier, our findings indicate more frequent occurrences of exacerbations among obese women (19/26) and late-onset asthmatics (17/24) [12]. Also, since no significant differences are observed in relation to patient therapy, it could be considered that the changes in miRNA expression are not due to the treatment. However, it is worth mentioning that, surprisingly, in contrast to what has been described in the literature in relation to lung function, that obese asthmatic patients have worse lung function than non-obese asthmatic patients. In our population the opposite is shown, obese asthma patients show more improved lung function than non-obese asthma subjects [32], which may be due, to some extent, to the fact that there are more non-obese than obese asthmatic patients and, within the group of non-obese asthmatic patients there was a high number of patients with eosinophilic asthma, which may cause these non-obese asthmatic patients to have lower pulmonary function since, subdividing obese and non-obese asthmatic patients into eosinophilic and non-eosinophilic categories, the eosinophilic asthmatic patient groups showed poorer lung function than the non-eosinophilic asthmatic patient groups, contributing to a decrease in the overall group's lung function. Moreover, according to the percentage of neutrophils in sputum in our study, the 27% of patients with obese asthma are individuals with a neutrophilic component. The differences in clinical and inflammatory findings between the four patient groups may be due to the combined influence of the different inflammation profiles in obese asthma [12] and the pathophysiological differences between eosinophilic and non-eosinophilic asthma [33].

Biomarkers are promising tools for categorising asthma patients into different phenotypes and endotypes, as they enable the identification of potential therapeutic targets and make it possible to deliver personalised therapy [17]; of these tools, miRNAs could be an effective biomarker [34]. In this study we found expression differences of the two miRNAs, hsa-miR-26a-1-3p and hsa-miR-376a-3p, in lung tissue samples from asthmatic individuals and healthy subjects. The same difference was previously observed in serum from patients with eosinophilic asthma and others with non-eosinophilic asthma [11]. This finding supports the involvement of these miRNAs in asthma, as they were detected in target tissue, and further indicates that the two miRNAs could serve as potential biomarkers to classify patients into either group. In terms of downstream effects, in silico experiments revealed that hsa-miR-26a-1-3p and hsa-miR-376a-3p significantly modulate two pathways associated with asthma pathogenesis that play a key role in the development of airway inflammation and asthma remodelling [35,36]. Also, of the target genes selected for study due to their relationship with differential expression of these two miRNAs, only *IGFBP3* displayed differential expression between lung biopsy samples from asthmatic individuals and healthy subjects, with higher values found in asthmatic individuals; the expression levels of the miRNA and its target gene were correlated, emphasising the involvement of this inflammation-related gene in asthma pathology, as has been previously described [37]. Moreover, we found differences in the expression of two soluble extracellular proteins, IGFBP-3 and IGF-1R, which are encoded by target genes of these miRNAs, in the serum of subjects with eosinophilic asthma and in individuals with non-eosinophilic asthma, further separated into obese and non-obese asthmatic patients.

Growth hormone (GH), which is deficient in obese patients, stimulates the secretion of insulin-like growth factor I (IGF-1) in most tissues. Jointly, GH and IGF-1 exert strong actions on fat, protein, and glucose metabolism [38]. The actions of IGF-1 are typically mediated through interactions with its receptor (IGF-1R), resulting in the activation of intracellular signalling pathways, including MAP kinase and PI3 kinase/AKT [39]. In addition, the bioavailability, half-life, and actions of IGF-1 in circulation are regulated through its binding to one of the members of the IGF binding proteins, of which IGFBP-3 is the most abundant in plasma and the one with the highest affinity for IGF-1 [40]. Alterations in the IGF axis may be implicated in a variety of pathological conditions, including obesity [41]. Furthermore, a link between asthma and obesity has been recognised alongside other markers of the metabolic syndrome, such as insulin resistance and hypertension [42]. It has also been proposed that insulin resistance may have an effect on asthma risk [43]. Peters and co-workers observed that, among asthma patients participating in a severe asthma research programme, patients with insulin resistance had poorer asthma-control test scores and greater deterioration in lung-function test responses [44]. Finally, it is well known that both insulin and IGF-1 can mutually activate insulin and IGF-1R receptors, sharing downstream signalling pathways but with different biological effects [45].

Regarding our study on obese asthma molecular mechanisms, on the one hand, hsa-miR-26a-1-3p showed higher expression in subjects with eosinophilic asthma and in the subgroups of obese individuals, while the concentration of the soluble extracellular protein IGFBP-3, which is encoded by the target gene of this miRNA, as expected in terms of miRNA-gene interaction, was lower in the non-obese patient groups. Also, since IGFBP-3 is a molecule that displays anti-inflammatory action [46], a greater concentration was expected in subjects with no additional inflammatory condition such as obesity. As reported in other studies [47], hsa-miR-376a-3p showed greater expression in non-obese subjects and in asthmatic patients without additional comorbid conditions, with expression increasing in eosinophilic asthmatics. As in the previous case, there is an inverse miRNA-gene relation, such that the concentration of IGF-1R, which is encoded by the target gene of hsa-miR-376a-3p, was elevated in obese individuals, and specifically in obese patients with eosinophilic asthma. This result supports previous reports indicating that the higher expression of this molecule results in more prominent inflammatory responses [48].

Since IGFBP-3 and IGF-1R are involved in the IGF axis, which is closely related to metabolism, we studied the correlation of the expression of hsa-miR-26a-1-3p and hsa-miR-376a-3p, in addition to these two soluble extracellular proteins encoded by their respective target genes, and established their correlation with clinical parameters. The results suggested that these molecules could be relevant in phenotypically distinguishing patients based on the inflammatory component and thus be used in asthma diagnosis.

Lastly, based on the AUC values calculated, hsa-miR-26a-1-3p could be a highly valuable biomarker in differentiating obese from non-obese subjects among patients with non-eosinophilic asthma or in discriminating individuals with no inflammatory condition (non-obese, non-eosinophilic asthma subjects) from individuals with combined inflammatory states (obese eosinophilic asthma subjects). Hsa-miR-26a-1-3p is also effective in distinguishing non-eosinophilic from eosinophilic individuals within the group of non-obese asthma patients. Therefore, this miRNA can discern the inflammatory conditions of obesity or eosinophilia (individually) and can distinguish the presence of both inflammatory processes from cases with no inflammatory condition. Based on the AUC score found, hsa-miR-376a-3p could be an acceptable biomarker to differentiate non-eosinophilic from eosinophilic subjects within the group of non-obese asthma patients (i.e., eosinophilia in the absence of obesity) or to separate individuals with one inflammatory state (i.e., non-obese, non-eosinophilic patients) from subjects with the other (non-obese eosinophilic patients).

We wish to point out that a limitation of this study is the low number of samples from obese individuals, mainly patients with eosinophilic asthma, compared to the number of samples from non-obese subjects (patients with eosinophilic asthma and non-eosinophilic asthma). A more even sample size in the four groups could provide more consistent results. Another limitation of our study is the existence of significant differences in age between the asthmatic and control groups of subjects, whose lung tissue samples were included for analysis, as this was not evaluated as a confounding factor. However, although there were significant age differences between these two groups, in the obese and non-obese groups (with a larger number of subjects), from which serum samples were included, or in the four groups after being subdivided, no significant age differences were observed; therefore, it is possible that age does not significantly affect miRNA expression. In addition, unfortunately, as the samples were from patients belonging to biobanks, we did not have all the clinical data available.

4. Materials and Methods

4.1. Study Subjects and Sample Collection

We obtained lung biopsy samples from 16 asthmatic patients and 20 control subject donors (population 1) to examine miRNAs and gene expression. Biopsy samples were provided by the CIBERES Pulmonary Biobank Consortium (PT13/0010/0030), a network currently made up of 12 tertiary Spanish hospitals (www.ciberes.org; accessed on 5 June 2023) listed in the Acknowledgements section and integrated in the Spanish National Biobanks Network. Lung biopsies were processed in accordance with standard operating procedures, and subsequent approval was granted for processing by the local ethics and scientific committees (ref. B.0000471 Registro Nacional de Biobancos—ISCIII; CIBERES Pulmonary Biobank Consortium started its activity in 2008 and has been ISO9001:2015 (formerly ISO9001:2008) certified since 2012.).

Lung tissue samples were preserved in RNAlater stabilisation solution at −80 °C until use.

A cohort of 77 subjects with diagnosed asthma (population 2) was selected for confirmation studies. This cohort was recruited from the allergy and pulmonology units of Fundación Jiménez Díaz Hospital in Madrid (Spain); 10 of these individuals were phenotyped as obese patients with eosinophilic asthma, 28 as non-obese patients with eosinophilic asthma, 16 as obese patients with non-eosinophilic asthma, and 23 as non-obese patients with non-eosinophilic asthma. Descriptive data representing the demographic, inflammatory, functional, and clinical characteristics of study subjects were collected.

The following scheme (Figure S1) shows the groups of the two studies (with the exact number of patients per group) and the material (lung tissue or serum) used from each patient in each phase of the study, facilitating the understanding of the design.

Serum samples were obtained by blood clotting in plain, additive-free tubes which were subsequently centrifuged at 3000 rpm for 10 min at 4 °C and stored at −80 °C until use.

The inclusion criteria were as follows: (i) acceptance to participate, providing signed informed consent; (ii) asthma diagnosis in accordance with the 2021 GINA criteria [49]; and (iii) age between 18 and 75 years. Eosinophilic asthma was determined by counts of $\geq$500 eosinophils/mm^3 in peripheral blood, while non-eosinophilic patients had $\leq$200 eosinophils/mm^3. Participants were considered obese if they had a body mass index (BMI) $\geq$30 kg/m^2, and non-obese if BMI was <30 kg/m^2.

The study was conducted in accordance with the tenets of the Declaration of Helsinki, and the protocol was approved by the CEIm of the participating hospital ethics committees (EOH2014/48, 4 December 2014).

4.2. miRNA and mRNA Analysis by RT-qPCR

Four µL of serum miRNA samples or 30 ng of miRNA-enriched lung tissue RNA were reverse-transcribed to cDNA using the miRCURY LNA RT Kit (Qiagen, Hilden, Germany) following the manufacturer's protocol. Synthetic Uni-Sp6 and miRNA cel-miR-39-3p (Qiagen, Hilden, Germany) were used to verify correct reverse transcription. The cDNA obtained was stored at −20 °C until use. Then, miRNA expression was evaluated by qPCR using the miRCURY LNA SYBR Green PCR Kit (Qiagen) as indicated in the instructions. For this purpose, we used 3 µL of cDNA from the serum or lung tissue miRNAs diluted in 1:30 or 1:60 in RNase-free water, respectively, according to the producer's recommendations. To ensure that differences in concentration, due to choosing the same volume in the case of serum samples, did not affect the miRNA expression measurements, the samples were normalised with endogenous control genes, which functioned as a control for the qPCR reaction. The following probes (Qiagen) were used for miRNA expression analysis in the serum and lung tissue: hsa-miR-26a-1-3p and hsa-miR-376a-3p. Additionally, hsa-miR-103a-3p, hsa-miR-191-5p, SP6, cel-miR-39-3p, and U6 (Qiagen) were selected as housekeeping miRNAs. The program GraphPad Prism® v6-8.0 was used to perform a homogeneity test of Ct means of the samples from each of the comparison groups to select endogenous and/or exogenous genes as internal control of the qPCR reaction. In addition, a normalisation analysis was carried out with the use of the Bestkeeper tool. All samples were run in triplicate, and reactions were performed in a LightCycler® 96 thermal cycler (Roche, Basel, Switzerland). Cycle threshold (Ct) values were analysed with LightCycler® 96 SW 1.1 (Roche) software, and miRNA expression was calculated by normalising levels of expression to the endogenous miRNA controls. Relative quantification of differences in expression ($RQ = 2^{-\Delta\Delta Ct}$; where $\Delta\Delta Ct = \Delta Ct_{eosinophilic\ asthmatics} - \Delta Ct_{non\text{-}eosinophilic\ asthmatics}$ and $\Delta Ct = Ct_{miRNA} - Ct_{housekeeping\ miRNAs}$) was carried out by means of the $\Delta\Delta Ct$ method [50].

For gene expression analysis in lung tissue, 500 ng of RNA quantified by Nanodrop® ND-1000 spectrophotometer (Thermo Fisher Scientific, Waltham, MA, USA) was reverse transcribed using the High-Capacity cDNA Reverse Transcription Kit (Applied Biosystems,

Foster City, CA, USA), followed by qPCR analysis according to manufacturer guidelines on a StepOne™ Real-Time PCR System (Applied Biosystems). TaqMan™ gene expression probes were purchased for *CDK6, CCND1, IGFBP3, COL3A1, COL6A2, COL1A1, COL3A3,* and *GAPDH* using TaqMan™ Gene Expression Master Mix (Applied Biosystems). Gene expression was calculated by normalising to the endogenous control gene *GAPDH* applying the $2^{-\Delta\Delta Ct}$ method as reported previously (RQ = $2^{-\Delta\Delta Ct}$; where $\Delta\Delta Ct = \Delta Ct_{asthmatics} - \Delta Ct_{healthy\ subjects}$) [50].

4.3. Pathway Enrichment Analyses

In order to identify the target genes related to asthma disease, hsa-miR-26-a-1-3p and hsa-miR-376a-3p, and differentially expressed miRNAs, a pathway enrichment analysis of the dysregulated miRNAs was performed using the DIANA-miRPath v3.0 online bioinformatic resource [51]. Relevant pathways for asthma pathology were represented when a *p*-value < 0.05, and the involved genes were analysed by RT-PCR, as previously mentioned.

4.4. Protein Detection by ELISA

Zehavi et al., identified the *IGF1R* gene as a target of hsa-miR-376a and hsa-miR-376c miRNAs. Since the protein encoded by the *IGF1R* gene is involved in the IGF axis, it was chosen, together with IGFBP-3, for ELISA evaluation [52].

For a determination of the soluble extracellular proteins IGFBP-3 and IGF-1R (genes encoded which are targeted by both hsa-miR-26a-1-3p and hsa-miR-376a-3p), 10 μL (dilution 1:100) or 100 μL of serum samples, in that order, were used for ELISA, performed with the Human IGFBP-3 Quantikine ELISA kit (R&D Systems, Minneapolis, MN, USA) and Human IGF-I R/IGF-1R DuoSet ELISA Kit (Bio-Techne, Minneapolis, MN, USA), respectively, following the manufacturers' instructions.

The optical density of each well was determined using a Tecan Infinite® F200 microplate reader (TECAN, Männedorf, Switzerland) adjusted to 450 nm, with the wavelength correction set to 570 nm.

4.5. Statistical Analysis

Statistical analyses and graphs were created with GraphPad Prism® v6-8.0 (GraphPad Software Inc., San Diego, CA, USA).

The results are shown as mean (standard deviation, SD) or median (interquartile range, IQR) values. For all statistical analyses, differences showing $p < 0.05$ were considered significant. Normality was analysed by means of the Shapiro–Wilk test. For continuous variables, comparisons of normally distributed data among non-paired groups were performed via an unpaired t test (equal to SD) and t test with Welch's correction (different SD), and non-normally distributed data and non-paired groups were compared by Mann–Whitney test.

Multiple comparisons between groups with data following a normal distribution were carried out with one-way ANOVA test (equal to SD) and Brown–Forsythe and Welch ANOVA test (different SD) without Welch's correction test or with Kruskal–Wallis test without Dunn's correction for non-normally distributed data.

Additionally, correlations among miRNA expression levels (ΔCt), soluble extracellular protein concentration, and clinical parameters (quantitative variables) were estimated by Spearman (non-normally distributed data) or Pearson (normally distributed data) correlations; one-way ANOVA and Chi-square test (contingency table) were applied to test the null hypothesis of the independence of groups and some demographic, inflammatory, functional, and clinical (quantitative and qualitative) variables. Finally, the expression profile (ΔCt) of each differentially expressed miRNA was used to create receiver operator characteristic (ROC) curves to evaluate miRNA performance as biomarkers; an area under the curve (AUC) of 0.7 indicated an acceptable biomarker.

5. Conclusions

In summary, this is the first study to describe differences in the expression of the previously validated serum miRNAs, hsa-miR-26a-1-3p and hsa-miR-376a-3p, in lung tissue samples. Findings from this study support their potential use as biomarkers to distinguish between obese and non-obese non-eosinophilic asthma patients, and in non-obese asthma patients, and to catalogue eosinophilic and non-eosinophilic patients. Finally, the findings can be used to discriminate non-eosinophilic asthma patients without obesity and eosinophilic asthma patients with obesity.

Supplementary Materials: The following supporting information can be downloaded at: https://www.mdpi.com/article/10.3390/ijms241411620/s1.

Author Contributions: V.d.P. conceived the manuscript and designed the study. M.G.-M., C.L.-S., J.M.R.-M. and S.N. performed experiments and data analysis. M.G.-M. drafted and revised the manuscript. Z.G.-d.C. assisted with the experiments. J.S., M.V.-M., S.Q., M.L.C. and J.M.O. recruited patients and obtained samples and clinical data. All authors contributed to the article and approved the submitted version. All authors have read and agreed to the published version of the manuscript.

Funding: This work was supported by ISCIII—Instituto de Salud Carlos III, FIS (Fondo de Investigación Sanitaria—Spanish Health Research Fund) grants PI21/00896 and FI19/00067; Ciber de Enfermedades Respiratorias (CIBERES); Comunidad de Madrid grant PEJ-2021-AI_BMD-22320; and FEDER funds (Fondo Europeo de Desarrollo Regional).

Institutional Review Board Statement: The study protocol was approved by the ethics committees of each hospital, and the study was conducted in accordance with the principles set forth in the Declaration of Helsinki.

Informed Consent Statement: Signed informed consent was obtained from all patients. The study was conducted in accordance with the tenets of the Declaration of Helsinki, and the protocol was approved by the participating hospitals' ethics committees.

Data Availability Statement: The data that support the findings of this study are available from the corresponding author, V.d.P., upon reasonable request.

Acknowledgments: We want to particularly acknowledge the patients and the CIBERES Pulmonary Biobank Consortium (PT13/0010/0030), on behalf of Hospital Universitario 12 de Octubre, Hospital del Mar-IMIM, Hospital Universitario de Getafe, Consorcio Universitario Hospital General de Valencia, Fundación Jiménez-Díaz, Hospital Clínic de Barcelona, Hospital Universitario de Bellvitge, Hospital Universitario Joan XXIII, Hospital Universitario Germans Trias i Pujol, Hospital Universitario Virgen del Rocío, Hospital Vall d'Hebron, and Hospital Universitario Son Espases, integrated in the Spanish National Biobanks Network, for their collaboration. The authors also acknowledge Oliver Shaw for editorial support.

Conflicts of Interest: J.S.: having served as a consultant to Thermo Fisher, MEDA, Novartis, Sanofi, Leti, Faes Farma, Mundipharma, and GSK; having been paid lecture fees by Novartis, GSK, Stallergenes, Leti, and Faes Farma; having received grant support for research from Thermo Fisher, Sanofi, and ALK. S.Q.: personal fees outside of the submitted manuscript received from AstraZeneca, Novartis, Sanofi, Boehringer Ingelheim, Teva, ALK, Mundipharma, GSK, Chiesi, and Leti. J.M.O.: grants from Sanofi during performance of the study; personal fees outside of the submitted work from AstraZeneca and Mundipharma. V.d.P.: honoraria (advisory board, speaker) and/or institutional grant/research support from AstraZeneca and GSK. The remaining authors declare that they have no conflict of interest.

References

1. Varricchi, G.; Ferri, S.; Pepys, J.; Poto, R.; Spadaro, G.; Nappi, E.; Paoletti, G.; Virchow, J.C.; Heffler, E.; Canonica, W.G. Biologics and airway remodeling in severe asthma. *Allergy* **2022**, *77*, 3538–3552. [CrossRef] [PubMed]
2. Kuruvilla, M.E.; Lee, E.-H.; Lee, G.B. Understanding Asthma Phenotypes, Endotypes, and Mechanisms of Disease. *Clin. Rev. Allergy Immunol.* **2019**, *56*, 219–233. [CrossRef]
3. Chung, K.F.; Dixey, P.; Abubakar-Waziri, H.; Bhavsar, P.; Patel, P.H.; Guo, S.; Ji, Y. Characteristics, phenotypes, mechanisms and management of severe asthma. *Chin. Med. J.* **2022**, *135*, 1141–1155. [CrossRef] [PubMed]

4. Porsbjerg, C.; Menzies-Gow, A. Co-morbidities in severe asthma: Clinical impact and management. *Respirology* **2017**, *22*, 651–661. [CrossRef] [PubMed]
5. Fasola, S.; Ferrante, G.; Cilluffo, G.; Malizia, V.; Alfano, P.; Montalbano, L.; Cuttitta, G.; La Grutta, S. Asthma Comorbidities: Frequency, Risk Factors, and Associated Burden in Children and Adolescents. *Children* **2022**, *9*, 1001. [CrossRef]
6. Peters, U.; Dixon, A.; Forno, E. Obesity and Asthma. *J. Allergy Clin. Immunol.* **2018**, *141*, 1169–1179. [CrossRef]
7. Roshan Lal, T.; Cechinel, L.R.; Freishtat, R.; Rastogi, D. Metabolic Contributions to Pathobiology of Asthma. *Metabolites* **2023**, *13*, 212. [CrossRef]
8. Kaplan, A.G.; Kim, J.W. Asthma Exacerbations and Glucagon-Like Peptide-1 Receptor Agonists: A Review of the Current Evidence. *Pulm. Ther.* **2022**, *8*, 343–358. [CrossRef]
9. Bantulà, M.; Roca-Ferrer, J.; Arismendi, E.; Picado, C. Asthma and Obesity: Two Diseases on the Rise and Bridged by Inflammation. *J. Clin. Med.* **2021**, *10*, 169. [CrossRef]
10. Wang, Y.; Hu, C. Leptin and Asthma: What Are the Interactive Correlations? *Biomolecules* **2022**, *12*, 1780. [CrossRef]
11. Gil-Martínez, M.; Rodrigo-Muñoz, J.M.; Sastre, B.; Cañas, J.A.; García-Latorre, R.; Redondo, N.; de la Fuente, L.; Mínguez, P.; Mahíllo-Fernández, I.; Sastre, J.; et al. Serum microRNAs Catalog Asthma Patients by Phenotype. *J. Investig. Allergol. Clin. Immunol.* **2022**, *32*, 471–478. [CrossRef]
12. Condrat, C.E.; Thompson, D.C.; Barbu, M.G.; Bugnar, O.L.; Boboc, A.; Cretoiu, D.; Suciu, N.; Cretoiu, S.M.; Voinea, S.C. miRNAs as Biomarkers in Disease: Latest Findings Regarding Their Role in Diagnosis and Prognosis. *Cells* **2020**, *9*, 276. [CrossRef] [PubMed]
13. O'Brien, J.; Hayder, H.; Zayed, Y.; Peng, C. Overview of MicroRNA Biogenesis, Mechanisms of Actions, and Circulation. *Front. Endocrinol. (Lausanne)* **2018**, *9*, 402. [CrossRef] [PubMed]
14. Gil-Martínez, M.; Lorente-Sorolla, C.; Rodrigo-Muñoz, J.M.; Lendínez, M.Á.; Núñez-Moreno, G.; de la Fuente, L.; Mínguez, P.; Mahíllo-Fernández, I.; Sastre, J.; Valverde-Monge, M.; et al. Analysis of Differentially Expressed MicroRNAs in Serum and Lung Tissues from Individuals with Severe Asthma Treated with Oral Glucocorticoids. *Int. J. Mol. Sci.* **2023**, *24*, 1611. [CrossRef] [PubMed]
15. Scichilone, N.; Barnes, P.J.; Battaglia, S.; Benfante, A.; Brown, R.; Canonica, G.W.; Caramori, G.; Cazzola, M.; Centanni, S.; Cianferoni, A.; et al. The Hidden Burden of Severe Asthma: From Patient Perspective to New Opportunities for Clinicians. *J. Clin. Med.* **2020**, *9*, 2397. [CrossRef] [PubMed]
16. Svenningsen, S.; Nair, P. Asthma Endotypes and an Overview of Targeted Therapy for Asthma. *Front. Med.* **2017**, *4*, 158. [CrossRef]
17. Carr, T.F.; Kraft, M. Use of biomarkers to identify phenotypes and endotypes of severe asthma. *Ann. Allergy Asthma Immunol.* **2018**, *121*, 414–420. [CrossRef]
18. Porpodis, K.; Tsiouprou, I.; Apostolopoulos, A.; Ntontsi, P.; Fouka, E.; Papakosta, D.; Vliagoftis, H.; Domvri, K. Eosinophilic Asthma, Phenotypes-Endotypes and Current Biomarkers of Choice. *J. Pers. Med.* **2022**, *12*, 1093. [CrossRef]
19. Badar, A.; Salem, A.M.; Bamosa, A.O.; Qutub, H.O.; Gupta, R.K.; Siddiqui, I.A. Association between FeNO, Total Blood IgE, Peripheral Blood Eosinophil and Inflammatory Cytokines in Partly Controlled Asthma. *J. Asthma Allergy* **2020**, *13*, 533–543. [CrossRef]
20. Gao, J.; Wu, F. Association between fractional exhaled nitric oxide, sputum induction and peripheral blood eosinophil in uncontrolled asthma. *Allergy Asthma Clin. Immunol.* **2018**, *14*, 21. [CrossRef]
21. Al Ghobain, M.O.; Alsubaie, A.S.; Aljumah, W.A.; Alrumayh, F.M.; Aldawsari, K.F.; Alqahtani, A.M.; Alotaibi, S.N. The Correlation between Fractional Exhaled Nitric Oxide (FeNO), Blood Eosinophil Count, Immunoglobulin E Levels, and Spirometric Values in Patients with Asthma. *Cureus* **2023**, *15*, e35289. [CrossRef]
22. McDowell, P.J.; Heaney, L.G. Different endotypes and phenotypes drive the heterogeneity in severe asthma. *Allergy* **2020**, *75*, 302–310. [CrossRef]
23. Kaplan, A.; Szefler, S.J.; Halpin, D.M.G. Impact of comorbid conditions on asthmatic adults and children. *NPJ Prim. Care Respir. Med.* **2020**, *30*, 36. [CrossRef] [PubMed]
24. Maule, M.; Olivieri, B.; Guarnieri, G.; De Franceschi, L.; Martinelli, N.; Vaia, R.; Argentino, G.; Vianello, A.; Senna, G.; Caminati, M. Hidden Comorbidities in Asthma: A Perspective for a Personalized Approach. *J. Clin. Med.* **2023**, *12*, 2294. [CrossRef]
25. Yawn, B.P.; Han, M.L.K. Practical Considerations for the Diagnosis and Management of Asthma in Older Adults. *Mayo Clin. Proc.* **2017**, *92*, 1697–1705. [CrossRef] [PubMed]
26. Ban, G.Y.; Trinh, T.H.K.; Ye, Y.M.; Park, H.S. Predictors of asthma control in elderly patients. *Curr. Opin. Allergy Clin. Immunol.* **2016**, *16*, 237–243. [CrossRef] [PubMed]
27. Tashiro, H.; Shore, S.A. Obesity and severe asthma. *Allergol. Int.* **2019**, *68*, 135–142. [CrossRef] [PubMed]
28. Oudjedi, A.; Said Aissa, K. Associations between obesity, asthma and physical activity in children and adolescents. *Apunts Sports Med.* **2020**, *55*, 39–48. [CrossRef]
29. Lin, X.; Li, H. Obesity: Epidemiology, Pathophysiology, and Therapeutics. *Front. Endocrinol.* **2021**, *12*, 706978. [CrossRef]
30. Juel, C.T.B.; Ulrik, C.S. Obesity and asthma: Impact on severity, asthma control, and response to therapy. *Respir. Care* **2013**, *58*, 867–873. [CrossRef]

31. Gonzalez-Barcala, F.-J.; Nieto-Fontarigo, J.-J.; Lourido-Cebreiro, T.; Rodríguez-García, C.; San-Jose, M.-E.; Carreira, J.-M.; Calvo-Alvarez, U.; Cruz, M.-J.; Facal, D.; Garcia-Sanz, M.-T.; et al. Obesity Does Not Increase the Risk of Asthma Readmissions. *J. Clin. Med.* **2020**, *9*, 221. [CrossRef]
32. Bermúdez Barón, N.; Kankaanranta, H.; Hedman, L.; Andersson, M.; Stridsman, C.; Lindberg, A.; Rönmark, E.; Backman, H. Body mass index increase: A risk factor for forced expiratory volume in 1 s decline for overweight and obese adults with asthma. *ERJ Open Res.* **2022**, *8*, 00110–02022. [CrossRef]
33. Heaney, L.G.; Perez de Llano, L.; Al-Ahmad, M.; Backer, V.; Busby, J.; Canonica, G.W.; Christoff, G.C.; Cosio, B.G.; FitzGerald, J.M.; Heffler, E.; et al. Eosinophilic and Noneosinophilic Asthma: An Expert Consensus Framework to Characterize Phenotypes in a Global Real-Life Severe Asthma Cohort. *Chest* **2021**, *160*, 814–830. [CrossRef]
34. Kyyaly, M.A.; Vorobeva, E.V.; Kothalawala, D.M.; Fong, W.C.G.; He, P.; Sones, C.L.; Al-Zahrani, M.; Sanchez-Elsner, T.; Arshad, S.H.; Kurukulaaratchy, R.J. MicroRNAs—A Promising Tool for Asthma Diagnosis and Severity Assessment: A Systematic Review. *J. Pers. Med.* **2022**, *12*, 543. [CrossRef] [PubMed]
35. Taghadosi, M.; Adib, M.; Jamshidi, A.; Mahmoudi, M.; Farhadi, E. The p53 status in rheumatoid arthritis with focus on fibroblast-like synoviocytes. *Immunol. Res.* **2021**, *69*, 225–238. [CrossRef]
36. Joseph, C.; Tatler, A.L. Pathobiology of Airway Remodeling in Asthma: The Emerging Role of Integrins. *J. Asthma Allergy* **2022**, *15*, 595–610. [CrossRef] [PubMed]
37. Veraldi, K.L.; Gibson, B.T.; Yasuoka, H.; Myerburg, M.M.; Kelly, E.A.; Balzar, S.; Jarjour, N.N.; Pilewski, J.M.; Wenzel, S.E.; Feghali-Bostwick, C.A. Role of insulin-like growth factor binding protein-3 in allergic airway remodeling. *Am. J. Respir. Crit. Care Med.* **2009**, *180*, 611–617. [CrossRef]
38. Asgharihanjani, N.; Vafa, M. The role of IGF-1 in obesity, cardiovascular disease, and cancer. *Med. J. Islam Repub. Iran* **2019**, *33*, 56. [CrossRef] [PubMed]
39. Bach, L.A. IGF-binding proteins. *J. Mol. Endocrinol.* **2018**, *61*, T11–T28. [CrossRef]
40. Zhang, W.B.; Aleksic, S.; Gao, T.; Weiss, E.F.; Demetriou, E.; Verghese, J.; Holtzer, R.; Barzilai, N.; Milman, S. Insulin-like Growth Factor-1 and IGF Binding Proteins Predict All-Cause Mortality and Morbidity in Older Adults. *Cells* **2020**, *9*, 1368. [CrossRef]
41. Kim, H.; Fu, Y.; Hong, H.J.; Lee, S.G.; Lee, D.S.; Kim, H.M. Structural basis for assembly and disassembly of the IGF/IGFBP/ALS ternary complex. *Nat. Commun.* **2022**, *13*, 4434. [CrossRef] [PubMed]
42. Singh, S.; Prakash, Y.S.; Linneberg, A.; Agrawal, A. Insulin and the lung: Connecting asthma and metabolic syndrome. *J. Allergy (Cairo)* **2013**, *2013*, 627384. [CrossRef] [PubMed]
43. Carr, T.F.; Granell, R.; Stern, D.A.; Guerra, S.; Wright, A.; Halonen, M.; Henderson, J.; Martinez, F.D. High Insulin in Early Childhood Is Associated with Subsequent Asthma Risk Independent of Body Mass Index. *J. Allergy Clin. Immunol. Pract.* **2022**, *10*, 785–792.e5. [CrossRef] [PubMed]
44. Peters, M.C.; Schiebler, M.L.; Cardet, J.C.; Johansson, M.W.; Sorkness, R.; DeBoer, M.D.; Bleecker, E.R.; Meyers, D.A.; Castro, M.; Sumino, K.; et al. The Impact of Insulin Resistance on Loss of Lung Function and Response to Treatment in Asthma. *Am. J. Respir. Crit. Care Med.* **2022**, *206*, 1096–1106. [CrossRef] [PubMed]
45. Nagao, H.; Cai, W.; Wewer Albrechtsen, N.J.; Steger, M.; Batista, T.M.; Pan, H.; Dreyfuss, J.M.; Mann, M.; Kahn, C.R. Distinct signaling by insulin and IGF-1 receptors and their extra- and intracellular domains. *Proc. Natl. Acad. Sci. USA* **2021**, *118*, e2019474118. [CrossRef] [PubMed]
46. Mohanraj, L.; Kim, H.S.; Li, W.; Cai, Q.; Kim, K.E.; Shin, H.-J.; Lee, Y.-J.; Lee, W.J.; Kim, J.H.; Oh, Y. IGFBP-3 inhibits cytokine-induced insulin resistance and early manifestations of atherosclerosis. *PLoS ONE* **2013**, *8*, e55084. [CrossRef]
47. Pescador, N.; Pérez-Barba, M.; Ibarra, J.M.; Corbatón, A.; Martínez-Larrad, M.T.; Serrano-Ríos, M. Serum circulating microRNA profiling for identification of potential type two diabetes and obesity biomarkers. *PLoS ONE* **2013**, *8*, e77251. [CrossRef]
48. Wang, Z.; Li, W.; Guo, Q.; Wang, Y.; Ma, L.; Zhang, X. Insulin-Like Growth Factor-1 Signaling in Lung Development and Inflammatory Lung Diseases. *Biomed. Res. Int.* **2018**, *2018*, 6057589. [CrossRef]
49. Global Initiative for Asthma—GINA. Available online: https://ginasthma.org/ (accessed on 10 May 2023).
50. Livak, K.J.; Schmittgen, T.D. Analysis of relative gene expression data using real-time quantitative PCR and the 2(-Delta Delta C(T)) Method. *Methods* **2001**, *25*, 402–408. [CrossRef]
51. Vlachos, I.S.; Zagganas, K.; Paraskevopoulou, M.D.; Georgakilas, G.; Karagkouni, D.; Vergoulis, T.; Dalamagas, T.; Hatzigeorgiou, A.G. DIANA-miRPath v3.0: Deciphering microRNA function with experimental support. *Nucleic Acids Res.* **2015**, *43*, W460–W466. [CrossRef]
52. Zehavi, L.; Avraham, R.; Barzilai, A.; Bar-Ilan, D.; Navon, R.; Sidi, Y.; Avni, D.; Leibowitz-Amit, R. Silencing of a large microRNA cluster on human chromosome 14q32 in melanoma: Biological effects of mir-376a and mir-376c on insulin growth factor 1 receptor. *Mol. Cancer* **2012**, *11*, 44. [CrossRef] [PubMed]

International Journal of
Molecular Sciences

MDPI

Article

Circular Intronic RNA circTTN Inhibits Host Gene Transcription and Myogenesis by Recruiting PURB Proteins to form Heterotypic Complexes

Nini Ai [1], Zonggang Yu [1], Xueli Xu [1], Sui Liufu [1], Kaiming Wang [1], Shengqiang Huang [1], Xintong Li [1], Xiaolin Liu [1], Bohe Chen [1], Haiming Ma [1,2,*] and Yulong Yin [1,3,*]

1 College of Animal Science and Technology, Hunan Agricultural University, Changsha 410128, China; 15291270712@163.com (N.A.); study236@163.com (Z.Y.); xxl1632501152@163.com (X.X.); liufusui0816@163.com (S.L.); m15116529648@163.com (K.W.); hsq07@126.com (S.H.); p13548614212@163.com (X.L.); lxl810711@163.com (X.L.); chenhe0213914@163.com (B.C.)

2 Guangdong Laboratory for Lingnan Modern Agriculture, Guangzhou 510642, China

3 Institute of Subtropical Agriculture, Chinese Academy of Sciences, Changsha 410125, China

* Correspondence: mahaiming@hunau.edu.cn (H.M.); yinyulong@isa.ac.cn (Y.Y.)

Abstract: Muscle cell growth plays an important role in skeletal muscle development. Circular RNAs (circRNAs) have been proven to be involved in the regulation of skeletal muscle growth and development. In this study, we explored the effect of circTTN on myoblast growth and its possible molecular mechanism. Using C2C12 cells as a functional model, the authenticity of circTTN was confirmed by RNase R digestion and Sanger sequencing. Previous functional studies have showed that the overexpression of circTTN inhibits myoblast proliferation and differentiation. Mechanistically, circTTN recruits the PURB protein on the *Titin* (*TTN*) promoter to inhibit the expression of the *TTN* gene. Moreover, PURB inhibits myoblast proliferation and differentiation, which is consistent with circTTN function. In summary, our results indicate that circTTN inhibits the transcription and myogenesis of the host gene *TTN* by recruiting PURB proteins to form heterotypic complexes. This work may act as a reference for further research on the role of circRNA in skeletal muscle growth and development.

Keywords: circTTN; C2C12; myogenesis; proliferation; PURB; *TTN*

1. Introduction

Skeletal muscle is an important component of the mammalian body, which plays an important role in energy transformation and posture maintenance [1]. The skeletal muscle development process includes three stages: (1) mesenchymal stem cells from mesoderm proliferate and differentiate into myoblasts, (2) myoblasts further differentiate to form primary myotubes and secondary myotubes, and (3) myotubes then fuse into muscle fibers [2]. This process is also known as myogenesis. Muscle fibers are the basic contractile units of skeletal muscle. They are surrounded by connective tissue layers and are formed into bundles to form skeletal muscle [3]. Skeletal muscle development is subject to myogenic regulatory factors (*MRFs*), paired box family (*PAX*), SIX family, and myocyte enhancer factor 2 (*MEF2*) [4,5]. In recent years, with the upsurge of research on non-coding RNAs (ncRNAs), the mechanism of circular RNAs (circRNAs) in skeletal muscle growth and development is also attracting researchers' interest.

circRNAs are a class of non-coding RNA molecules that do not have a 5′ cap and 3′ poly(A) tail. Dicer [6] first discovered the potato spindle tuber "virus" in 1971 and called it viroid. In 1993, circular transcription of the *Sry* gene was found in mice, but the circRNAs discovered were thought to be low-abundance RNA molecules formed by the missplicing of exon transcripts [7]. With the rapid development of high-throughput sequencing, more circRNAs have been discovered by researchers [8]. Currently, circRNAs have been shown to be associated with myogenesis, the transformation of muscle fiber types, and muscle diseases [9–11]. For example, circZfp609 can act as a molecular sponge to adsorb miR-194-5p and isolate its inhibition of *BCLAF1*, thus inhibiting myogenic differentiation [12]. In addition, circRNAs can regulate the variable splicing of transcripts and parental gene transcription and translation by binding to RBPs [13–15]. However, the previous research on this mechanism has mainly been focused on the aspect of cancer treatment and less on the aspect of skeletal muscle development, which requires further exploration.

We previously found a circular intronic RNA circTTN produced by the *TTN* (*Titin*) gene through high-throughput sequencing [16]. circTTN was significantly differentially expressed in the skeletal muscle of Ningxiang pig at 30 d, 90 d, 150 d, and 210 d after birth, suggesting that circTTN may be involved in the regulation of skeletal muscle myogenesis. In vitro, functional validation models (C2C12 cells) showed that circTTN is a negative regulator of myogenesis and that circTTN recruits PURB protein to inhibit the transcription of the host gene *TTN*. Our study revealed the molecular regulatory mechanism of circTTN and provided a reference for further research on skeletal muscle growth and development.

2. Results

2.1. Characterization Analysis of circTTN

circTTN is derived from intron15 of the *TTN* gene; its length is 339 bases, and it is located in chromosome 15 (Chr15_85761324_85761662) (Figure 1A). First, the divergent primer and convergent primer were designed to perform qRT-PCR experiments using the cDNA and gDNA of muscle tissue as templates, and DNA gel electrophoresis was performed on the products. The convergent primer and divergent primer could amplify bands using cDNA as a template. Only the convergent primer could amplify a band when gDNA was used as the template (Figure 1B). The products of the divergent primer were analyzed using Sanger sequencing (Figure 1A). After treatment with RNase R, *TTN* and *β-actin* genes had no bands after electrophoresis, while circTTN had bands (Figure 1C). The above experimental results can confirm the authenticity and stability of circTTN.

Then, the expression of circTTN in different tissues of Ningxiang pig at four periods after birth were detected using qRT-PCR assay. The results showed that the expression of circTTN in the heart and muscle of Ningxiang pig at four stages was the highest (Figure 1D). In addition, circTTN was differentially expressed during the proliferation and differentiation of C2C12 cells (Figure 1E,F).

2.2. circTTN Inhibits Proliferation and Differentiation of C2C12 Cells and Host Gene TTN

To explore the function of circTTN, the circTTN overexpression vector was transfected into C2C12 cells, and the transfection efficiency of circTTN was more than 3000 times higher ($p < 0.01$) (Figure 2A), which can be used for subsequent experiments. First, qRT-PCR and Western blot were used to detect the expression of proliferation biomarker genes. The overexpression of circTTN significantly reduced the mRNA levels of *CCND* and *CDK4* ($p < 0.05$); the mRNA expression level of *PCNA* was markedly decreased ($p < 0.01$) (Figure 2B). The protein expression levels of proliferating genes CCND and PCNA in the overexpression group were markedly lower than those in the control group ($p < 0.01$); the CDK4 protein expression level was significantly decreased ($p < 0.05$) (Figure 2C,D). The results of the CCK8 experiments displayed that there were no significant difference in the proliferative activity between the overexpressed group and the control group ($p > 0.05$) (Figure 2E); the EdU results showed that the number of positive proliferating cells in the overexpressed group was markedly decreased ($p < 0.01$) (Figure 2F,G). In addition, flow cytometry was used to detect circTTN expression during cell cycle process, and the number of cells in the overexpressed group was significantly reduced in the S phase ($p < 0.01$) (Figure 2H,I). In summary, circTTN inhibited the proliferation of C2C12 cells.

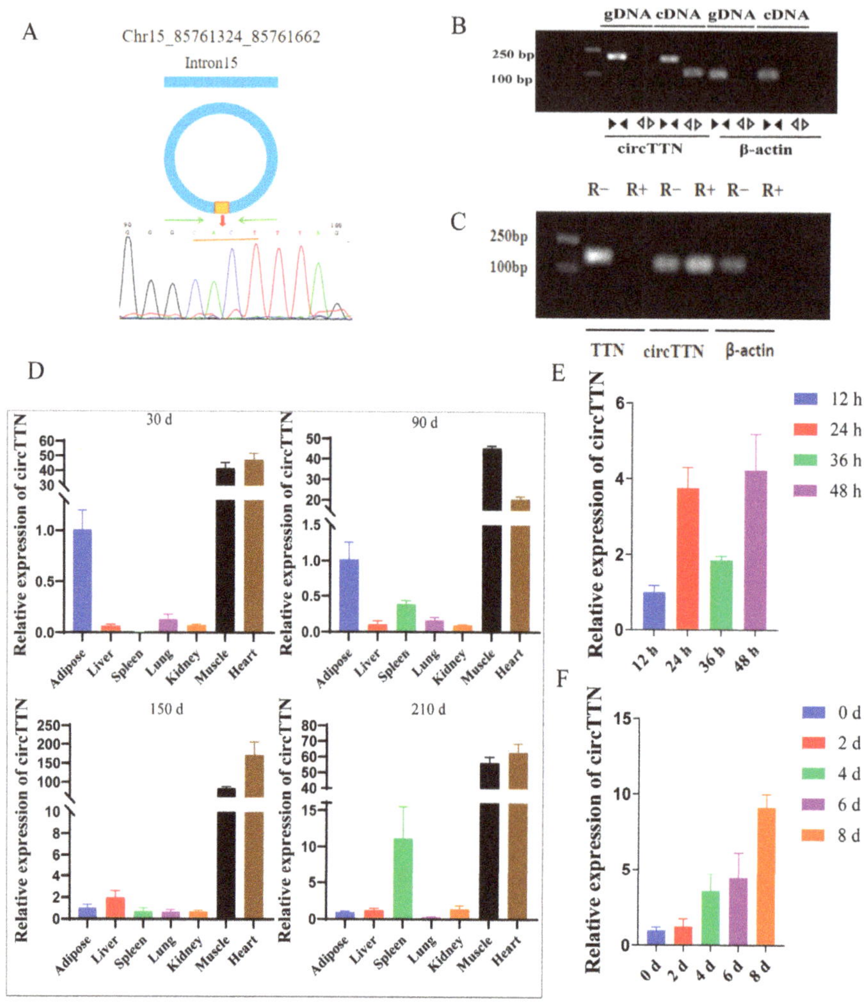

Figure 1. Characterization analysis of circTTN. (**A**) circTTN source diagram and the joint sequence by sanger sequencing. (**B**) The authenticity of circTTN was detected by divergent primer and convergent primer. (**C**) The stability of circTTN was determined by RNase R digestion. (**D**) The expression of circTTN in different tissues of Ningxiang pig at 30 d, 90 d, 150 d, and 210 d after birth was detected. (**E**) circTTN expression in C2C12 cells during proliferation period was detected. (**F**) circTTN expression during C2C12 cells differentiation was detected.

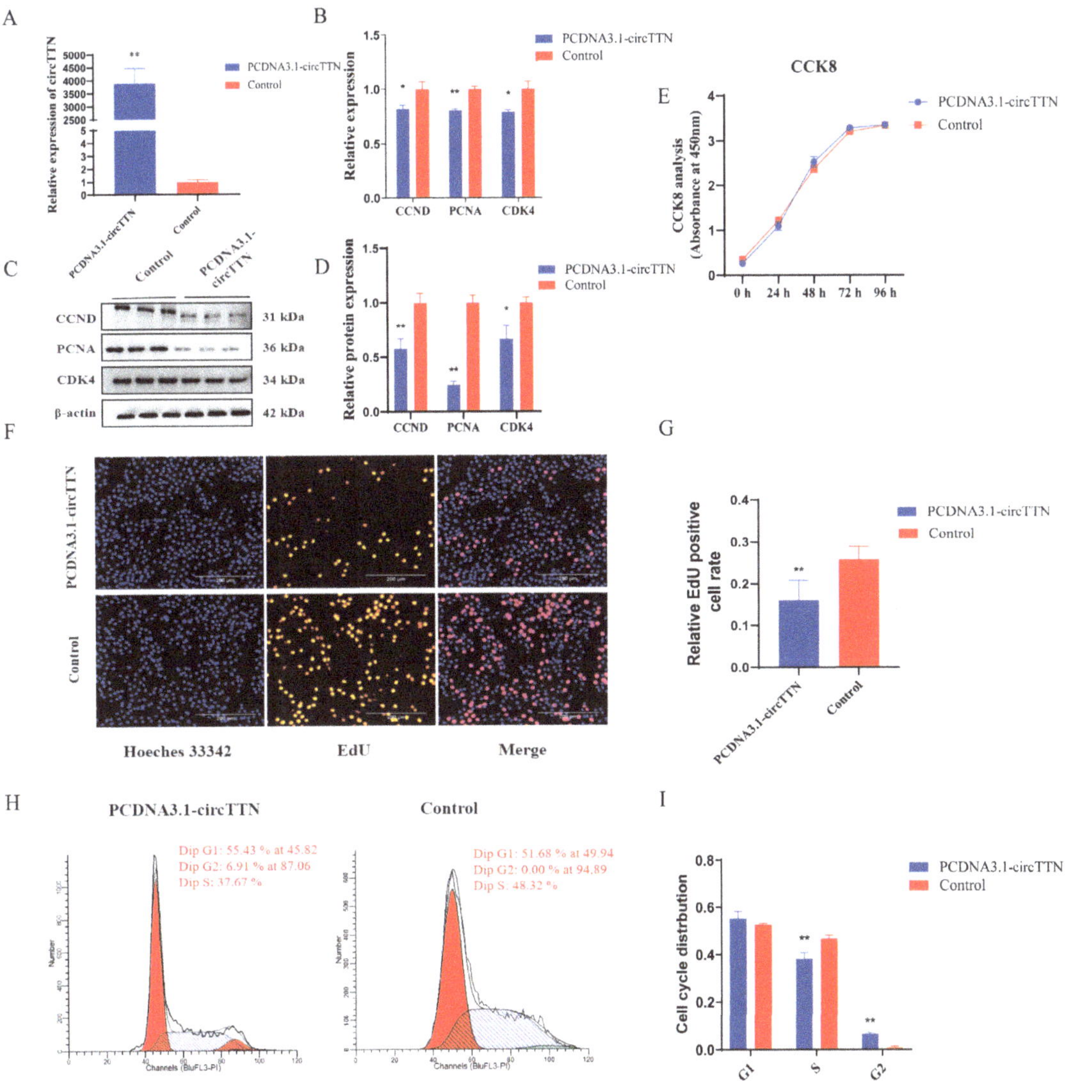

Figure 2. Effect of overexpression of circTTN on proliferation of C2C12 cells. (**A**) Transfection efficiency was screened by qRT-PCR 24 h after transfection with circTTN overexpression (PCDNA3.1-circTTN) and empty vector (control). The transfection condition is 4.0 μg PCDNA3.1-circTTN/control and 10 μL Lipofectamine 2000 reagent. (**B**) The effect of circTTN overexpression was detected by qRT-PCR on proliferation biomarker genes at the mRNA level. (**C**) The effect of circTTN overexpression was detected by Western blot on proliferation biomarker genes at the protein level. (**D**) Gray scale analysis of protein band in B. (**E**) Detection of cell proliferative activity using CCK8 after transfection with PCDNA3.1-circTTN and control. (**F**) Cell proliferation was detected by EdU. Scale bars indicate 200 μm. (**G**) EdU-positive cells were counted. (**H**) Cell cycle changes were detected using flow cytometry. (**I**) Count the number of proliferating cells during the cell cycle. * $p < 0.05$ and ** $p < 0.01$. Results are presented as mean ± S.E.M. n = 3.

We further explored the effect of circTTN on the differentiation of C2C12 cells; the overexpressed circTTN vector was transfected into cells for induction differentiation culture, and then, qRT-PCR and a Western blot assay were used to detect the expression of differentiation biomarker genes. The overexpression of circTTN significantly inhibited the expression of differentiation biomarker genes (*MyoD*, *MyHC*, and *MyoG*) at the mRNA level (Figure 3A) ($p < 0.01$). At the protein level, the expression of *MyoD* and *MyoG* were markedly repressed ($p < 0.01$), significantly inhibiting the expression of *MyHC* ($p < 0.05$) (Figure 3B,C). In addition, an immunofluorescence assay was used to observe the changes in phenotypic myotube differentiation, and the number in differentiated myotubes in the circTTN overexpression group was significantly reduced compared with the control group ($p < 0.05$) (Figure 3D). These results indicate that circTTN inhibited the differentiation of C2C12 cells.

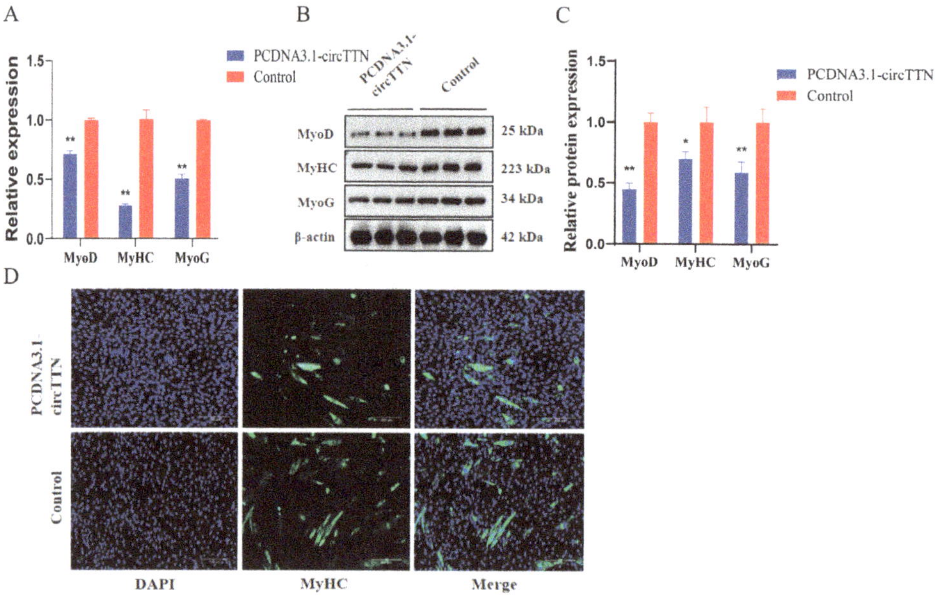

Figure 3. Effect of overexpression of circTTN on differentiation of C2C12 cells. (**A**) At 72 h after transfection and differentiation, qRT-PCR was used to detect the effect of circTTN overexpression on the mRNA level of differentiation biomarker genes. (**B**) Effect of overexpression of circTTN on differentiation biomarker genes at the protein level. (**C**) Gray scale analysis of protein band in B. (**D**) The changes in circTTN overexpressed myotubes were detected by immunofluorescence at sixth day of differentiation. Scale bars indicate 200 μm. * $p < 0.05$ and ** $p < 0.01$. Results are presented as mean ± S.E.M. n = 3.

TTN is a coding gene that plays an important role in skeletal muscle. We intended to discover if circTTN acts on host gene *TTN*. The qRT-PCR results implied that circTTN overexpression significantly downregulated *TTN* gene expression during the proliferation period (Figure 4A) ($p < 0.01$). At the protein level, the overexpression of circTTN was markedly lower during proliferation compared with the control group (Figure 4B,C) ($p < 0.01$). Similarly, the effect of circTTN overexpression on the *TTN* gene during differentiation was consistent with that during proliferation, and the mRNA and protein levels of circTTN were significantly decreased compared with those of the control group ($p < 0.01$) (Figure 4D–F).

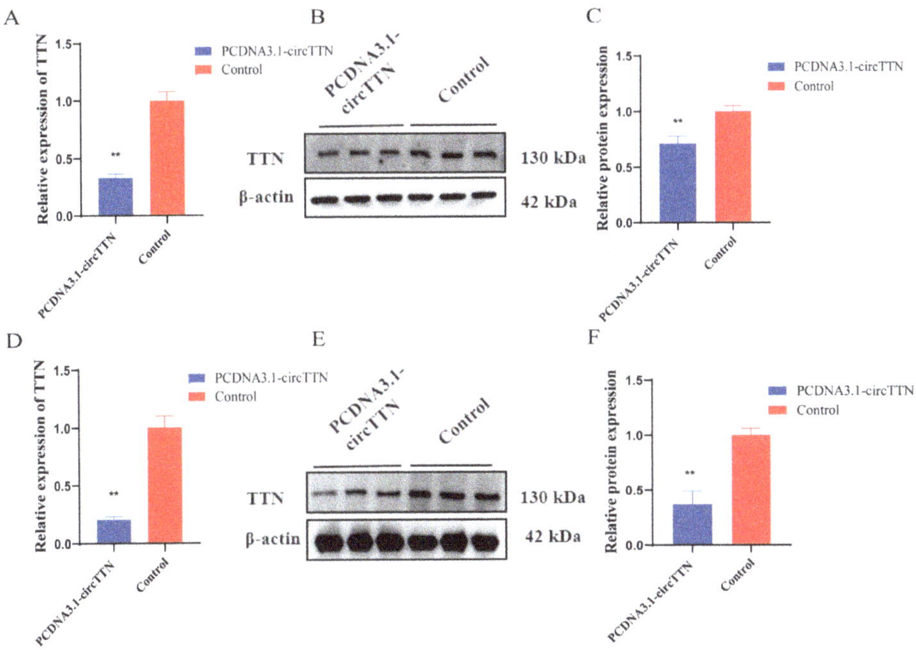

Figure 4. Effect of overexpression of circTTN on *TTN* gene. (**A**) qRT-PCR assay was used to detect the effect of overexpressed circTTN on the proliferation stage of *TTN*. (**B**) Western blot assay was used to detect the effect of overexpressed circTTN on *TTN*. (**C**) Gray scale analysis of protein bands in figure B. (**D**) The effect of circTTN overexpression on *TTN* differentiation stage was detected by qRT-PCR. (**E**) Western blot assay was used to detect the effect of overexpressed circTTN on *TTN* at differentiation stage. (**F**) Protein band gray analysis statistics in figure (**E**). ** $p < 0.01$. Results are presented as mean ± S.E.M. n = 3.

2.3. circTTN Interacts with PURB Protein

First, we determined the subcellular localization of circTTN by nucleoplasmic isolation and RNA fluorescence in situ hybridization experiments. *GAPDH* and *U6* were used as reference genes in the cytoplasm and nucleus, respectively. The results showed that circTTN and *TTN* mRNA existed in the nucleus and cytoplasm, but most of them were located in the nucleus (Figure 5A). The results of the FISH experiment were consistent with the results of nucleoplasmic separation. (Figure 5B). These results suggest that circTTN may be regulated in the nucleus. circRNA can interact with RNA-binding proteins to influence the expression of parent genes. We then examined circTTN-bound proteins using an RNA pull-down assay. The circTTN group showed obvious bands of difference compared with the control NC group (Figure 5C). Subsequently, a total of 356 proteins were identified using mass spectrometry (MS) analysis, among which 218 different proteins were found in the circTTN group (RPD group) (Figure 5D, Table 1). Gene ontology (GO) functional enrichment analysis and Kyoto Encyclopedia of Genes and Genomes (KEGG) analysis were conducted, respectively. GO functional enrichment showed that these proteins were mainly enriched in cellular processes, protein complexes, regulatory translation activities, biological regulation, developmental processes, and binding. KEGG suggested that some proteins were enriched in transcription, translation, and metabolism (Figure 5F). These results implied that circTTN may bind to these proteins and play a certain function. Next, the expression of Pur-beta (PURB) protein in the RNA pull-down eluent was detected using a Western blot assay (Figure 5G), and it was found that the expression of PURB protein could be only detected in the circTTN group's eluent. In addition, we reversely demonstrated an interaction between PURB protein and circTTN through an RIP experiment (Figure 5H). The results showed that the PURB antibody could be significantly enriched to circTTN ($p < 0.01$), further supporting the idea that circTTN interacts with PURB.

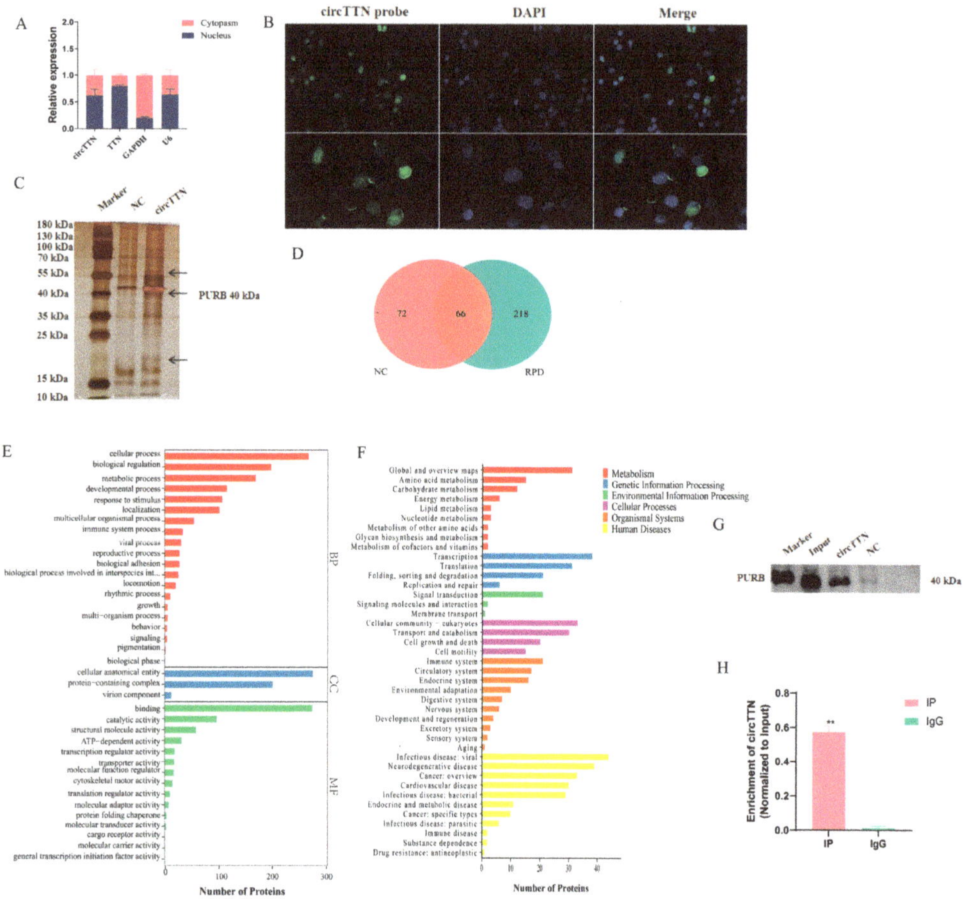

Figure 5. circTTN interacts with PURB protein. (**A**) The location of circTTN was determined by nucleoplasmic separation. (**B**) We determined the location of circTTN using FISH. 400× (**upper** images) and 1000× (**bottom** images) (**C**) In the RNA pull-down experimental silver glue diagram, arrows indicate differential proteins. (**D**) Protein statistics were identified by mass spectrometry. (**E**) Functional enrichment of protein GO was identified by mass spectrometry. (**F**) Protein KEGG analysis was identified by mass spectrometry. (**G**) The expression of Pur-beta (PURB) protein in RNA pull-down eluent was detected using WB. (**H**) RIP experiments reverse demonstrated that PURB protein interacts with circTTN. ** $p < 0.01$. Results are presented as mean ± S.E.M. n = 3.

Table 1. Possible circTTN interaction proteins identified by RNA pull-down RPD group (top 30).

Accessions	Names	Unused	Description
Q6QAQ1	ACTB	191.97	Actin, cytoplasmic 1
P02543	VIM	149.62	Vimentin
A0A287AU59	ACACA	133.63	Acetyl-CoA carboxylase 1
A0A287BNK7	PLEC	133	Uncharacterized protein
A0A287BHM1	TPM1	91.59	Tropomyosin alpha-1 chain
A0A5G2RI50	MYO1C	83.87	Unconventional myosin-Ic isoform b
Q2XVP4	TUBA1B	61.55	Tubulin alpha-1B chain
P02554	Tubulin beta	60.54	Tubulin beta chain
F6Q364	ACTA2	59.23	Uncharacterized protein
P02540	DES	52.77	Desmin
A0A287BQP1	RBM25	36.79	Uncharacterized protein
A0A288CG57	EEF1A1	36.1	Elongation factor 1-alpha
A0A286ZVS0	SF3B3	35.56	Splicing factor 3B subunit 3
F1SSA6	MYH10	33.34	Myosin motor domain-containing protein
A0A287BSP1	MYO5A	32.65	Uncharacterized protein
A0A287A4H2	PURB	32.47	Transcriptional activator protein Pur-beta
P80021	ATPATP5F1A	31.56	ATP synthase subunit alpha, mitochondrial
F1RUL1	RBM17	29.87	Splicing factor 45
A0A287B1F6	MPRIP	29.21	Uncharacterized protein
K9J4V0	SNRNP200	29	Protogenin isoform X1
I3L920	CORO1C	27.2	Coronin
I3L650	CALD1	26.54	Uncharacterized protein
A0A287BIL8	GN = HSPA5	26.3	78 kDa glucose-regulated protein
K7GLT8	ATP5F1B	24.2	ATP synthase subunit beta
A0A5G2R8T3	MYO1E	23.9	Uncharacterized protein
A0A286ZHW0	SF3B1	21.89	Splicing factor 3B subunit 1 isoform 1
F1S5A8	DHX15	21.49	RNA helicase
A0A286ZWK2	HSPA8	20.98	Uncharacterized protein
I3LLT2	CAD	20.86	Aspartate carbamoyltransferase
A0A286ZXU2	FLNA	19.38	Uncharacterized protein

2.4. PURB Knockdown Promotes Proliferation and Differentiation of C2C12 Cells and Increase TTN Expression

Previously, we confirmed the interaction between PURB protein and circTTN. Next, we detected the expression of PURB in C2C12 cells using a Western blot assay (Figure 6A), and it was found that PURB was highly expressed in C2C12 cells. To further explore the effect of PURB on C2C12 cells, we designed 3 siRNA sequences, as shown in Figure 6B; qRT-PCR was used to select the one with the best knockdown effect, and siRNA2 had the best knockdown efficiency ($p < 0.01$) (5′-GUGGACUCCAAGCGUUUCUTT-3′), which can be used for subsequent experiments. siRNA-PURB was transfected into C2C12 cells, and siRNA-PURB significantly promoted the expression of *CDK4* and *PCNA* genes at the mRNA level (Figure 6C) ($p < 0.01$). The protein expression level of CDK4 was markedly higher than that of the control group ($p < 0.01$), and the protein expression level of PCNA was significantly increased (Figure 6D,E) ($p < 0.05$). These results indicated that the knockdown PURB promoted the proliferation of C2C12. Similarly, siRNA-PURB was transfected into cells and induced differentiation was performed until myotubes were grown. As shown in Figure 6F, siRNA-PURB significantly upregulated the expression of differentiation biomarker genes (*MyoD, MyHC,* and *MyoG*) at the mRNA level ($p < 0.01$). siRNA-PURB markedly elevated the protein expression of MyHC ($p < 0.01$) and significantly promoted the protein expression of MyoD and MyoG ($p < 0.05$) (Figure 6G,H). In conclusion, PURB inhibited the proliferation and differentiation of C2C12 cells.

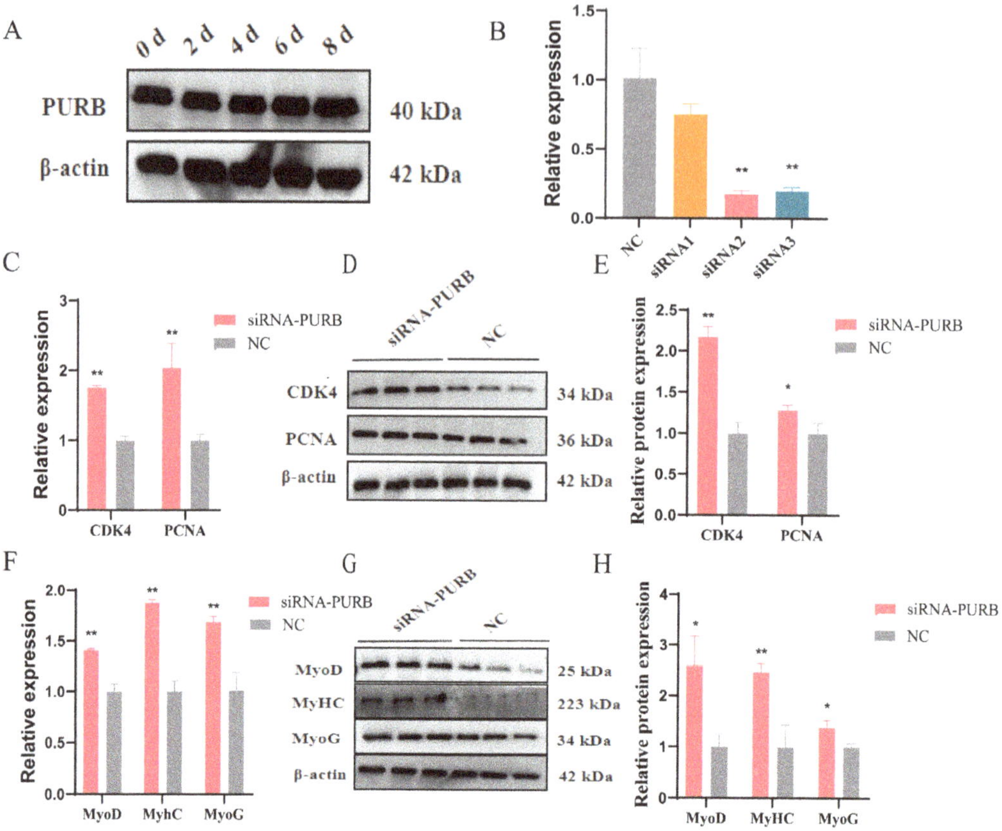

Figure 6. Knockdown effect of PURB on proliferation and differentiation of C2C12 cells. (**A**) PURB protein expression in naturally differentiated C2C12 cells. (**B**) Screening of siRNA sequences by qRT-PCR 24h after transfection with siRNA-PURB and NC. The transfection conditions are 100 pmol siRNA-PURB/NC and 5 μL Lipofectamine 2000 reagent. (**C**) The effect of siRNA-PURB proliferation biomarker genes at mRNA level. (**D**) The effect of siRNA-PURB on proliferation biomarker genes at the protein level. (**E**) Gray scale analysis of protein bands in (**D**). (**F**) At 72 h after transfection and differentiation, qRT-PCR was used to detect the effect of siRNA-PURB on the mRNA level of differentiation biomarker genes (**G**) The effect of siRNA-PURB on differentiation biomarker genes at the protein level. (**H**) Gray scale analysis of protein bands in Figure (**G**). * $p < 0.05$ and ** $p < 0.01$. Results are presented as mean ± S.E.M. n = 3.

To explore the effect of PURB on the *TTN* gene, siRNA-PURB was transfected into C2C12 cells, and the effects of PURB on *TTN* during proliferation and differentiation were detected, respectively. The results show that the mRNA and protein levels of siRNA-PURB were significantly higher than those of the control group at the proliferation stage ($p < 0.01$) (Figure 7A–C). siRNA-PURB were markedly increased compared with the control group at the differentiation stage ($p < 0.01$) (Figure 7D), and the protein level was significantly elevated ($p < 0.05$) (Figure 7E,F). These results indicate that PURB inhibited *TTN* gene expression.

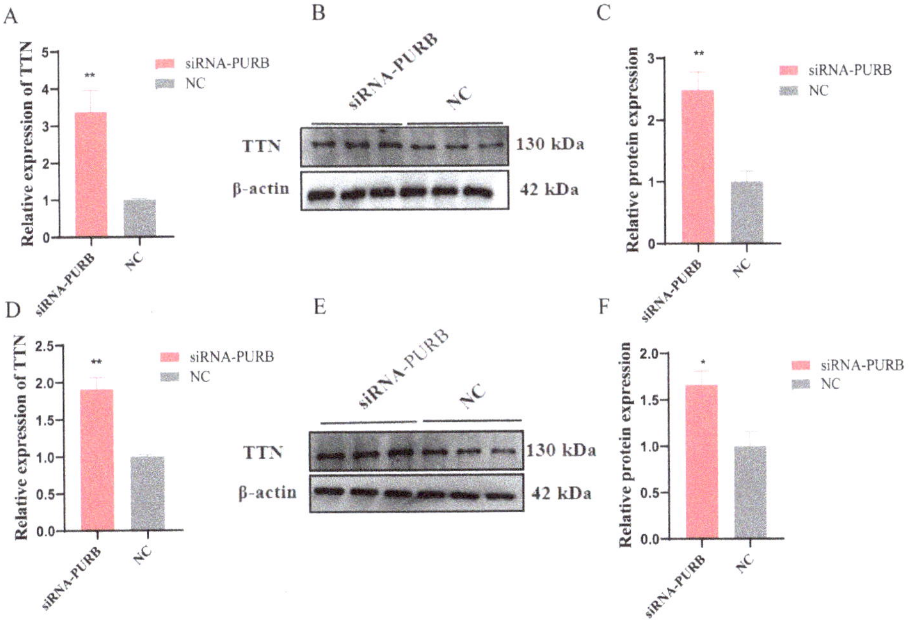

Figure 7. Effect of PURB knockdown on *TTN* gene. (**A**) qRT-PCR was used to detect the effect of PURB knockdown on *TTN* during proliferation stage. (**B**) Western blot assay was used to detect the effect of PURB knockdown on *TTN* during proliferation stage. (**C**) Gray scale analysis of protein bands in (**B**). (**D**) qRT-PCR assay was used to detect the effect of PURB knockdown on *TTN* during differentiation. (**E**) Western blot assay was used to detect the effect of PURB knockdown on *TTN* during differentiation. (**F**) Protein band gray analysis statistics (**E**). * $p < 0.05$ and ** $p < 0.01$. Results are presented as mean ± S.E.M. n = 3.

2.5. circTTN Inhibits TTN Transcription by Recruiting PURB

We intended to verify whether PURB are regulated by circTTN. After circTTN overexpression, PURB induced no significant changes in mRNA and protein levels (Figure 8A–C) ($p < 0.05$). This indicated that circTTN overexpression did not affect the PURB expression level. In a previous study, the effects of circTTN and PURB on the TTN gene were explored, respectively. Then, we utilized the UCSC Genome Browser (http://genome.ucsc.edu/ accessed on 25 August 2022) and PROMO database (https://alggen.lsi.upc.es/cgi-bin/promo_v3/promo/promoinit.cgi?dirDB=TF_8.3 accessed on 25 August 2022) for bioinformatics analysis. We found that the PURB binding motif was present at the *TTN* promoter (Figure 8D), and then, ChIP-qPCR was used to further verify the binding of PURB and the *TTN* promoter. PURB protein significantly bound to the *TTN* promoter (Figure 8E,F) ($p < 0.01$). After circTTN overexpression, we found that PURB protein bound to the *TTN* promoter was significantly higher than in the control group ($p < 0.01$) (Figure 8G). Our study suggested that circTTN inhibited the transcription of the host gene *TTN* by recruiting PURB protein.

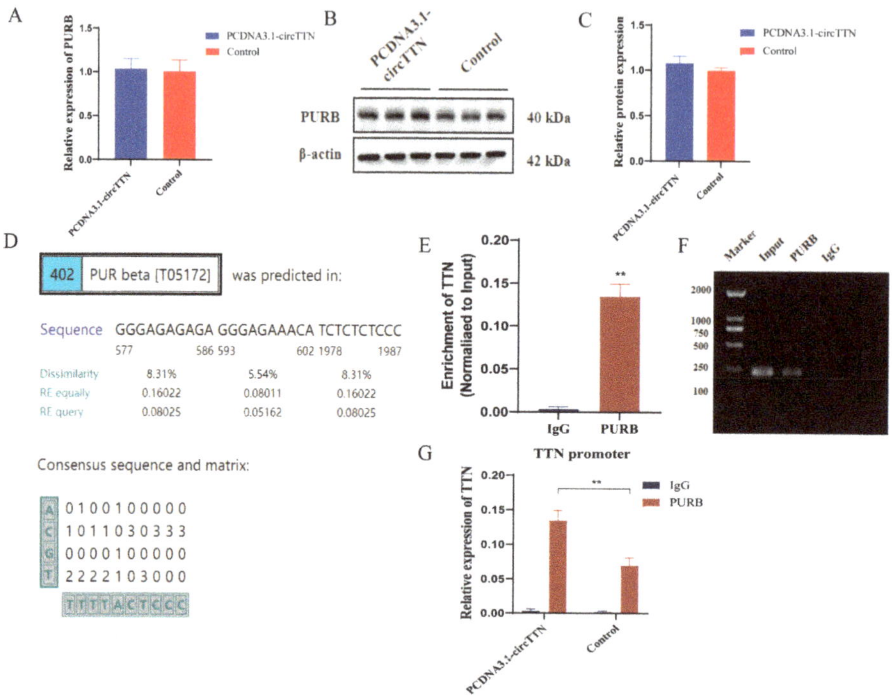

Figure 8. circTTN inhibits *TTN* gene expression by recruiting PURB. (**A**) The effect of overexpression of circTTN on PURB mRNA level. (**B**) The effect of overexpression of circTTN on PURB at the protein level. (**C**) Gray scale analysis of protein bands in (**B**). (**D**) PROMO database predicted PURB binding to *TTN* promoter. (**E**) ChIP-qPCR showed that PURB antibody was enriched to *TTN* promoter. (**F**) Glue map of ChIP-qPCR demonstrated that PURB antibody was enriched to *TTN* promoter. (**G**) ChIP-qPCR showed that circTTN overexpression promoted PURB and *TTN* promoter aggregation. ** $p < 0.01$. Results are presented as mean ± S.E.M. n = 3.

3. Discussion

Skeletal muscle plays a role in coordinating movement and maintaining normal metabolism in organisms [17]. At present, circRNA is involved in skeletal muscle proliferation and differentiation as a positive or negative regulator of skeletal muscle myogenesis [18]. The C2C12 cell line is a valuable in vitro model, originally monocytes, which can rapidly integrate into multinucleated myotubes after culture with an induced differentiation solution [19]. Due to its unique biochemical characteristics, researchers can better understand the metabolism and differentiation of skeletal muscle at cellular and molecular levels [20]. Sun et al. [21] identified circCSDE with a stable and high expression in porcine skeletal muscle and studied the function of circCSDE on C2C12 cells, where they found that circCSDE1 and its target miR-21-3p jointly regulated the proliferation and differentiation of myoblasts. In their study, the mechanism of circCSDE regulating the growth and development of skeletal muscle in the porcine skeletal muscle injury model was verified. In our study, we first reported that circular intronic RNA circTTN acts as a negative regulator of myogenesis and inhibits the transcription of the host gene *TTN* by recruiting PURB protein in the nucleus (Figure 9).

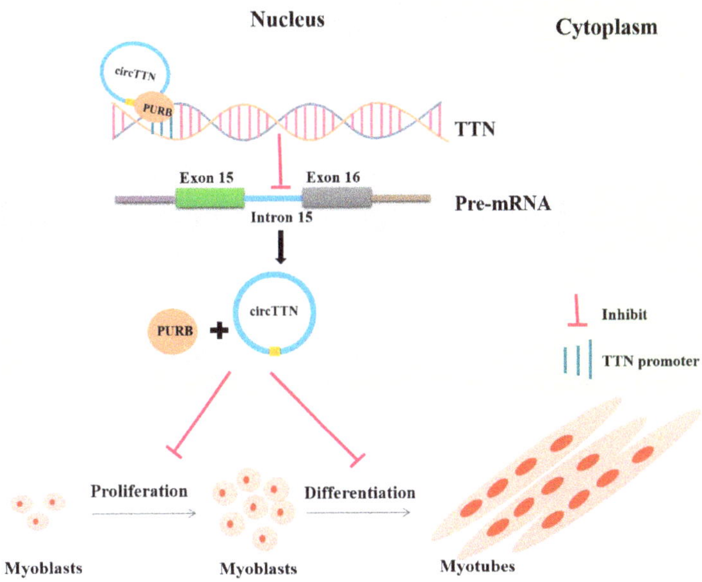

Figure 9. Molecular regulatory mechanism model of circTTN. Circular intronic RNA circTTN acts as a negative regulator of myogenesis and inhibits transcription of host gene TTN by recruiting PURB protein in the nucleus.

circRNAs are usually expressed in a tissue-specific manner and perform different biological functions [22]. circRNAs regulate gene expression through the use of a variety of mechanisms. Some studies have shown that circRNA can act as a miRNA sponge or an effective competing endogenouse RNA (ceRNA) molecule [23,24], which is also one of the most classical mechanisms of circRNAs. circRNAs can also bind RBPs. Although the RBPs' binding site of circRNA is less than that of the corresponding linear mRNA [25], there is still a lot of evidence that RBPs interacts with circRNAs; moreover, they are involved in almost every aspect of the circRNA life cycle, including generation, transcriptional regulation, post-transcriptional regulation, functional execution, specific modifications, and potential extracellular transport pathways [26–28]. In this study, we identified the authenticity of circTTN through the use of RNase R digestion and other methods, and then, detected the expression level of circTTN in Ningxiang pig tissues of different days of age and proliferative and differentiated C2C12 cells. It was found that circTTN was highly expressed in the heart and muscle tissues of Ningxiang pig at different periods. It was also differentially expressed in proliferating and differentiated C2C12 cells. This suggested that circTTN may be involved in the regulation of skeletal muscle development. Then, we demonstrated that circTTN inhibits the proliferation and differentiation of C2C12 cells. A previous study has shown that circTTN promotes the proliferation and differentiation of bovine myoblasts [29], suggesting that circTTN may have different effects on the proliferation and differentiation of myoblasts in different mammals. The reason may be that circRNA plays different functions due to different subcellular localization. Our study showed that circTTN is a negative regulator of muscle regeneration in skeletal muscle. Interestingly, we found that circTTN inhibited the expression of the host gene *TTN* during both proliferation and differentiation. In addition, we found that circTTN is mainly located in the nucleus and interacts with PURB protein. Our study showed that PURB appears to be involved in circTTN expression and function.

PURB is a highly conserved protein whose conserved region includes ssDNA and RNA domains. It is a DNA and RNA binding protein involved in the regulation of DNA replication and transcription [30] and can inhibit or promote gene transcription [31]. PURB has been reported to play different roles in many physiological and pathological processes [32]. It can interact with gonadotropin-releasing hormone1 (*GnRH1*) to control fish reproduction by stimulating the pituitary to release gonadotropin-releasing hormone1 [33]. Hariharan et al. [34] demonstrated that PURB protein negatively regulates the trans-activation of smooth muscle *actin* genes in myoblasts. Based on previous studies on PURB, in our study, we confirmed that PURB inhibits the proliferation and differentiation of C2C12 cells and the expression of the *TTN* gene, which was consistent with the function of circTTN.

Previous studies showed that PURB, as a transcription factor, can be recruited by linc-HOXA1 to mediate the transcriptional regulation of lincHOXA1 in embryonic stem cells [35]. A recent study indicated that LncCMRR functionally interacts with the transcriptional suppressor PURB to inhibit its binding potential in the *Flk1* promoter region, thereby protecting *Flk1* transcription during cardiac mesoderm formation [36]. By binding to PURB protein, Circ-Calm4 promoted the transcription of autophagy-related protein *Beclin1*, thereby regulating the hypoxia-induced autophagy of pulmonary artery smooth muscle [37]. circRNAs can not only act as miRNA molecular sponges or translation proteins to regulate muscle growth and development [38] but they also regulate muscle generation by interacting with RBPs. Pandey et al. [14] determined that PURB was distributed in the nucleus and cytoplasm of C2C12 cells, and it combined with circSamd4 located in the nucleus to participate in the transcription regulation of *MHC*, thus promoting myogenesis, which was also the first revelation that circRNA could interact with RBPs to regulate the growth and development of skeletal muscle. Chen et al. [39] pointed out that circMYBPC1 can directly bind MyIIC protein, promote the differentiation of myoblasts, and possibly promote skeletal muscle regeneration. One study showed that LncMGPF promoted the expression of HuR in C2C12 cytoplasm, enhancing the stability of *MyoD* and *MyoG* mRNA, and thus promoting myogenesis [40,41]. In this study, we found that PURB and the *TTN* promoter have a binding motif through the use of the UCSC genome browser and PROMO database, and we found that circTTN overexpression promoted the binding of PURB and the *TTN* promoter and inhibited the generation of the *TTN* gene. The protein encoded by the *TTN* gene is the largest known protein and plays a key structural, developmental, and regulatory role in heart and skeletal muscle [42,43]. circTTN inhibits *TTN* transcription through PURB recruitment, which also indicates that circTTN affects skeletal muscle development through PURB recruitment. In addition, our study showed that circTTN or PURB negatively regulated the proliferation and differentiation of C2C12 cells; however, it was unclear whether circTTN inhibited the proliferation and differentiation of myoblasts by recruiting PURB. Furthermore, the role of circTTN in muscle regeneration was not explored in vivo, which is also a limitation of our study. These are something that needs to be addressed in future studies.

4. Materials and Methods

4.1. Experimental Animals and Samples

All the animals were obtained from Dalong Animal Husbandry Technology Co., LTD., Ningxiang City, Hunan Province. In total, 7 tissue samples (n = 3) were taken from Ningxiang pigs at 30 d, 90 d, 150 d, and 210 d after birth, including heart, liver, spleen, lung, kidney, longissimus dorsi muscle, and adipose. The sampling process strictly followed the asepsis rules. The samples were placed in liquid nitrogen where they were then brought back to the laboratory and finally stored in a freezer at −80 °C.

4.2. Culture and Differentiation of C2C12 Cells

C2C12 cells were purchased from Anweisci (Shanghai, China). The resuscitated cells were cultured with growth medium (CM) in 10 cm dishes. CM was composed of 89% DMEM (Gibco, Waltham, MA, USA), 10% fetal bovine serum (Gibco, Waltham, MA, USA), and 1% penicillin–streptomycin (Gibco, Waltham, MA, USA). When the cell confluence reached 80–90%, trypsin was used for digestion (Gibco, Waltham, MA, USA), and the medium was changed to differentiation medium (DM), which was composed of 2% horse serum and 98% DMEM. C2C12 cells were cultured in a 37 °C incubator containing 5% CO_2.

4.3. RNA Extraction and cDNA Synthesis

An RNA Extraction kit (TIANGEN, Beijing, China) was used to extract total RNA. A nucleic acid concentration was used for the measurements using NanoDrop ND-2000 (Thermo Scientific, Waltham, MA, USA). The samples that needed to be processed with RNase R were digested using RNase R (JISAI, Guangzhou, China). Then, cDNA synthesis was performed using a RevertAid First Strand cDNA Synthesis Kit (Thermo Scientific, Waltham, MA, USA) according to the manufacturer's instructions.

4.4. Quantitative Real-Time PCR

Quantitative real-time PCR (qRT-PCR) primer sequences were designed from the NCBI (https://www.ncbi.nlm.nih.gov/tools/primer-blast/ accessed on 10 November 2021) database, and primers were synthesized from Tsingke Biotechnology (Beijing, China). The primer sequences are shown in Table 2. qRT-PCR was performed using a CFX connect real-time system (Bio-Rad, Hercules, CA, USA), and qRT-PCR data were analyzed using the *β-actin* gene as the reference gene. The relative expression was calculated using the $2^{-\Delta\Delta Ct}$ method.

Table 2. Primer sequences information.

Genes	Primer (5′→3′)	Length
circTTN (convergent primer)	F: TGTCTCACTGCCTTGTCTGATG R: TTAGCAGCTGGTTCAGTCACC	241 bp
circTTN (divergent primer)	F: AGCTGCTAAAGTGCCCATTCC R: AAACAAAGAACATCAGACAAGGCA	139 bp
β-actin	F: GGACTGTTACTGAGCTGCGTT R: CGCCTTCACCGTTCCAGTT	290 bp
CDK4	F: CGAGCGTAAGGCTGATGGAT R: CCAGGCCGCTTAGAAACTGA	177 bp
PCNA	F: GCCGAGACCTTAGCCACATT R: GTAGGAGACAGTGGAGTGGC	229 bp
CCND	F: TCAAGTGTGACCCGGACTG R: GCTCCTTCCTCTTTGCGGG	235 bp
MyoD	F: AGTGAATGAGGCCTTCGAGA R: GCATCTGAGTCGCCACTGTA	169 bp
MyoG	F: CAATGCACTGGAGTTCGGT R: CTGGGAAGGCAACAGACAT	134 bp
MyHC	F: CGGTCGAAGTTGCATCCCT R: GAGCCTCGATTCGCTCCTTT	141 bp
TTN	F: GTCCTCCATCTCCTCCTGGT R: GCGTTTTGTTGACCCTCACC	173 bp
PURB	F: GGAGCGACAGAGGGATAAGC R: TAGCTCACTGGGGACGAGAA	181 bp
TTN-promoter	F: TGCCCACAAGTGTCTTCAACT .R: ACCTGTTCCTCTGATTGGCAG	179 bp

4.5. Vector Construction and siRNA

The circTTN overexpression vector (PCDNA3.1-circTTN) and empty vector (control) were synthesized in JTS Scientific Biotechnology (Wuhan, China). The siRNA and NC of PURB were synthesized in GenePharma Biotechnology (Suzhou, China). Transfection was performed using a Lipofectamine 2000 reagent (Invitrogen, Waltham, MA, USA) according to the manufacturer's instructions.

4.6. CCK8 and EdU Assays

Cell proliferation was detected through Cell Counting Kit-8 (CCK8). The cells were cultured after transfection and inoculated into 96-well plates when the confluence reached 90%. Then, they were cultured for 0 h, 24 h, 48 h, 72 h, and 96 h, respectively, and 10 μL of CCK8 reagent was added 4 h before each assay (Beyotime Biotechnology, Shanghai, China). The absorbance at 450 nm was measured using a microplate reader (Multiskan FC, Thermo Scientific, Waltham, MA, USA). Cell proliferation was detected using the EdU kit (Meilunbio Biotechnology, Shanghai, China). Similarly, the cells were inoculated into 96-well plates and cultured. The EdU solution was diluted with cell growth medium at a ratio of 500:1 to prepare 20 μM of EdU medium. Each well was added with 100 μL of EdU solution, incubated for 2 h, and then fixed with 4% paraformaldehyde, followed by 0.5% tritonX-100 osmotic solution, and finally, stained with EdU and Hoeches 33342.

4.7. Flow Cytometry

The cells were transfected when the cells grew to about 60%. After transfection, the cells were mixed with 70% ethanol and fixed at 4 °C for 2 h. Each tube of the sample was stained with 0.5 mL of staining solution (0.5 mL of staining buffer, 10 μL of propidium iodide storage, and 10 μL of RNaseA) and incubated at 37 °C for 30 min in the absence of light. Then, the treated cell suspension was measured using Cytek DxP Athena flow cytometry (Cytek, Fremont, CA, USA).

4.8. Immunofluorescence Staining

The cells were fixed with 4% paraformaldehyde for 30 min, incubated with 0.5% triton X-100 at room temperature for 20 min, and closed with 5% BSA (Bio Froxx, Frankfurt, Germany) for 2 h. Anti-MyHC monoclonal antibody was used for diluting 5% BAS (1:200, DSHB, Iowa, IA, USA) and was incubated overnight at 4 °C. Then, it was incubated with anti-mouse IgG (1:1000, Abbkine, Wuhan, China) for 2 h. Next, nuclei were stained using DAPI (Solarbio, Beijing, China) for 10 min without light. Finally, photographs were taken and analyzed under an inverted microscope (Axio Vert A1, ZEISS, Oberkochen, Germany).

4.9. Western Blot Analysis

The cell proteins were extracted with RIPA Lysis Buffer (Beyotime Biotechnology, Shanghai, China), and the protein concentration was quantified using a BCA protein quantitative/concentration assay kit (Meilunbio Biotechnology, Shanghai, China). The denatured protein samples were electrophoreted in Tris/Glycine/SDS Buffer (Epizyme, Biotechnology, Shanghai, China), and then, the proteins were transferred to the PDVF membrane for electrotransfer in the electrotransfer solution (Yamei, Biotechnology, Shanghai, China). The membranes were sealed with the sealing liquid for 2 h, and then β-actin (1:1000, TransGen, Beijing, China), CDK4 (1:1000, AiFang, Changsha, China), PCNA (1:2000, ZENGBIO, Chengdu, China), CCND (1:1000, CST, Boston, MA, USA), MyoD (1:1000, HUABIO, Hangzhou, China), MyoG (1:200, DSHB, Iowa, IA, USA), MyHC (1:1000, DSHB, Iowa, IA, USA), Titin (1:1000, Bioword, Minnesota, MN, USA), and PURB (1:1000, Proteintech, San Diego, CA, USA) antibodies were incubated at 4 °C overnight. Finally, the membranes were incubated with the secondary antibody (1:10,000, ZENBIO, Chengdu, China) for 2 h, and the protein bands were developed using an ECL chemiluminescent substrate kit (Biosharp, Guangzhou, China). All the protein levels were analyzed by β-actin protein reference normalization.

4.10. Nucleoplasmic Separation and RNA Fluorescence In Situ Hybridization

Nuclear and cytoplasmic RNA was extracted using a cytoplasmic and nuclearRNA purification kit (Norgen Biotek, Thorold, ON, Canada) and was quantified using qRT-PCR analysis. We made cell creep tablets, fixed them with 4% paraformaldehyde, and sent them to Pinuofei Biotechnology (Wuhan, China) for a RNA fluorescence in situ hybridization

(FISH) experiment. The probe sequence of circTTN is 5′-UGAGGCUGCCAACCGGAACA AUGACG-3′.

4.11. RNA Pull-Down and Mass Spectrometry

Biotin-labeled circTTN probe and NC probe were provided by Ribo Biotechnology (Guangzhou, China). Pull-down protein experiments were performed using an RNA pull-down kit (BersinBio, Guangzhou, China). First, the circTTN probe and NC probe formed a secondary structure and were incubated with streptavidin magnetic beads, respectively. Then, 2×10^7 cell samples were collected for protein extraction and the nucleic acid of protein samples were removed, and then, the probe–magnetic bead complex was rotatively incubated with cell extracts. Finally, the histone samples of the circTTN group and NC group were collected, and silver dyeing was performed with a silver dyeing kit (Beyotime, Shanghai, China). First, 30% ethanol was used for fixing, and then, silver sensitizing solution, silver solution, silver dyeing solution, and silver dyeing stop solution were added successively. The silver dyed adhesive strips were photographed and the results were analyzed. RNA pulldown protein samples were determined and analyzed using mass spectrometry in Genecreate Biotechnology (Wuhan, China).

4.12. RNA Immunoprecipitation (RIP) Analysis

An RNA Immunoprecipitation Kit (BersinBio, Guangzhou, China) was used for RIP analysis. First, 2×10^7 cells were lysed to remove the DNA, and then the cell lysate samples were divided into 3 parts according to 0.8 mL (IP), 0.8 mL (IgG), and 0.1 mL (Input). IP and IgG samples were added with the corresponding experimental antibodies PURB (ProteintechSan Diego, CA, USA) and IgG (BersinBio, Guangzhou, China), respectively, and were incubated at 4 °C for 16 h. Then, they were rotatively incubated with balanced protein A/G beads for 1 h, followed by RNA eluting using an elution buffer. Finally, RNA was extracted with a mixture of phenol–chloroform–isoamyl alcohol, which was detected and analyzed using qRT-PCR according to the percentage input method.

4.13. ChIP-qPCR Analysis

A SimpleChIP® Plus Enzymatic Chromatin IP Kit (CST, Boston, MA, USA) was used for ChIP analysis. In brief, the cells were crosslinked with 1% formaldehyde for 10 min and incubated with 10× glycine for 5 min, followed by the addition of protease enzyme (PIC); then, chromatin fragment and diluted chromatin were added to the corresponding IP (PURB) and IgG immunoprecipitation antibody. They were then incubated overnight by rotation at 4 °C, followed by the addition of protein G magnetic beads. After incubation at 4 °C for 2 h, the DNA complex was cleaned with high-salt and low-salt washing solution, and chromatin was eluted with a ChIP Elution buffer and Proteinase K was added. Finally, DNA purification was performed. The combination of PURB and the *TTN* promoter were analyzed using qRT-PCR according to the percentage input method.

4.14. Statistical Analysis

We used Microsoft Excel to sort the data. We analyzed all the data using *t*-tests and a one-way ANOVA and calculated qRT-PCR data using the $2^{-\Delta\Delta Ct}$ method. In addition, we analyzed RIP-qPCR and ChIP-qPCR using the percentage input method, counted the number of cells using Image J software (version 1.49), and mapped these results using GraphPad Prism 8.0 software. Data analysis was expressed as mean ± SEM, n = 3. * $p < 0.05$ and ** $p < 0.01$ denote a significant and extremely significant mean, respectively.

5. Conclusions

In conclusion, we verified a novel circular intronic RNA circTTN and validated its function in C2C12 cells. Our study shows that circTTN inhibits myoblast proliferation and differentiation and influences skeletal muscle development through the regulation of transcription using the circTTN/PURB/TTN axis. This study provides new insights and references for future comprehensive analyses of skeletal muscle growth and development.

Author Contributions: Data curation, H.M. and N.A.; formal analysis, K.W., B.C. and S.L.; methodology, X.L. (Xintong Li) and Z.Y.; supervision, Y.Y. and H.M.; validation, X.X. and X.L. (Xiaolin Liu); writing—original draft, N.A.; writing—review and editing, H.M., S.H. and Y.Y. All authors have read and agreed to the published version of the manuscript.

Funding: This research was funded by the Laboratory of Lingnan Modern Agriculture Project (NT2021005), Hunan Provincial Natural Science Joint Foundation (2022JJ50018), Changsha Municipal Natural Science Foundation (kq2202230), and Major science and technology projects in Yunnan Province (202202AE090032).

Institutional Review Board Statement: The Institutional Animal Care and Use Committee of Hunan Agricultural University evaluated and approved the experimental protocols under approval number 2013-06.

Informed Consent Statement: Not applicable.

Data Availability Statement: The data analyzed during the current study are available from the corresponding author on reasonable request.

Acknowledgments: We express thanks to all the members of the lab for their help.

Conflicts of Interest: The authors declare no conflict of interest.

References

1. Wolfe, R.R. The underappreciated role of muscle in health and disease. *Am. J. Clin. Nutr.* **2006**, *84*, 475–482. [CrossRef] [PubMed]
2. Bentzinger, C.F.; Wang, Y.X.; Rudnicki, M.A. Building muscle: Molecular regulation of myogenesis. *Cold Spring Harb. Perspect. Biol.* **2012**, *4*, a008342. [CrossRef] [PubMed]
3. Charge, S.B.; Rudnicki, M.A. Cellular and molecular regulation of muscle regeneration. *Physiol. Rev.* **2004**, *84*, 209–238. [CrossRef] [PubMed]
4. Hernandez-Hernandez, J.M.; Garcia-Gonzalez, E.G.; Brun, C.E.; Rudnicki, M.A. The myogenic regulatory factors, determinants of muscle development, cell identity and regeneration. *Semin. Cell Dev. Biol.* **2017**, *72*, 10–18. [CrossRef]
5. Li, H.; Zhu, C.; Tao, Z.; Xu, W.; Song, W.; Hu, Y.; Zhu, W.; Song, C. MyoD and Myf6 gene expression patterns in skeletal muscle during embryonic and posthatch development in the domestic duck (*Anas platyrhynchos domestica*). *J. Anim. Breed. Genet.* **2014**, *131*, 194–201. [CrossRef]
6. Diener, T.O. Potato spindle tuber "virus". IV. A replicating, low molecular weight RNA. *Virology* **1971**, *45*, 411–428. [CrossRef]
7. Capel, B.; Swain, A.; Nicolis, S.; Hacker, A.; Walter, M.; Koopman, P.; Goodfellow, P.; Lovell-Badge, R. Circular transcripts of the testis-determining gene Sry in adult mouse testis. *Cell* **1993**, *73*, 1019–1030. [CrossRef]
8. Jeck, W.R.; Sorrentino, J.A.; Wang, K.; Slevin, M.K.; Burd, C.E.; Liu, J.; Marzluff, W.F.; Sharpless, N.E. Circular RNAs are abundant, conserved, and associated with ALU repeats. *RNA* **2013**, *19*, 141–157. [CrossRef]
9. Li, B.; Yin, D.; Li, P.; Zhang, Z.; Zhang, X.; Li, H.; Li, R.; Hou, L.; Liu, H.; Wu, W. Profiling and Functional Analysis of Circular RNAs in Porcine Fast and Slow Muscles. *Front. Cell Dev. Biol.* **2020**, *8*, 322. [CrossRef]
10. Wei, X.; Li, H.; Yang, J.; Hao, D.; Dong, D.; Huang, Y.; Lan, X.; Plath, M.; Lei, C.; Lin, F.; et al. Circular RNA profiling reveals an abundant circLMO7 that regulates myoblasts differentiation and survival by sponging miR-378a-3p. *Cell Death Dis.* **2017**, *8*, e3153. [CrossRef]
11. Zheng, S.; Zhang, X.; Odame, E.; Xu, X.; Chen, Y.; Ye, J.; Zhou, H.; Dai, D.; Kyei, B.; Zhan, S.; et al. CircRNA-Protein Interactions in Muscle Development and Diseases. *Int. J. Mol. Sci.* **2021**, *22*, 3262. [CrossRef]
12. Legnini, I.; Di Timoteo, G.; Rossi, F.; Morlando, M.; Briganti, F.; Sthandier, O.; Fatica, A.; Santini, T.; Andronache, A.; Wade, M.; et al. Circ-ZNF609 Is a Circular RNA that Can Be Translated and Functions in Myogenesis. *Mol. Cell* **2017**, *66*, 22–37. [CrossRef] [PubMed]
13. Yu, Y.; Fang, L. CircRPAP2 regulates the alternative splicing of PTK2 by binding to SRSF1 in breast cancer. *Cell Death Discov.* **2022**, *8*, 152. [CrossRef] [PubMed]
14. Pandey, P.R.; Yang, J.H.; Tsitsipatis, D.; Panda, A.C.; Noh, J.H.; Kim, K.M.; Munk, R.; Nicholson, T.; Hanniford, D.; Argibay, D.; et al. circSamd4 represses myogenic transcriptional activity of PUR proteins. *Nucleic Acids Res.* **2020**, *48*, 3789–3805. [CrossRef] [PubMed]
15. Abdelmohsen, K.; Panda, A.C.; Munk, R.; Grammatikakis, I.; Dudekula, D.B.; De, S.; Kim, J.; Noh, J.H.; Kim, K.M.; Martindale, J.L.; et al. Identification of HuR target circular RNAs uncovers suppression of PABPN1 translation by CircPABPN1. *RNA Biol.* **2017**, *14*, 361–369. [CrossRef]

16. Li, B.; Yang, J.; Yang, J.; He, J.; Gong, Y.; Xiao, Y.; Zeng, Q.; Xu, K.; Duan, Y.; He, J.; et al. Spatiotemporal Regulation and Functional Analysis of Circular RNAs in Skeletal Muscle and Subcutaneous Fat during Pig Growth. *Biology* **2021**, *10*, 841. [CrossRef]
17. Yin, L.; Li, N.; Jia, W.; Wang, N.; Liang, M.; Yang, X.; Du, G. Skeletal muscle atrophy: From mechanisms to treatments. *Pharmacol. Res.* **2021**, *172*, 105807. [CrossRef]
18. Chen, R.; Lei, S.; Jiang, T.; Zeng, J.; Zhou, S.; She, Y. Roles of lncRNAs and circRNAs in regulating skeletal muscle development. *Acta Physiol.* **2020**, *228*, e13356. [CrossRef]
19. Ding, H.; Heng, B.; He, W.; Shi, L.; Lai, C.; Xiao, L.; Ren, H.; Mo, S.; Su, Z. Chronic reactive oxygen species exposure inhibits glucose uptake and causes insulin resistance in C2C12 myotubes. *Biochem. Biophys. Res. Commun.* **2016**, *478*, 798–803. [CrossRef]
20. Wong, C.Y.; Al-Salami, H.; Dass, C.R. C2C12 cell model: Its role in understanding of insulin resistance at the molecular level and pharmaceutical development at the preclinical stage. *J. Pharm. Pharmacol.* **2020**, *72*, 1667–1693. [CrossRef]
21. Sun, D.; An, J.; Cui, Z.; Li, J.; You, Z.; Lu, C.; Yang, Y.; Gao, P.; Guo, X.; Li, B.; et al. CircCSDE1 Regulates Proliferation and Differentiation of C2C12 Myoblasts by Sponging miR-21-3p. *Int. J. Mol. Sci.* **2022**, *23*, 12038. [CrossRef] [PubMed]
22. Patop, I.L.; Wust, S.; Kadener, S. Past, present, and future of circRNAs. *EMBO J.* **2019**, *38*, e100836. [CrossRef]
23. Dhawan, A.; Harris, A.L.; Buffa, F.M.; Scott, J.G. Endogenous miRNA sponges mediate the generation of oscillatory dynamics for a non-coding RNA network. *J. Theor. Biol.* **2019**, *481*, 54–60. [CrossRef] [PubMed]
24. Panda, A.C. Circular RNAs Act as miRNA Sponges. *Adv. Exp. Med. Biol.* **2018**, *1087*, 67–79.
25. You, X.; Vlatkovic, I.; Babic, A.; Will, T.; Epstein, I.; Tushev, G.; Akbalik, G.; Wang, M.; Glock, C.; Quedenau, C.; et al. Neural circular RNAs are derived from synaptic genes and regulated by development and plasticity. *Nat. Neurosci.* **2015**, *18*, 603–610. [CrossRef] [PubMed]
26. Wang, Z.; Lei, X. A web server for identifying circRNA-RBP variable-length binding sites based on stacked generalization ensemble deep learning network. *Methods* **2022**, *205*, 179–190. [CrossRef]
27. Yang, Y.; Hou, Z.; Ma, Z.; Li, X.; Wong, K.C. iCircRBP-DHN: Identification of circRNA-RBP interaction sites using deep hierarchical network. *Brief. Bioinform.* **2021**, *22*, bbaa27. [CrossRef]
28. Wang, Z.; Lei, X. Matrix factorization with neural network for predicting circRNA-RBP interactions. *BMC Bioinform.* **2020**, *21*, 229. [CrossRef]
29. Wang, X.; Cao, X.; Dong, D.; Shen, X.; Cheng, J.; Jiang, R.; Yang, Z.; Peng, S.; Huang, Y.; Lan, X. Circular RNA TTN Acts As a miR-432 Sponge to Facilitate Proliferation and Differentiation of Myoblasts via the IGF2/PI3K/AKT Signaling Pathway. *Mol. Ther. Nucleic Acids* **2019**, *18*, 966–980. [CrossRef]
30. Kelm, R.J., Jr.; Wang, S.X.; Polikandriotis, J.A.; Strauch, A.R. Structure/function analysis of mouse Purbeta, a single-stranded DNA-binding repressor of vascular smooth muscle alpha-actin gene transcription. *J. Biol. Chem.* **2003**, *278*, 38749–38757. [CrossRef]
31. Ji, J.; Tsika, G.L.; Rindt, H.; Schreiber, K.L.; McCarthy, J.J.; Kelm, R.J., Jr.; Tsika, R. Puralpha and Purbeta collaborate with Sp3 to negatively regulate beta-myosin heavy chain gene expression during skeletal muscle inactivity. *Mol. Cell. Biol.* **2007**, *27*, 1531–1543. [CrossRef] [PubMed]
32. Johnson, E.M. The Pur protein family: Clues to function from recent studies on cancer and AIDS. *Anticancer Res.* **2003**, *23*, 2093–2100. [PubMed]
33. Kelm, R.J., Jr.; Lamba, G.S.; Levis, J.E.; Holmes, C.E. Characterization of purine-rich element binding protein B as a novel biomarker in acute myelogenous leukemia prognostication. *J. Cell. Biochem.* **2018**, *119*, 2073–2083. [CrossRef] [PubMed]
34. Hariharan, S.; Kelm, R.J., Jr.; Strauch, A.R. The Puralpha/Purbeta single-strand DNA-binding proteins attenuate smooth-muscle actin gene transactivation in myofibroblasts. *J. Cell. Physiol.* **2014**, *229*, 1256–1271. [CrossRef] [PubMed]
35. Maamar, H.; Cabili, M.N.; Rinn, J.; Raj, A. linc-HOXA1 is a noncoding RNA that represses Hoxa1 transcription in cis. *Genes Dev.* **2013**, *27*, 1260–1271. [CrossRef] [PubMed]
36. Yang, Y.; Wang, X.; Tan, Y.; Xu, Y.; Guo, X.; Wu, Y.; Wang, W.; Jing, R.; Zhu, F.; Ye, D.; et al. LncCMRR Plays an Important Role in Cardiac Differentiation by Regulating the Purb/Flk1 Axis. *Stem Cells* **2023**, *41*, 11–25. [CrossRef]
37. Zhang, J.; Li, Y.; Chen, Y.; Yu, X.; Yu, X.; Wang, S.; Sun, H.; Zheng, X.; Zheng, X.; Zhang, L.; et al. Circ-calm4 regulates hypoxia-induced pulmonary artery smooth muscle autophagy by binding Purb. *J. Mol. Cell. Cardiol.* **2023**, *176*, 41–54. [CrossRef]
38. Li, L.; Chen, Y.; Nie, L.; Ding, X.; Zhang, X.; Zhao, W.; Xu, X.; Kyei, B.; Dai, D.; Zhan, S.; et al. MyoD-induced circular RNA CDR1as promotes myogenic differentiation of skeletal muscle satellite cells. *Biochim. Biophys. Acta Gene Regul. Mech.* **2019**, *1862*, 807–821. [CrossRef]
39. Chen, M.; Wei, X.; Song, M.; Jiang, R.; Huang, K.; Deng, Y.; Liu, Q.; Shi, D.; Li, H. Circular RNA circMYBPC1 promotes skeletal muscle differentiation by targeting MyHC. *Mol. Ther. Nucleic Acids* **2021**, *24*, 352–368. [CrossRef]
40. Lv, W.; Zhao, S.; Hou, Y.; Tong, Q.; Peng, Y.; Li, J.; Xu, Z.; Zuo, B. Single Nucleotide Polymorphisms of Porcine lncMGPF Regulate Meat Production Traits by Affecting RNA Stability. *Front. Cell Dev. Biol.* **2021**, *9*, 731712. [CrossRef]
41. Lv, W.; Jin, J.; Xu, Z.; Luo, H.; Guo, Y.; Wang, X.; Wang, S.; Zhang, J.; Zuo, H.; Bai, W.; et al. lncMGPF is a novel positive regulator of muscle growth and regeneration. *J. Cachexia Sarcopenia Muscle* **2020**, *11*, 1723–1746. [CrossRef] [PubMed]

42. Chauveau, C.; Rowell, J.; Ferreiro, A. A rising titan: TTN review and mutation update. *Hum. Mutat.* **2014**, *35*, 1046–1059. [CrossRef] [PubMed]
43. Gautel, M.; Djinovic-Carugo, K. The sarcomeric cytoskeleton: From molecules to motion. *J. Exp. Biol.* **2016**, *219*, 135–145. [CrossRef] [PubMed]

International Journal of
Molecular Sciences

MDPI

Article

Serum microRNA Levels as a Biomarker for Diagnosing Non-Alcoholic Fatty Liver Disease in Chinese Colorectal Polyp Patients

Lui Ng [1,*], Ryan Wai-Yan Sin [1], David Him Cheung [1], Wai-Keung Leung [2], Abraham Tak-Ka Man [1], Oswens Siu-Hung Lo [1], Wai-Lun Law [1] and Dominic Chi-Chung Foo [1,*]

1 Department of Surgery, School of Clinical Medicine, Li Ka Shing Faculty of Medicine, The University of Hong Kong, Hong Kong SAR, China; ryansin@connect.hku.hk (R.W.-Y.S.); cheung96@hku.hk (D.H.C.); tkman@hku.hk (A.T.-K.M.); oswenslo@yahoo.com (O.S.-H.L.); lawwl@hku.hk (W.-L.L.)

2 Department of Medicine, School of Clinical Medicine, Li Ka Shing Faculty of Medicine, The University of Hong Kong, Hong Kong SAR, China; waikleung@hku.hk

* Correspondence: luing@hku.hk (L.N.); ccfoo@hku.hk (D.C.-C.F.)

Abstract: Non-alcoholic fatty liver disease (NAFLD) is one of the most common chronic liver diseases and its prevalence is increasing worldwide. It is reported that NAFLD is associated with colorectal polyps. Since identifying NAFLD in its early stages could prevent possible disease progression to cirrhosis and decrease the risk of HCC by early intervention, patients with colorectal polyp may thus be considered a target group for screening NAFLD. This study aimed to investigate the potential of serum microRNAs (miRNAs) in identifying NAFLD for colorectal polyp patients. Serum samples were collected from 141 colorectal polyp patients, of which 38 had NAFLD. The serum level of eight miRNAs was determined by quantitative PCR and delta Ct values of different miRNA pairs which were compared between NAFLD and control groups. A miRNA panel was formulated from candidate miRNA pairs by multiple linear regression model and ROC analysis was performed to evaluate its diagnostic potential for NAFLD. Compared to the control group, the NAFLD group showed significantly lower delta Ct values of miR-18a/miR-16 (6.141 vs. 7.374, p = 0.009), miR-25-3p/miR-16 (2.311 vs. 2.978, p = 0.003), miR-18a/miR-21-5p (4.367 vs. 5.081, p = 0.021) and miR-18a/miR-92a-3p (8.807 vs. 9.582, p = 0.020). A serum miRNA panel composed of these four miRNA pairs significantly identified NAFLD in colorectal polyp patients with an AUC value of 0.6584 (p = 0.004). The performance of the miRNA panel was further improved to an AUC value of 0.8337 (p < 0.0001) when polyp patients with other concurrent metabolic disorders were removed from the analysis. The serum miRNA panel is a potential diagnostic biomarker for screening NAFLD in colorectal polyp patients. This serum miRNA test could be performed for colorectal polyp patients for early diagnosis and for prevention of the disease from progressing into more advanced stages.

Keywords: NAFLD; colorectal polyp; serum microRNA (miRNA); diagnostic biomarker

1. Introduction

Non-alcoholic fatty liver disease (NAFLD) is one of the most common chronic liver diseases and its prevalence is increasing worldwide, being estimated to affect up to 25% of the global population [1]. Comparing to other metabolic disorders such as obesity and diabetes which have multiple well-known risk factors as predictors, a substantial portion of individuals with normal body mass index (BMI) have NAFLD [2]. Initiating from simple hepatic steatosis (NAFL, nonalcoholic fatty liver), and although remission can easily be achieved, NAFLD is difficult to diagnose due to a lack of characteristic symptoms [3]. Approximately 30% to 40% of NAFLD patients develop nonalcoholic steatohepatitis (NASH), which is characterized by excessive accumulation of fat coupled with the development of inflammation. Without proper treatment, patients are subjected to the risk of the disease developing into fibrosis, cirrhosis or hepatocellular carcinoma over time [4]. It is suggested that NAFLD or more specifically NASH might have already become the most common cause of liver diseases and liver-related deaths globally [5].

Colorectal polyp is a benign growth on the mucosal surface of the colon or rectum, and over a period of 5 to 15 years, can develop into colorectal cancer (CRC) [6], which is one of the most common cancers and a leading cause of cancer-related death worldwide [7]. Hence, screening of polyps is crucial for reducing the morbidity of CRC. Adenomatous polyps are considered the precursor lesions for the majority of CRCs through an adenoma-carcinoma sequence, whereas recent studies also proposed that hyperplastic polyps contribute to CRCs through serrated or microsatellite instable pathways [8]. The prevalence of colorectal polyp is rapidly increasing worldwide in recent years, which may be influenced by factors such as changes in diet and lifestyle habits [9]. A US study reported that the prevalence of colorectal polyp varies in different geographical locations and races, with the highest incidence in Western countries and the lowest in Africa and South-Central Asia [10]. Recently, more Asian studies investigating prevalence and risk factors for colorectal polyps and CRC have been published. The prevalence of colorectal polyps in an average risk population for Asian countries is around 20% and generally with higher prevalence in older age and male gender [8,11,12]. Recent studies have reported some risk factors for adenomatous polyps, including age, gender, metabolic syndrome, Helicobacter pylori infection, smoking, alcohol drinking, etc. [13–15].

Researches have been investigating whether NAFLD has a relationship to colorectal polyps, since metabolic syndrome and insulin resistance are associated with a higher risk of colon cancer, whereas NAFLD is regarded as a manifestation of metabolic syndrome in the liver [16]. In line with this hypothesis, epidemiological research reported that the occurrence of NAFLD is associated with colorectal polyps [16–18]. Though the underlying mechanism of the association between NAFLD and colorectal polyps is still unclear, the significant roles of insulin resistance, inflammatory response and obesity in the pathophysiological pathways of NAFLD and colorectal polyps have been suggested [19]. The levels of hormones are an important factor for the development of colorectal neoplasms in patients with NAFLD. For example, insulin resistance, which is a risk factor for NAFLD, is closely related to colorectal polyp and colorectal cancer [20]. Moreover, high serum levels of pro-inflammatory cytokines, such as those including tumor necrosis factor (TNF)-α, interleukin (IL)-6 and IL-8, are associated with both NAFLD, and colorectal neoplasm patients with NAFLD are significantly higher than those in the normal population, which have been found to be significantly associated with the risk of developing colorectal neoplasms [21]. Similarly, the high prevalence of a high-energy diet and obesity has played a role in the increase in the prevalence of both NAFLD and colorectal neoplasms [22]. On the other hand, serum albumin is a recognized biomarker for assessing the nutritional condition. Chronological decline in serum albumin concentration was found to be indicator of the occurrence of serious complications of NAFLD/NASH and development of colorectal polyps [8,23], yet the causality requires further investigations. The prevalence of NAFLD was higher in the colorectal polyp group when compared to the control group [16], and an increased risk for NAFLD was more evident in patients with a greater number of adenomatous polyps [16]. Since identifying NAFLD in its early stages could prevent possible disease progression to cirrhosis and decrease the risk of HCC by early intervention, patients with colorectal polyp may thus be considered a high-risk group, and screening of NAFLD will be highly beneficial.

The gold standard for diagnosing and monitoring NAFLD is liver biopsy [24], yet it is not feasible to all colorectal polyp patients within clinical practice due to its cost and invasiveness [25]. In addition, there remains a risk of serious complications despite liver biopsy being considered safe, including pain at the biopsy site, serious bleeding and, rarely, death [26]. Therefore, a practical screening biomarker is warranted for screening colorectal polyp patients with a high risk of NAFLD.

Circulating blood microRNAs (miRNAs) are endogenous, noncoding, small RNAs present in blood. Due to their size, abundance, tissue specificity, relative stability in peripheral circulation and dysregulated levels associated with different diseases, they offer great potential as noninvasive screening biomarkers [27]. There have been several studies demonstrating the clinical significance of serum miRNA in monitoring NAFLD progression. For instance, serum levels of certain miRNAs showed high accuracy for distinguishing NASH from simple steatosis or have the potential for steatohepatitis diagnosis [28–30]. Furthermore, serum levels of certain hepatic miRNAs were able to predict the percentage of liver fat with errors of less than 5% [31]. However, no studies have identified serum miRNAs for screening NAFLD specifically in colorectal polyp patients which are considered the high-risk group for the disease. Hence, this study aimed to identify serum miRNAs which show differential levels between colorectal polyp patients with or without NAFLD and evaluate the potential of serum miRNAs in identifying NAFLD in patients with colorectal polyps.

2. Results

2.1. Patient Characteristics

A total of 141 Chinese colorectal polyp patients, confirmed by colonoscopy, were included in this study, including 38 patients diagnosed with NAFLD as the NAFLD group, and 103 patients confirmed with absence of fatty liver or other liver diseases as the control group. As shown in Table 1, among the 38 NAFLD patients, 57.9% were male and the mean age of these patients was 62.2 ± 8.50 years. Among the 103 control subjects, 53.8% were male and the mean age of these patients was 64.9 ± 9.96 years. Regarding the type of polyps, tubular adenoma was the most prevalent in both groups, followed by hyperplastic polyp. The presence of other concurrent metabolic disorders including diabetes and obesity/overweight (BMI ≥ 30) were obtained from the patients. Within the NAFLD patient group, 8 of them (20.0%) were diagnosed with diabetes, 5 (13.2%) with overweight/obesity and 11 (28.9%) with both diseases. Within the control group, 13 of them (12.6%) had diabetes, 14 (13.5%) with obesity/overweight and 9 (8.7%) had both diseases.

Table 1. Characteristics of colorectal polyp patients recruited.

	NAFLD Group (n = 38)	Control Group (n = 103)	*p*-Value
Male	22 (57.9%)	57 (55.3%)	
Female	16 (42.1%)	46 (44.7%)	0.936
Age	62.2 ± 8.50	64.9 ± 9.96	0.138
Type of polyp			
Hyperplastic	6	11	
Tubular adenoma	25	81	
Others	3	5	
Mixed types	4	6	0.470
Complications:			
Diabetes only	8	13	
Obesity/overweight only	5	14	
Diabetes and obesity/overweight	11	9	0.366

2.2. Comparison of Different miRNA Pairs between Polyp Patients with or without NAFLD

We determined the serum levels of eight miRNAs including miR-16, miR-18a-5b, miR-21-5p, miR-25-3p, miR-92a-3p, miR-93b-5p, miR-106-5p and miR-1246. We found that the delta Ct values of certain miRNA pairs were significantly lower in colorectal polyp patients with NAFLD (Figure 1). Compared to colorectal polyp patients without NAFLD, those with NAFLD showed significantly lower delta Ct values of miR-18a/miR-16 (6.141 vs. 7.374, $p = 0.009$), miR-25-3p/miR-16 (2.311 vs. 2.978, $p = 0.003$), miR-18a/miR-21-5p (4.367 vs. 5.081, $p = 0.021$) and miR-18a/miR-92a-3p (8.807 vs. 9.582, $p = 0.020$).

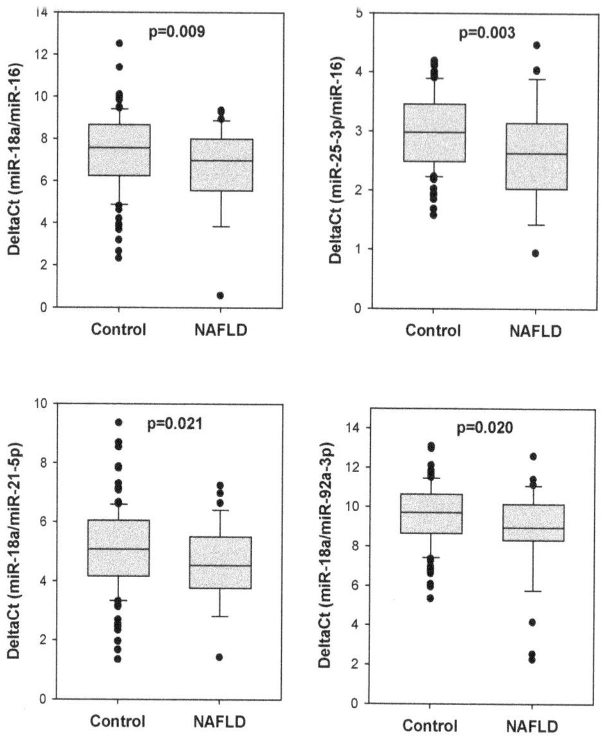

Figure 1. Four serum miRNA pairs showing differential delta Ct values between NAFLD and control groups. The delta Ct values of four serum miRNA pairs, including miR-18a/miR-16, miR-25-3p/miR-16, miR-18a/miR-21-5p and miR-18-3p/miR-92a-3p, were significantly lower in colorectal polyp patients with NAFLD (N = 38) when compared to NAFLD-free patients (control; N = 103).

Moreover, in addition to NAFLD, 60 polyp patients (24 from NAFLD group; 36 from control group) were diagnosed with other concurrent metabolic disorders including diabetes and obesity or overweight. To rule out the potential interference of concurrent metabolic diseases on serum miRNA levels [32], we refined polyp patients to the NAFLD group and control group without other concurrent metabolic diseases. Using such approach, there was generally a larger difference in delta Ct value of different miRNA pairs between the NAFLD group and control group (Figure 2): miR-18a/miR-16 (5.708 vs. 7.287, $p = 0.042$), miR-25-3p/miR-16 (1.820 vs. 2.940, $p = 0.002$), miR-18a/miR-21-5p (3.579 vs. 5.135, $p = 0.001$) and miR-18a/miR-92a-3p (8.378 vs. 9.678, $p = 0.011$).

2.3. Diagnostic Performance of Serum miRNA Panel for NAFLD in Colorectal Polyp Patients

We applied multiple linear regression analysis to formulate a combination of these four miRNA pairs for diagnosing NAFLD in colorectal polyp patients: serum miRNA panel score = 1.088 + (0.0302 × miR-18a/miR-16) − (0.118 × miR-25-3p/miR-16) + (0.00378 × miR-18a/miR-21-5p) − (0.0714 × miR-18a/miR-92a-3p). The miRNA panel score in the NAFLD group was significantly higher than that of the control group (0.365 vs. 0.268, $p < 0.0001$, Figure 3A). As shown in Figure 3B, the serum miRNA panel significantly identified NAFLD in colorectal polyp patients with an AUC value of 0.6584 (95% CI, 0.5578 To 0.7590; $p = 0.0040$).

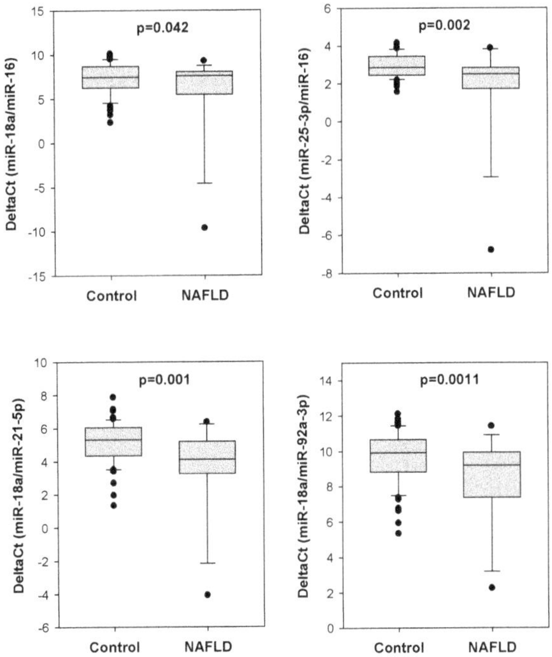

Figure 2. Delta Ct values of the four differential serum miRNA pairs in NAFLD and control groups without concurrent diabetes nor obesity/overweight. The four serum miRNA pairs—miR-18a/miR-16, miR-25-3p/miR-16, miR-18a/miR-21-5p and miR-18a/miR-92a-3p—generally showed a larger difference in delta Ct values between the NAFLD group (N = 14) and control group (N = 67) after colorectal polyp patients with other concurrent metabolic disorders (diabetes and obesity/overweight) were removed from the analysis.

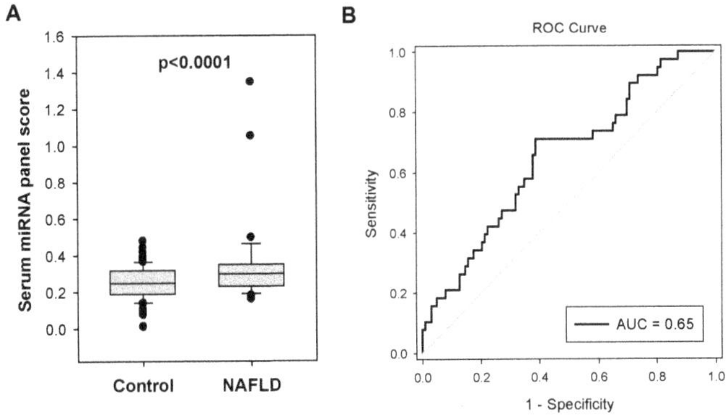

Figure 3. Serum miRNA panel as a potential biomarker for screening NAFLD in colorectal polyp patients. (**A**) A serum miRNA panel which was formulated from the 4 serum miRNA pairs using multiple linear regression 1.088 + (0.0302 × miR-18a/miR-16) − (0.118 × miR-25-3p/miR-16) + (0.00378 × miR-18a/miR-21-5p) − (0.0714 × miR-18a/miR-92a-3p) showed significantly higher value ($p < 0.0001$) in NAFLD group (N = 38) compared to the control group (N = 103). (**B**) ROC analysis showed that the serum miRNA panel significantly identify NAFLD in colorectal polyp patients with an AUC value of 0.66.

Furthermore, when colorectal polyp patients with other concurrent metabolic diseases were removed, this further extended the significant difference between miRNA panel scores in NAFLD and control patients (0.438 vs. 0.263, $p < 0.0001$, Figure 4A). The AUC of this miRNA panel for NAFLD diagnosis was increased to 0.8337 (95% CI, 0.7329 To 0.9345; $p < 0.0001$, Figure 4B). The best combination of sensitivity and specificity which was at the cut-off value of 0.3066 was 92.9% and 73.1%, respectively.

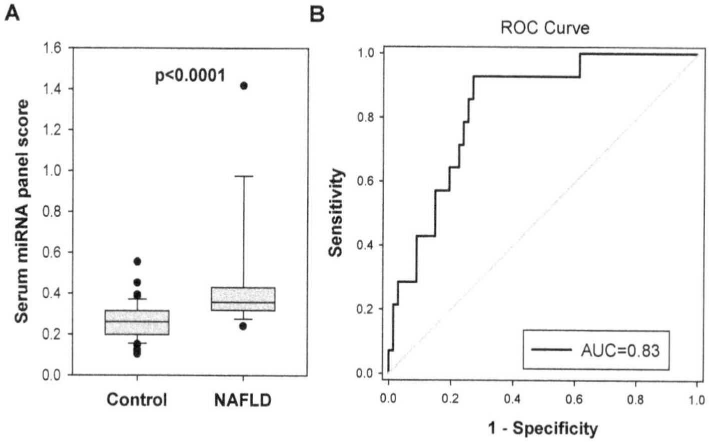

Figure 4. Serum miRNA panel showed better performance in identifying NAFLD in colorectal polyp patients without concurrent diabetes nor obesity/overweight. (**A**) The serum miRNA panel showed an improved significant difference ($p < 0.0001$) between the NAFLD group (N = 14) and control group after colorectal polyp patients (N = 67) with other concurrent metabolic disorders (diabetes and obesity/overweight) were removed from the analysis. (**B**) ROC analysis showed that the AUC of such miRNA panel for NAFLD diagnosis was increased to 0.8337 after colorectal polyp patients with concurrent metabolic disorders (diabetes and obesity/overweight) were removed from the analysis.

2.4. Validation of Serum miRNA Panel for Diagnosing Probable NAFLD in Another Small Cohort of Colorectal Polyp Patients

Finally, we examined the performance of the serum miRNA panel in a validation cohort consisting of 22 Chinese colorectal polyp patients who had been scanned for fatty liver by ultrasound. Twelve of them showed no sign of fatty liver, whereas 10 of them were probable NAFLD patients with fatty liver or liver cyst. The characteristics of these patients are shown in Table 2. We determined the values of miRNA pairs for these samples. In line with the above findings, probable NAFLD patients showed significantly lower delta Ct values of miR-18a/miR-16 (Figure 5, 5.823 vs. 7.758, $p = 0.043$), miR-25-3p/miR-16 (2.304 vs. 2.785, $p = 0.042$), miR-18a/miR-21-5p (4.035 vs. 5.719, $p = 0.046$) and miR-18a/miR-92a-3p (8.288 vs. 10.592, $p = 0.007$).

Table 2. Characteristics of colorectal polyp patients recruited in the validation cohort.

	Probable NAFLD Group (n = 10)	Control Group (n = 12)	p-Value
Male	7	9	
Female	3	3	1.000
Age	61.3 ± 5.95	63.8 ± 9.31	0.481
Type of polyp			
Hyperplastic	2	2	
Tubular adenoma	6	8	
Mixed types	2	2	0.949

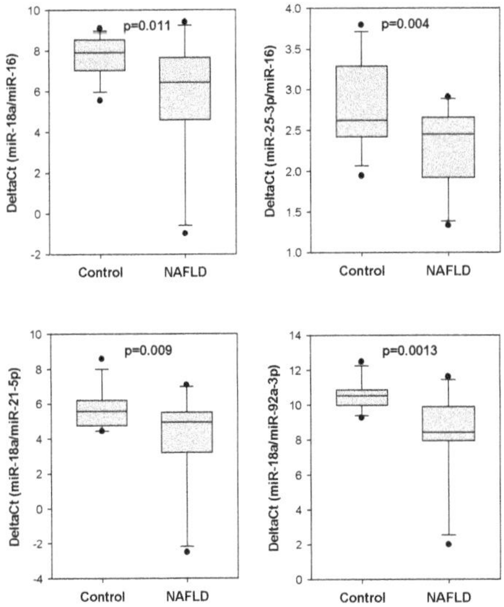

Figure 5. Four serum miRNA pairs showing differential delta Ct values between NAFLD and control groups in an independent cohort of colorectal polyp patients. The delta Ct values of four serum miRNA pairs, including miR-18a/miR-16, miR-25-3p/miR-16, miR-18a/miR-21-5p and miR-18a/miR-92a-3p, were significantly lower in colorectal polyp patients with NAFLD (N = 10) when compared to those without NAFLD (control; N = 12).

The serum miRNA panel score for these patients was determined. As shown In Figure 6, the score was significantly higher in the probable NAFLD group (0.393) when compared to the control group (0.230, $p = 0.002$). We examined the performance of the serum miRNA test in identifying polyp patients with probable NAFLD disease based on the cut-off value of 0.3066. Eight out of ten patients were correctly identified with probable NAFLD (sensitivity = 80.0%), whereas nine out of twelve normal subjects were correctly identified as normal (specificity = 75.0%). These findings confirmed the serum miRNA panel showing consistent performance in diagnosing NAFLD patients.

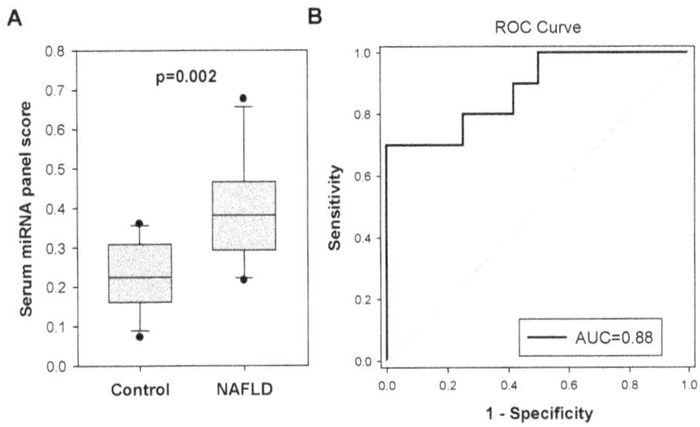

Figure 6. Serum miRNA panel showed consistent performance in identifying probable NAFLD in an independent cohort of colorectal polyp patients. (**A**) The serum miRNA panel score was significantly higher in the probable NAFLD group (median = 0.393; N = 10) when compared to the control group (median = 0.230, $p = 0.002$; N = 12). (**B**) ROC analysis showed that the AUC of the serum miRNA panel for NAFLD diagnosis was 0.8800.

3. Discussion

NAFLD is significantly associated with the presence of colorectal polyps. Therefore, it is suggested that polyp patients should be screened for the presence of NAFLD to prevent disease progression to cirrhosis and decrease the risk of HCC by early intervention. One major advantage of our serum miRNA approach as a screening biomarker is that serum samples are routinely collected from colorectal polyp patients when they have a follow-up at hospitals. Our results showed that the delta Ct values of four miRNA pairs in serum samples, namely miR-18a/miR-16, miR-25-3p/miR-16, miR-18a/miR-21-5p and miR-18a/miR-92a-3p, were significantly lower in colorectal polyp patients with NAFLD when compared to those without NAFLD. A miRNA panel formulated from these four miRNA pairs was able to identify NAFLD in colorectal polyp patients with an AUC value of 0.6584. These findings suggested that our serum miRNA panel is a potential biomarker for screening NAFLD in colorectal polyp patients. More importantly, when patients with other metabolic disorders (i.e., diabetes and obesity or overweightness) were removed from the analysis, the AUC value was further increased to 0.8337. This finding suggests that colorectal polyp patients who measured with high serum miRNA panel scores (i.e., above 0.3066 which showed the best combination of sensitivity and specificity in our patient cohort), in particular for those without diabetes nor obesity/overweight, can be classified as the high-risk group, and further verification of the occurrence of NAFLD is recommended. We further validated these findings using an independent cohort of colorectal polyps patients who potentially develop a fatty liver or cyst, but have not tested for other metabolic diseases. In line with the aforementioned findings, our data suggested that the delta Ct level of the four miRNA pairs was lower in probable NAFLD patients when compared to control patients. More importantly, the serum miRNA panel score was able to identify probable NAFLD patients with 80.0% sensitivity and 75% specificity, indicating the serum miRNA panel exhibits consistent performance in diagnosing NAFLD patients.

In this study, we evaluated the potential of applying serum miRNA levels for identifying NAFLD in colorectal polyp patients. Due to the low level of circulating miRNAs in the peripheral blood, effective data normalization using RT-qPCR is somewhat challenging. We tested different normalization strategies including endogenous controls that display stable expression across all samples (e.g., U6, miR-16, miR-39) [33] or spike-in cel-miR-39. Though the normalized levels of certain miRNAs showed statistically significant differences between NAFLD and normal groups, the variation was unfavorably high among samples within the same group, which greatly affected the performance of the serum miRNA test. Alternatively, we found that the delta Ct values of certain miRNA pairs were significantly distinct between NAFLD and control groups, and they showed more consistent levels among study subjects within each group. A similar strategy has been adopted in several blood biomarker studies for different diseases or cancers [33–38]. Moreover, this miRNA-pair-based normalization method presents a better ability to identify more biomarkers as candidates, in comparison to those using internal controls or spike-in as the normalizing agent [33]. In line with their findings, our data showed that for individual miRNA normalized by blood volume, only miR-16 and miR-21 showed significant difference between NAFLD and control groups (Supplementary File S1), whereas for miRNA-pair approach, there were four miRNA-pairs showing significant difference, and the p-value was also more significant ($p = 0.003$ to 0.021) than that of individual miR-16 or miR-21 normalized by blood volume ($p = 0.040$ and 0.045, respectively). Hence, our study provides further support to the use of miRNA pair in biomarker identification study, and hopefully more potential biomarkers can be identified with this approach.

Our serum miRNA panel is composed of miR-16, miR-18a, miR-21-5p, miR-25-3p and miR-92a-3p. These noncoding RNAs have been reported to be associated with NAFLD or liver disease. MiR-16 and miR-21 are commonly deregulated in liver fibrosis and hepatocellular carcinoma [39]. Serum miR-16 level was elevated in NAFLD patients with simple steatosis (NAFLD-SS) when compared to the control group [39] and its upregulated level was associated with NAFLD/NASH progression [40], but another study indicated that the serum level of miR-16-5p was lower in the group of patients with steatosis [29]. MiR-18a was significantly downregulated in the liver of mice with NAFLD [41]. Serum miR-21 levels in NAFLD patients, when compared to controls, varied depending on the studies [42]. One study suggested that the serum miR-21 level was lower in 25 NAFLD patients than the expressions in 12 healthy controls [43], whereas another study claimed that the serum level of miR-21 was higher in patients with NAFLD [44]. Serum miR-25-3p expression is significantly suppressed in NASH patients with significant fibrosis compared to NAFLD patients [45], indicating its association with NAFLD progression. Moreover, a decrease in serum miR-25-3p level was also associated with poor survival in patients with liver disease [46]. MiR-92a-3p was downregulated in the liver tissues in aged NAFLD rat model [47], and serum level of such miRNA was associated with different stages of liver steatosis [48].

One main limitation of this study is the relatively small sample size. Due to logistic issues, serum samples were collected from all patients prior to the colonoscopy examination. After confirmation of their colonoscopy result and diagnostic result for the presence/absence of NAFLD or other metabolic diseases including diabetes and obesity/overweight, the refined number of the NAFLD group and control group was 38 and 103, respectively. To validate our finding, we retrieved an independent cohort of 22 colorectal polyp patients whose serum samples were collected in 2013. Among them, 12 were confirmed to be free of fatty liver disease, whereas 10 of them were probable NAFLD patients who carried fatty liver disease or liver cyst. Though the sample size of this independent cohort was small, we were able to validate our findings. The four serum miRNA pairs showed a significant difference between probable NAFLD group and the control group, and the values for both groups were comparable to that obtained in the previous cohort. Moreover, the serum miRNA panel was able to effectively identify colorectal polyp patients with probable NAFLD in this validation cohort with the sensitivity of 80% and specificity of 75.0%. Another limitation of this study is that all NAFLD patients recruited were at stage 1 (simple fatty liver or steatosis). Although it is an advantage for the serum miRNA panel to detect patients with NAFLD at early stage, it is warranted to investigate whether higher stage disease would affect the performance of serum miRNA panel. The last limitation of this study is that certain variables that might affect the profile of serum miRNA were not investigated and might affect the accuracy of the test in some patients. As we demonstrated in our study, the precision of the serum miRNA test for identifying NAFLD in colorectal polyp patients was reduced when the patients had other concurrent metabolic diseases including diabetes and obesity/overweight. This is mainly caused by the serum miRNA profiles being affected by these metabolic disorders [49]. However, NAFLD is a multisystem disease and is closely associated with many complications which are not investigated in this study, including viral infections (HCV, HIV) [50,51], bacterial infection [52,53], and diseases in various systems such as cardiovascular disease, chronic kidney disease, psoriasis, polycystic ovarian syndrome, chronic intermittent hypoxia of obstructive sleep apnea syndrome, and hypothyroidism [52,54].

To summarize, this study demonstrated that our serum miRNA panel, which is made up of miR-18a/miR-16, miR-25-3p/miR-16, miR-18a/miR-21-5p and miR-18a/miR-92a-3p, is a potential non-invasive biomarker for identifying NAFLD in colorectal polyp patients. Since patients with colorectal polyp have a higher risk of concurrent NAFLD, we suggest serum miRNA test could be performed for these patients for early diagnosis and to prevent the disease from progressing to cirrhosis by early intervention, hence reducing the risk of HCC. To improve the clinical value of this serum miRNA panel, in the future we plan to recruit a larger cohort of colorectal polyp patients with different stages of NAFLD for constructing a more comprehensive nomogram by including other variables including complete blood count, liver enzyme level, the complication of other infections and diseases to improve the performance of the serum miRNA panel in screening the presence of NAFLD in colorectal polyp patients.

4. Materials and Methods

4.1. Patient Specimens

This was a case-controlled study conducted in the departments of Surgery and Medicine of Queen Mary Hospital of Hong Kong. Serum samples were collected from all patients prior to the colonoscopy examination after obtaining written informed consent for participation in accordance with the Declaration of Helsinki through protocols approved by the Institutional Review Board of the University of Hong Kong/Hospital Authority Hong Kong West Cluster. One hundred forty-one colorectal polyp patients who were diagnosed with colorectal polyp by colonoscopy examination were recruited into this study. The presence of NAFLD was diagnosed by blood liver function test and ultrasound scan. The liver was scanned by ultrasound diagnostic instrument and conducted by an experienced sonographer who was blinded to the clinical details of the patients. Radiologists who performed the ultrasound evaluation were blinded to the laboratory data. According to 2010 guidelines for the diagnosis of NAFLD, patients with viral hepatitis, autoimmune hepatitis, drug-induced hepatitis, various liver cirrhosis and alcoholic liver diseases were excluded. The diagnosis of NAFLD was established if the ultrasonogram showed increased echogenicity when compared to the renal parenchyma [55]. Thirty-eight colorectal polyp patients diagnosed with NAFLD (simple fatty liver or steatosis) were sorted into the NAFLD group, whereas 103 colorectal polyp patients free of NAFLD based on their blood liver function test and ultrasound scan were sorted into the control group (Supplementary File S1). Other metabolic disorders including diabetes and obesity/overweight were retrieved from hospital's clinical database. Three millilitres of blood was taken by Serum Vacuum Blood Collection Tubes, followed by the extraction of serum. Blood tubes were first centrifuged at 3000 rpm for 10 min at 4 °C, and the upper layer of serum was collected. Further centrifugation at 10,000× *g* for 10 min at 4 °C was performed to remove any remaining clot before serum was transferred to a sterile polypropylene tube and was stored at −80 °C until use.

4.2. Serum miRNA Extraction

The serum miRNA was extracted using a combination of TRIzol™ LS Reagent (Invitrogen, Carlsbad, CA, USA) and mirVana™ miRNA Isolation Kit, without phenol (Invitrogen). In brief, 300 µL of serum was taken and centrifuged at 12,000× *g* for 10 min at 4 °C. Two hundred fifty microliters of supernatant was collected and homogenised with 750 µL of TRIzol™ LS Reagent. Two hundred microliters of chloroform was added to the homogenate, followed by a 3 min incubation at room temperature and subsequent centrifugation at 12,000× *g* for 15 min at 4 °C. Then, the upper aqueous layer was collected, followed by the addition of 650 µL of 100% ethanol. The mixture was then added to the filter cartilage provided in the mirVana™ miRNA Isolation Kit and was centrifuged at 10,000× *g* for 15 s. After washing once with 700 µL of Wash Solution 1 and twice with 500 µL of Wash Solution 2/3, the total RNA on the filter cartilage was eluted using 95 °C-preheated 30 µL RNase-free water.

4.3. Reverse Transcription (RT) and Quantitative Polymerase Chain Reaction (qPCR)

Total RNA was reversely transcribed using a combination of E. coli Poly(A) Polymerase (New England BioLabs, Ipswich, MA, USA) and PrimeScript RT reagent kit with gDNA eraser (Perfect Real Time) (Takara, Tokyo, Japan) to generate 20 µL of genomic DNA-free cDNA. To maximize the efficiency of miRNA detection, polyA tail was added to 2 µL of total RNA samples under 1X Reaction Buffer, 1 mM of ATP and 1.25 U of E. coli Poly(A) Polymerase, to generate 5 µL of the product after incubation at 37 °C for 30 min. Then, 1 µL of gDNA Eraser in 1X gDNA Eraser Buffer was added to the samples and was topped up to 10 µL with RNase-free water, followed by incubation at 42 °C for 2 min. Then, 1 µL of PrimeScript RT Enzyme Mix 1, 12.5 mM of the universal RT primer designed (5′-CAGGTCCAGTTTTTTTTTTTTTTTVN-3′) in 1X Buffer was used to generate 20 µL of cDNA after incubation at 37 °C for 15 min then 85 °C for 5 s.

To perform qPCR, TB Green Premix Ex Taq II (Tli RNaseH Plus) (Takara) was used together with miRNA-specific primers (Supplementary File S2), following the instructions of the manufacturer. In brief, 0.5 µL of gDNA-free cDNA and 0.4 µM of both forward and reverse primers were used per 10 µL reaction mix per well of a 96-well plate. Quantitative PCR was performed using ViiA7 Real-Time PCR System (ThermoFisher Scientific, Waltham, MA, USA) at 95 °C for 30 s, followed by 40 cycles at 95 °C for 5 s and 60 °C for 34 s.

4.4. Statistical Analysis

Descriptive statistics were summarized with mean and standard deviations. Inter-group comparisons were conducted by *t*-tests. All statistical analyses, including plotting of dot histograms, calculation of the area under the curve (AUC) of the receiver operating characteristic (ROC) curve for specific miRNA, and multiple linear regression, were conducted using SigmaPlot 10.0 (Systat Software Inc., San Jose, CA, USA). A *p*-value less than 0.05 (two-tailed) was considered statistically significant.

4.5. Sample Size Calculation

Sample size calculation is performed by G*power 3.1 software [56]. Based on the results obtained from the serum miRNA score for NAFLD group and control group of this study, the calculated effect size was 0.5514. To achieve 80% power at a 5% significance level, the minimum number of samples in NAFLD and control group were 38 and 102, respectively.

Supplementary Materials: The following supporting information can be downloaded at: https://www.mdpi.com/article/10.3390/ijms24109084/s1.

Author Contributions: Conceptualization, L.N., W.-L.L. and D.C.-C.F.; methodology, R.W.-Y.S., D.H.C., W.-K.L.; formal analysis, L.N., R.W.-Y.S., D.H.C. and A.T.-K.M.; resources, W.-K.L., O.S.-H.L., W.-L.L. and D.C.-C.F.; writing—original draft preparation, L.N.; writing—review and editing, L.N. and D.C.-C.F.; funding acquisition, W.-L.L. and D.C.-C.F. All authors have read and agreed to the published version of the manuscript.

Funding: This research was funded by the Health and Medical Research Fund of the Research Fund Secretariat, Food and Health Bureau, HKSAR, grant number 04151956.

Institutional Review Board Statement: The study was conducted according to the guidelines of the Declaration of Helsinki and approved by the Institutional Review Board (or Ethics Committee) of The University of Hong Kong (protocol code UW 17-037 and date 14 February 2017).

Informed Consent Statement: Informed consent was obtained from all subjects involved in the study.

Data Availability Statement: The data presented in this study are available on request from the corresponding author. The data are not publicly available due to privacy.

Conflicts of Interest: The authors declare no conflict of interest.

References

1. Pydyn, N.; Miekus, K.; Jura, J.; Kotlinowski, J. New therapeutic strategies in nonalcoholic fatty liver disease: A focus on promising drugs for nonalcoholic steatohepatitis. *Pharm. Rep.* **2020**, *72*, 1–12. [CrossRef] [PubMed]
2. Estes, C.; Chan, H.L.Y.; Chien, R.N.; Chuang, W.L.; Fung, J.; Goh, G.B.; Hu, T.H.; Huang, J.F.; Jang, B.K.; Jun, D.W.; et al. Modelling NAFLD disease burden in four Asian regions-2019–2030. *Aliment Pharm.* **2020**, *51*, 801–811. [CrossRef]
3. Kleiner, D.E.; Brunt, E.M.; Van Natta, M.; Behling, C.; Contos, M.J.; Cummings, O.W.; Ferrell, L.D.; Liu, Y.C.; Torbenson, M.S.; Unalp-Arida, A.; et al. Design and validation of a histological scoring system for nonalcoholic fatty liver disease. *Hepatology* **2005**, *41*, 1313–1321. [CrossRef] [PubMed]
4. Streba, L.A.; Vere, C.C.; Rogoveanu, I.; Streba, C.T. Nonalcoholic fatty liver disease, metabolic risk factors, and hepatocellular carcinoma: An open question. *World J. Gastroenterol.* **2015**, *21*, 4103–4110. [CrossRef] [PubMed]
5. Rinella, M.; Charlton, M. The globalization of nonalcoholic fatty liver disease: Prevalence and impact on world health. *Hepatology* **2016**, *64*, 19–22. [CrossRef] [PubMed]
6. Huck, M.B.; Bohl, J.L. Colonic Polyps: Diagnosis and Surveillance. *Clin. Colon Rectal Surg.* **2016**, *29*, 296–305. [CrossRef]
7. Xi, Y.; Xu, P. Global colorectal cancer burden in 2020 and projections to 2040. *Transl. Oncol.* **2021**, *14*, 101174. [CrossRef]

8. Pan, J.; Cen, L.; Xu, L.; Miao, M.; Li, Y.; Yu, C.; Shen, Z. Prevalence and risk factors for colorectal polyps in a Chinese population: A retrospective study. *Sci. Rep.* **2020**, *10*, 6974. [CrossRef]
9. Iyer, D.N.; Wan, T.M.H.; Man, J.H.W.; Sin, R.W.Y.; Li, X.; Lo, O.S.H.; Foo, D.C.C.; Pang, R.W.C.; Law, W.L.; Ng, L. Small RNA profiling of piRNAs in colorectal cancer identifies consistent overexpression of piR-24000 that correlates clinically with an aggressive disease phenotype. *Cancers* **2020**, *12*, 188. [CrossRef]
10. Polite, B.N.; Dignam, J.J.; Olopade, O.I. Colorectal cancer and race: Understanding the differences in outcomes between African Americans and whites. *Med. Clin. N. Am.* **2005**, *89*, 771–793. [CrossRef]
11. Siripongpreeda, B.; Mahidol, C.; Dusitanond, N.; Sriprayoon, T.; Muyphuag, B.; Sricharunrat, T.; Teerayatanakul, N.; Chaiwong, W.; Worasawate, W.; Sattayarungsee, P.; et al. High prevalence of advanced colorectal neoplasia in the Thai population: A prospective screening colonoscopy of 1404 cases. *BMC Gastroenterol.* **2016**, *16*, 101. [CrossRef]
12. Yang, M.H.; Rampal, S.; Sung, J.; Choi, Y.H.; Son, H.J.; Lee, J.H.; Kim, Y.H.; Chang, D.K.; Rhee, P.L.; Rhee, J.C.; et al. The prevalence of colorectal adenomas in asymptomatic Korean men and women. *Cancer Epidemiol. Biomark. Prev.* **2014**, *23*, 499–507. [CrossRef] [PubMed]
13. Nakai, K.; Watari, J.; Tozawa, K.; Tamura, A.; Hara, K.; Yamasaki, T.; Kondo, T.; Kono, T.; Tomita, T.; Ohda, Y.; et al. Sex differences in associations among metabolic syndrome, obesity, related biomarkers, and colorectal adenomatous polyp risk in a Japanese population. *J. Clin. Biochem. Nutr.* **2018**, *63*, 154–163. [CrossRef] [PubMed]
14. Mao, Y.; Du, J.; Xu, Y.; Zhu, Z.; Cao, H. Helicobacter pylori infection associated with an increased risk of colorectal adenomatous polyps in the Chinese population. *BMC Gastroenterol.* **2019**, *19*, 14. [CrossRef]
15. Colussi, D.; Fabbri, M.; Zagari, R.M.; Montale, A.; Bazzoli, F.; Ricciardiello, L. Lifestyle factors and risk for colorectal polyps and cancer at index colonoscopy in a FIT-positive screening population. *United Eur. Gastroenterol. J.* **2018**, *6*, 935–942. [CrossRef]
16. Hwang, S.T.; Cho, Y.K.; Park, J.H.; Kim, H.J.; Park, D.I.; Sohn, C.I.; Jeon, W.K.; Kim, B.I.; Won, K.H.; Jin, W. Relationship of non-alcoholic fatty liver disease to colorectal adenomatous polyps. *J. Gastroenterol. Hepatol.* **2010**, *25*, 562–567. [CrossRef] [PubMed]
17. Bhatt, B.D.; Lukose, T.; Siegel, A.B.; Brown, R.S., Jr.; Verna, E.C. Increased risk of colorectal polyps in patients with non-alcoholic fatty liver disease undergoing liver transplant evaluation. *J. Gastrointest. Oncol.* **2015**, *6*, 459–468. [CrossRef]
18. Chen, W.; Wang, M.; Jing, X.; Wu, C.; Zeng, Y.; Peng, J.; Cai, X. High risk of colorectal polyps in men with non-alcoholic fatty liver disease: A systematic review and meta-analysis. *J. Gastroenterol. Hepatol.* **2020**, *35*, 2051–2065. [CrossRef]
19. Li, Y.; Liu, S.; Gao, Y.; Ma, H.; Zhan, S.; Yang, Y.; Xin, Y.; Xuan, S. Association between NAFLD and Risk of Colorectal Adenoma in Chinese Han Population. *J. Clin. Transl. Hepatol.* **2019**, *7*, 99–105. [CrossRef]
20. Keku, T.O.; Lund, P.K.; Galanko, J.; Simmons, J.G.; Woosley, J.T.; Sandler, R.S. Insulin resistance, apoptosis, and colorectal adenoma risk. *Cancer Epidemiol. Biomark. Prev.* **2005**, *14*, 2076–2081. [CrossRef]
21. Yu, X.; Xie, L.; Zhou, Y.; Yuan, X.; Wu, Y.; Chen, P. Patients with non-alcoholic fatty liver disease may be a high-risk group for the development of colorectal polyps: A cross-sectional study. *World Acad. Sci. J.* **2020**, *2*, 1. [CrossRef]
22. Hernandez-Rodas, M.C.; Valenzuela, R.; Videla, L.A. Relevant Aspects of Nutritional and Dietary Interventions in Non-Alcoholic Fatty Liver Disease. *Int. J. Mol. Sci.* **2015**, *16*, 25168–25198. [CrossRef] [PubMed]
23. Kawaguchi, K.; Sakai, Y.; Terashima, T.; Shimode, T.; Seki, A.; Orita, N.; Takeshita, Y.; Shimakami, T.; Takatori, H.; Arai, K.; et al. Decline in serum albumin concentration is a predictor of serious events in nonalcoholic fatty liver disease. *Medicine* **2021**, *100*, e26835. [CrossRef] [PubMed]
24. Chalasani, N.; Younossi, Z.; Lavine, J.E.; Charlton, M.; Cusi, K.; Rinella, M.; Harrison, S.A.; Brunt, E.M.; Sanyal, A.J. The diagnosis and management of nonalcoholic fatty liver disease: Practice guidance from the American Association for the Study of Liver Diseases. *Hepatology* **2018**, *67*, 328–357. [CrossRef]
25. Rowe, I.A.; Parker, R. Liver biopsy for the selection of patients with nonalcoholic steatohepatitis for clinical trials. *Gastroenterology* **2015**, *148*, 262. [CrossRef]
26. Bedossa, P.; Dargere, D.; Paradis, V. Sampling variability of liver fibrosis in chronic hepatitis C. *Hepatology* **2003**, *38*, 1449–1457. [CrossRef]
27. Wang, Z.; Lu, Y.; Han, J. Peripheral blood microRNAs: A novel tool for diagnosing disease? *Intractable Rare Dis. Res.* **2012**, *1*, 98–102. [CrossRef] [PubMed]
28. Xin, S.; Zhan, Q.; Chen, X.; Xu, J.; Yu, Y. Efficacy of serum miRNA test as a non-invasive method to diagnose nonalcoholic steatohepatitis: A systematic review and meta-analysis. *BMC Gastroenterol.* **2020**, *20*, 186. [CrossRef]
29. Vulf, M.; Shunkina, D.; Komar, A.; Bograya, M.; Zatolokin, P.; Kirienkova, E.; Gazatova, N.; Kozlov, I.; Litvinova, L. Analysis of miRNAs Profiles in Serum of Patients With Steatosis and Steatohepatitis. *Front. Cell Dev. Biol.* **2021**, *9*, 736677. [CrossRef]
30. Kim, T.H.; Lee, Y.; Lee, Y.S.; Gim, J.A.; Ko, E.; Yim, S.Y.; Jung, Y.K.; Kang, S.; Kim, M.Y.; Kim, H.; et al. Circulating miRNA is a useful diagnostic biomarker for nonalcoholic steatohepatitis in nonalcoholic fatty liver disease. *Sci. Rep.* **2021**, *11*, 14639. [CrossRef]
31. Quintas, G.; Caiment, F.; Rienda, I.; Perez-Rojas, J.; Pareja, E.; Castell, J.V.; Jover, R. Quantitative Prediction of Steatosis in Patients with Non-Alcoholic Fatty Liver by Means of Hepatic MicroRNAs Present in Serum and Correlating with Hepatic Fat. *Int. J. Mol. Sci.* **2022**, *23*, 9298. [CrossRef] [PubMed]
32. Landrier, J.F.; Derghal, A.; Mounien, L. MicroRNAs in Obesity and Related Metabolic Disorders. *Cells* **2019**, *8*, 859. [CrossRef] [PubMed]
33. Fang, R.; Zhu, Y.; Hu, L.; Khadka, V.S.; Ai, J.; Zou, H.; Ju, D.; Jiang, B.; Deng, Y.; Hu, X. Plasma MicroRNA Pair Panels as Novel Biomarkers for Detection of Early Stage Breast Cancer. *Front. Physiol.* **2018**, *9*, 1879. [CrossRef] [PubMed]

34. Boeri, M.; Verri, C.; Conte, D.; Roz, L.; Modena, P.; Facchinetti, F.; Calabro, E.; Croce, C.M.; Pastorino, U.; Sozzi, G. MicroRNA signatures in tissues and plasma predict development and prognosis of computed tomography detected lung cancer. *Proc. Natl. Acad. Sci. USA* **2011**, *108*, 3713–3718. [CrossRef] [PubMed]
35. Hennessey, P.T.; Sanford, T.; Choudhary, A.; Mydlarz, W.W.; Brown, D.; Adai, A.T.; Ochs, M.F.; Ahrendt, S.A.; Mambo, E.; Califano, J.A. Serum microRNA biomarkers for detection of non-small cell lung cancer. *PloS ONE* **2012**, *7*, e32307. [CrossRef]
36. Matthaei, H.; Wylie, D.; Lloyd, M.B.; Dal Molin, M.; Kemppainen, J.; Mayo, S.C.; Wolfgang, C.L.; Schulick, R.D.; Langfield, L.; Andruss, B.F.; et al. miRNA biomarkers in cyst fluid augment the diagnosis and management of pancreatic cysts. *Clin. Cancer Res. Off. J. Am. Assoc. Cancer Res.* **2012**, *18*, 4713–4724. [CrossRef]
37. Sheinerman, K.S.; Umansky, S. Universal screening test based on analysis of circulating organ-enriched microRNAs: A novel approach to diagnostic screening. *Expert Rev. Mol. Diagn.* **2015**, *15*, 329–338. [CrossRef]
38. Srinivasan, S.; Treacy, R.; Herrero, T.; Olsen, R.; Leonardo, T.R.; Zhang, X.; DeHoff, P.; To, C.; Poling, L.G.; Fernando, A.; et al. Discovery and Verification of Extracellular miRNA Biomarkers for Non-invasive Prediction of Pre-eclampsia in Asymptomatic Women. *Cell Rep. Med.* **2020**, *1*, 10003. [CrossRef]
39. Cermelli, S.; Ruggieri, A.; Marrero, J.A.; Ioannou, G.N.; Beretta, L. Circulating microRNAs in patients with chronic hepatitis C and non-alcoholic fatty liver disease. *PloS ONE* **2011**, *6*, e23937. [CrossRef]
40. Gori, M.; Arciello, M.; Balsano, C. MicroRNAs in nonalcoholic fatty liver disease: Novel biomarkers and prognostic tools during the transition from steatosis to hepatocarcinoma. *BioMed Res. Int.* **2014**, *2014*, 741465. [CrossRef]
41. Liang, T.; Liu, C.; Ye, Z. Deep sequencing of small RNA repertoires in mice reveals metabolic disorders-associated hepatic miRNAs. *PloS ONE* **2013**, *8*, e80774. [CrossRef] [PubMed]
42. Zhang, T.; Yang, Z.; Kusumanchi, P.; Han, S.; Liangpunsakul, S. Critical Role of microRNA-21 in the Pathogenesis of Liver Diseases. *Front. Med.* **2020**, *7*, 7. [CrossRef]
43. Sun, C.; Huang, F.; Liu, X.; Xiao, X.; Yang, M.; Hu, G.; Liu, H.; Liao, L. miR-21 regulates triglyceride and cholesterol metabolism in non-alcoholic fatty liver disease by targeting HMGCR. *Int. J. Mol. Med.* **2015**, *35*, 847–853. [CrossRef]
44. Yamada, H.; Suzuki, K.; Ichino, N.; Ando, Y.; Sawada, A.; Osakabe, K.; Sugimoto, K.; Ohashi, K.; Teradaira, R.; Inoue, T.; et al. Associations between circulating microRNAs (miR-21, miR-34a, miR-122 and miR-451) and non-alcoholic fatty liver. *Clin. Chim. Acta Int. J. Clin. Chem.* **2013**, *424*, 99–103. [CrossRef] [PubMed]
45. Johnson, K.; Leary, P.J.; Govaere, O.; Barter, M.J.; Charlton, S.H.; Cockell, S.J.; Tiniakos, D.; Zatorska, M.; Bedossa, P.; Brosnan, M.J.; et al. Increased serum miR-193a-5p during non-alcoholic fatty liver disease progression: Diagnostic and mechanistic relevance. *JHEP Rep. Innov. Hepatol.* **2022**, *4*, 100409. [CrossRef] [PubMed]
46. Cisilotto, J.; do Amaral, A.E.; Rosolen, D.; Rode, M.P.; Silva, A.H.; Winter, E.; da Silva, T.E.; Fischer, J.; Matiollo, C.; Rateke, E.C.M.; et al. MicroRNA profiles in serum samples from Acute-On-Chronic Liver Failure patients and miR-25-3p as a potential biomarker for survival prediction. *Sci. Rep.* **2020**, *10*, 100. [CrossRef]
47. Zhang, Y.; Xiang, D.; Hu, X.; Ruan, Q.; Wang, L.; Bao, Z. Identification and study of differentially expressed miRNAs in aged NAFLD rats based on high-throughput sequencing. *Ann. Hepatol.* **2020**, *19*, 302–312. [CrossRef]
48. Pratama, M.Y.; Pascut, D.; Tamini, S.; Minocci, A.; Tiribelli, C.; Grugni, G.; Sartorio, A. Circulating microRNA Associated to Different Stages of Liver Steatosis in Prader-Willi Syndrome and Non-Syndromic Obesity. *J. Clin. Med.* **2020**, *9*, 1123. [CrossRef]
49. Pescador, N.; Perez-Barba, M.; Ibarra, J.M.; Corbaton, A.; Martinez-Larrad, M.T.; Serrano-Rios, M. Serum circulating microRNA profiling for identification of potential type 2 diabetes and obesity biomarkers. *PloS ONE* **2013**, *8*, e77251. [CrossRef]
50. Iacob, S.A.; Iacob, D.G. Non-Alcoholic Fatty Liver Disease in HIV/HBV Patients—A Metabolic Imbalance Aggravated by Antiretroviral Therapy and Perpetuated by the Hepatokine/Adipokine Axis Breakdown. *Front. Endocrinol.* **2022**, *13*, 814209. [CrossRef]
51. Adinolfi, L.E.; Rinaldi, L.; Guerrera, B.; Restivo, L.; Marrone, A.; Giordano, M.; Zampino, R. NAFLD and NASH in HCV Infection: Prevalence and Significance in Hepatic and Extrahepatic Manifestations. *Int. J. Mol. Sci.* **2016**, *17*, 803. [CrossRef] [PubMed]
52. Adenote, A.; Dumic, I.; Madrid, C.; Barusya, C.; Nordstrom, C.W.; Rueda Prada, L. NAFLD and Infection, a Nuanced Relationship. *Can. J. Gastroenterol. Hepatol.* **2021**, *2021*, 5556354. [CrossRef] [PubMed]
53. Nseir, W.; Taha, H.; Khateeb, J.; Grosovski, M.; Assy, N. Fatty liver is associated with recurrent bacterial infections independent of metabolic syndrome. *Dig. Dis. Sci.* **2011**, *56*, 3328–3334. [CrossRef]
54. Parameswaran, M.; Hasan, H.A.; Sadeque, J.; Jhaveri, S.; Avanthika, C.; Arisoyin, A.E.; Dhanani, M.B.; Rath, S.M. Factors That Predict the Progression of Non-alcoholic Fatty Liver Disease (NAFLD). *Cureus* **2021**, *13*, e20776. [CrossRef] [PubMed]
55. Lesmana, C.R.; Pakasi, L.S.; Inggriani, S.; Aidawati, M.L.; Lesmana, L.A. Development of non-alcoholic fatty liver disease scoring system among adult medical check-up patients: A large cross-sectional and prospective validation study. *Diabetes Metab. Syndr. Obes. Targets Ther.* **2015**, *8*, 213–218. [CrossRef]
56. Faul, F.; Erdfelder, E.; Buchner, A.; Lang, A.G. Statistical power analyses using G*Power 3.1: Tests for correlation and regression analyses. *Behav. Res. Methods* **2009**, *41*, 1149–1160. [CrossRef]

International Journal of *Molecular Sciences*

Article

Circulating miRNA-451a and miRNA-328-3p as Potential Markers of Coronary Artery Aneurysmal Disease

Sylwia Iwańczyk [1,*], Tomasz Lehmann [2], Artur Cieślewicz [3], Katarzyna Malesza [3], Patrycja Woźniak [1], Agnieszka Hertel [2], Grzegorz Krupka [1], Paweł P. Jagodziński [2], Marek Grygier [1], Maciej Lesiak [1] and Aleksander Araszkiewicz [1]

[1] 1st Department of Cardiology, Poznan University of Medical Sciences, 61-848 Poznań, Poland
[2] Department of Biochemistry and Molecular Biology, Poznan University of Medical Sciences, 60-781 Poznań, Poland
[3] Clinical Pharmacology, Poznan University of Medical Sciences, 61-848 Poznań, Poland
* Correspondence: sylwia.iwanczyk@skpp.edu.pl; Tel.: +48-662-712-627

Abstract: MicroRNAs (miRNAs) are currently investigated as crucial regulatory factors which may serve as a potential therapeutic target. Reports on the role of miRNA in patients with coronary artery aneurysmal disease (CAAD) are limited. The present analysis aims to confirm the differences in the expression of previously preselected miRNAs in larger study groups and evaluate their usefulness as potential markers of CAAD. The study cohort included 35 consecutive patients with CAAD (Group 1), and two groups of 35 patients matched Group 1 regarding sex and age from the overall cohort of 250 patients (Group 2 and Group 3). Group 2 included patients with angiographically documented coronary artery disease (CAD), while Group 3 enrolled patients with normal coronary arteries (NCA) assessed during coronary angiography. We applied the RT-qPCR method using the custom plates for the RT-qPCR array. We confirmed that the level of five preselected circulating miRNAs was different in patients with CAAD compared to Group 2 and Group 3. We found that miR-451a and miR-328 significantly improved the CAAD prediction. In conclusion, miR-451a is a significant marker of CAAD compared to patients with CAD. In turn, miR-328-3p is a significant marker of CAAD compared to patients with NCA.

Keywords: microRNAs; coronary aneurysm; coronary artery disease; atherosclerosis

1. Introduction

Circulating non-coding RNAs have been proposed as a new class of biomarkers and are investigated for their potential in many areas of cardiology, including coronary artery disease (CAD) [1]. As critical regulatory factors, microRNAs (miRNAs) may serve as a potential therapeutic target [2,3]. Reports on the role of miRNA in patients with coronary artery aneurysmal disease (CAAD) are limited and relate mainly to patients with Kawasaki disease.

CAAD is a rare condition affecting epicardial arteries. It occurs in 1.2% to 4.9% of patients undergoing coronary angiography [4]. CAAD includes two phenotypes: coronary artery aneurysm (CAA) and coronary artery ectasia (CAE) [5]. CAA is defined as a focal dilatation of the artery with a diameter of 1.5 times the adjacent normal coronary artery, whereas CAE describes more diffuse dilatation, often involving the predominant part of the artery and more than one vessel [6]. Moreover, CAA is classified into fusiform type if the longitudinal dimension is greater than the transverse or saccular type when the transverse dimension exceeds the longitudinal [7].

In symptomatic patients, the most common clinical presentation of CAAD is stable angina pectoris. It is significantly less frequent that myocardial infarction (MI) and aneurysm complications occur, i.e., thrombus formation following distal embolization, development of a fistula, and aneurysm rupture [6–8]. Moreover, CAAD is associated with poor long-term outcomes, regardless of concomitant atherosclerotic coronary disease [9,10].

The most common cause of abnormal dilatation of the coronary arteries is the atherosclerotic process followed by positive remodeling of the vessel. Approximately 20 to 30% of CAADs are considered inflammatory or congenital [11,12]. However, pathogenesis is multifactorial and can be influenced by the disturbance of physiological processes at the molecular level. Understanding these pathomechanisms is crucial for diagnosing and preventing CAAD complications. According to a growing body of literature, miRNAs significantly regulate human physiological processes, including gene expression in the cardiovascular system [13].

Based on our previous study [14], we selected the potential miRNA markers to validate them in a larger independent patient cohort. The present analysis aims to confirm the differences in the expression of five preselected miRNA markers (miR-125b-5p, miR-210-3p, miR-328-3p, miR-425-3p, and miR-483-5p), and three reference miRNAs (miR-16-5p, miR-30d-5p, miR-320d) in larger study groups and evaluate their usefulness as potential markers of CAAD occurrence.

2. Results

2.1. Clinical Characteristics of the Study Population

We presented baseline clinical characteristics in Table 1. Patients were predominantly male at the mean age of 65.9 ± 9.4, 67.1 ± 8.3, and 64.6 ± 12.3 in Group 1, Group 2, and Group 3, respectively. The prevalence of major cardiovascular risk factors did not differ significantly between the study groups. Despite the comparable incidence of renal failure, serum creatinine levels were significantly higher in Groups 1 and 2 compared to Group 3. In addition, Group 3 patients had no history of myocardial infarction (MI) or percutaneous coronary intervention (PCI). Therefore, a significantly lower group percentage was treated with aspirin and P2Y12 inhibitor. Moreover, beta-blocker and statin use were significantly less frequent in Group 3 than in Group 2. There were no differences in pharmacotherapy between Groups 1 and 2.

Table 1. Baseline clinical characteristics.

Variable	Group 1 (n = 33)	Group 2 (n = 35)	Group 3 (n = 35)	*p*-Value		
				Group 1 vs. Group 2	Group 1 vs. Group 3	Group 2 vs. Group 3
Sex, male	28 (84.8)	28 (80.0)	27 (77.1)	0.41	0.69	0.18
Age, years	65.9 ± 9.4	67.1 ± 8.3	64.6 ± 12.3	0.88	0.85	0.55
BMI, kg/m^2	29.7 (26.5–32.9)	28.9 (25.3–31.9)	30.1 (27.2–33.7)	0.72	0.72	0.63
Hypertension	29 (87.9)	30 (85.7)	31 (88.6)	0.81	0.90	0.72
Hyperlipidemia	20 (60.6)	29 (82.9)	27 (77.1)	0.09	0.26	0.55
Diabetes mellitus	10 (30.3)	10 (28.6)	11 (31.4)	0.97	0.77	0.79
HF, NYHA ≥ 2	6 (18.2)	9 (25.7)	4 (11.4)	0.12	0.40	0.12
LVEF, %	57.5 (55–60)	60 (50–60)	60 (55–60)	0.87	0.87	0.87
Stroke	4 (12.1)	3 (8.6)	1 (2.9)	0.90	0.30	0.60
Smoking	11 (33.3)	11 (31.4)	8 (22.8)	0.80	0.30	0.42
CKD, eGFR ≤ 60 mL/min	2 (6.1)	7 (20)	9 (25.7)	0.45	0.88	0.20
Previous MI	10 (33.3)	13 (37.1)	0	0.61	0.01	<0.01
Previous PCI	10 (33.3)	20 (57.1)	0	0.03	<0.01	<0.01
Previous CABG	1 (3.0)	3 (8.6)	0	0.67	0.46	0.24

Table 1. Cont.

Variable	Group 1 (n = 33)	Group 2 (n = 35)	Group 3 (n = 35)	p-Value		
				Group 1 vs. Group 2	Group 1 vs. Group 3	Group 2 vs. Group 3
Drugs administration						
Aspirin	25 (75.8)	26 (74.3)	22 (62.9)	0.60	0.01	0.05
P2Y12 inhibitor	14 (42.4)	20 (57.1)	1 (2.9)	0.24	<0.01	<0.01
Beta-blocker	23 (69.7)	25 (71.4)	22 (62.9)	0.84	0.15	0.04
Statin	26 (78.8)	28 (80.0)	30 (85.7)	0.95	0.12	0.02
ACEI/ARB	25 (75.8)	26 (74.3)	31 (88.6)	0.78	0.82	1.00
OHA	7 (21.2)	6 (17.1)	12 (34.3)	0.96	0.33	0.29
Insulin	2 (6.1)	1 (2.9)	3 (8.6)	0.90	0.74	1.00
Laboratory test						
WBC, 10^9/L	7.2 (5.9–9.1)	7.5 (6.1–9.6)	6.5 (5.5–7.7)	0.58	0.15	0.05
HGB, mmol/L	9.2 (8.3–9.9)	9 (8.1–9.5)	9 (8.4–9.8)	0.45	0.85	0.45
PLT, 10^9/L	223.5 (188.5–257)	204 (180–239)	201 (176.5–249.5)	0.91	0.91	0.91
LDL-C, mmol/L	2.1 (1.6–2.6)	1.9 (1.6–3.2)	2.8 (2.0–3.3)	0.78	0.33	0.44
Creatinine, mmol/L	85.7 (77.6–99.5)	93 (80–99.5)	80.9 (69.4–86.0)	0.45	0.04	<0.01
Glucose, mmol/L	5.6 (5.1–6.3)	5.8 (5.2–6.6)	5.8 (5.3–6.4)	0.86	0.86	0.86

We presented continuous variables as a mean ± s.d. or median (interquartile range), and categorical variables as a number (percentage). Abbreviations: ACEI, angiotensin-converting enzyme inhibitor; ARB, angiotensin II receptor blocker; BMI, body mass index; CABG, coronary artery bypass graft; CCB, calcium channel blocker; CKD, chronic kidney disease; HGB, hemoglobin; HF, heart failure; LDL-C, low-density lipoprotein cholesterol; LVEF, left ventricle ejection fraction; MI, myocardial infarction; NYHA, New York Heart Association; OHA, oral hypoglycemic agents; PCI, percutaneous coronary intervention; PLT, platelet count; s.d., standard deviation; WBC, white blood cells; significant difference = $p < 0.05$.

2.2. Expression Profile of miRNAs in the Plasma of Group 1 Versus Group 2 and Group 3

We confirmed that the concentration of five preselected plasma miRNAs was different in patients with CAAD (Group 1) compared to patients with CAD (Group 2) and patients without significant coronary artery disease (Group 3) (Table 2). We identified a specific signature of plasma miRNAs up- or down-regulated in individual groups.

Table 2. Deregulated miRNAs in the circulation of all studied groups.

Symbol	Up-Down Regulation					
	Group 2 Compared to Group 1		Group 3 Compared to Group 1		Group 3 Compared to Group 2	
	Fold Regulation	p-Value	Fold Regulation	p-Value	Fold Regulation	p-Value
hsa-miR-451a	−2.18	0.03	−2.64	0.01	−1.51	0.39
hsa-miR-23a-3p	1.54	0.01	2.11	<0.01	1.10	0.59
hsa-miR-125b-5p	−1.51	0.28	−1.78	0.07	−1.51	0.14
hsa-miR-210-3p	−2.07	0.22	−4.71	<0.01	−1.50	0.03
hsa-miR-328-3p	−1.08	0.53	−2.25	0.03	−1.88	0.03
hsa-miR-425-3p	−1.96	0.17	−3.30	0.01	−1.63	0.14
hsa-miR-483-5p	−1.28	0.45	−1.66	0.30	−1.47	0.02
hsa-miR-16-5p	−1.34	0.07	−1.23	0.15	−1.15	0.49
hsa-miR-30d-5p	−1.18	0.96	−1.49	0.15	−1.59	0.18
hsa-miR-320d	−1.38	0.27	−1.46	0.04	−1.38	0.10

The calculation does not include data for the spike-in controls (UniSP2-6 and cel-miR-39-3p) or any blank spots. MiRNA markers selected for further analysis are marked in red.

2.3. Comparison of Group 1 with Group 2

The univariate logistic regression analysis revealed three independent variables potentially increasing the risk of CAAD (Table 3).

Table 3. The univariate and multivariate logistic regression analysis of clinical variables and miRNA expression profile in Group 1 compared to Group 2.

Variable	Univariate Analysis		Multivariate Analysis	
	OR [95% CI]	*p*-Value	OR [95% CI]	*p*-Value
Previous PCI	0.3 [0.1; 0.9]	0.03	0.2 [0.1; 0.7]	0.01
Dyslipidemia	0.2 [0.1; 1.1]	0.06	0.2 [0.1; 0.8]	0.02
hsa-miR-451a	0.7 [0.5; 0.9]	0.03	0.7 [0.5; 0.99]	0.049

Abbreviations: OR, odds ratio; PCI, percutaneous coronary intervention.

We included all of them in the multivariate logistic regression analysis. Model 1 included two independent clinical variables, previous percutaneous coronary intervention (PCI) and dyslipidemia. Despite the low model fit (R2Pseudo = 0.11), it was statistically significant ($p = 0.004$). Both previous PCI and dyslipidemia significantly reduced the risk of CAAD. In Model 2, miR-451a improved the model fit (R2Pseudo = 0.16) and was associated with a significant reduction of the risk of CAAD (Table 3).

Comparative analysis of Model 1 and Model 2 based on R2Pseudo and the Likelihood-Ratio Test (LRT) ($p = 0.038$) revealed a significantly better fit of Model 2, including miR-451a, than of Model 1 (AIC = 83.6, AIC = 85.9, for Model 2 and Model 1, respectively) (Figure 1).

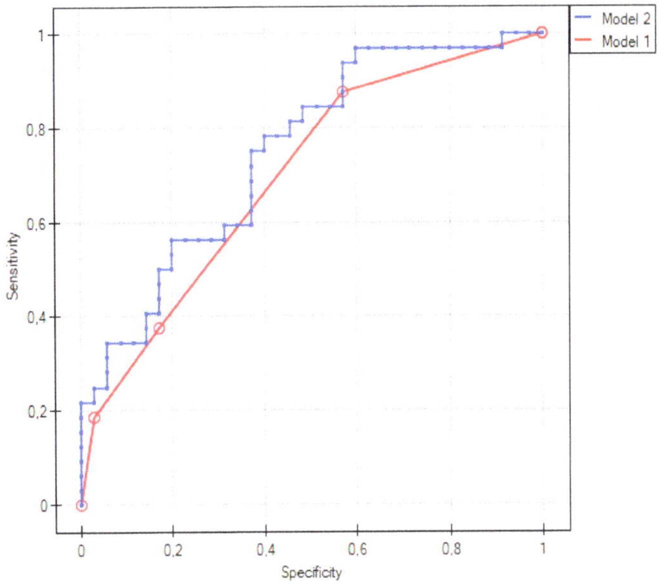

Figure 1. Comparison of logistic regression models between Group 1 and Group 2.

2.4. Comparison between Group 1 and Group 3

The univariate logistic regression analysis revealed six independent clinical variables and five miRNAs potentially increasing the risk of CAAD (Table 4). We included all of them in the multivariate logistic regression analysis.

Table 4. The univariate logistic regression analysis of clinical variables and miRNA expression profile in Group 1 compared to Group 3.

Variable	Univariate Analysis	
	OR [95% CI]	*p*-Value
Previous MI	7.5 [1.5; 37.6]	0.01
P2Y12 inhibitor	36.3 [4.4; 229.0]	<0.01
Aspirin	5.5 [1.4; 21.8]	0.01
WBC, 10^9/L	1.2 [0.9; 1.6]	0.06
RBC, 10^9/L	3.2 [1.1; 10.0]	0.04
Creatinine, mmol/L	1.02 [1.0; 1.1]	0.03
hsa-miR-328-3p	0.7 [0.5; 0.9]	<0.01
hsa-miR-451a	0.6 [0.4; 0.9]	<0.01
hsa-miR-23a-3p	2.6 [1.5; 4.7]	<0.01
hsa-miR-210-3p	0.7 [0.5; 0.8]	<0.01
hsa-miR-425-3p	0.8 [0.6; 0.9]	0.01

Abbreviations: MI, myocardial infarction; RBC, red blood cells; WBC, white blood cells; OR, odds ratio.

Model 1 included all independent clinical variables, such as previous myocardial infarction (MI), P2Y12 inhibitor and aspirin taking, white blood cell (WBC) count, red blood cell (RBC) count, and serum creatinine level. It was significantly associated with the risk of CAAD compared to patients without significant coronary artery lesions ($p < 0.01$). In addition, the goodness-of-fit was quite high (R2Pseudo = 0.46). Among the independent variables included in the model, only taking the P2Y12 inhibitor significantly increased the risk of CAAD (OR [95% CI] 80.3 [5.8; 1108.4, $p < 0.01$). In Model 2, miRNAs were added separately to Model 1. Only miR-328-3p improved the model fit (R2Pseudo = 0.62) and was associated with a significant reduction in the risk of CAAD (OR [95% CI] 0.21 [0.07; 0.67], $p = 0.01$) (Table 5).

Table 5. The multivariate logistic regression analysis of clinical variables and miRNA expression profile in Group 1 compared to Group 3 (Model 2).

Variable	Multivariate Analysis	
	OR [95% CI]	*p*-Value
Previous MI	7.0 [0.9; 51.7]	0.06
P2Y12 inhibitor	80.3 [5.8; 1108.4]	<0.01
Aspirin	3.7 [0.4; 29.8]	0.22
WBC, 10^9/L	1.3 [0.9; 1.9]	0.08
RBC, 10^9/L	4.5 [0.8; 24.6]	0.11
Creatinine, mmol/L	1.0 [0.97; 1.1]	0.54
hsa-miR-328-3p	0.2 [0.1; 0.7]	<0.01

Abbreviations: MI, myocardial infarction; RBC, red blood cells; WBC, white blood cells; OR, odds ratio.

Comparative analysis of Model 1 and Model 2 based on R2Pseudo and LRT ($p < 0.01$) revealed a significantly better fit of Model 2, including miR-328-3p, than of Model 1 (AIC = 49.0, AIC = 61.4, for Model 2 and Model 1 respectively) (Figure 2).

2.5. Functional Enrichment Analysis

Recent research reports that exogenous, circulating miRNAs can enter recipient cells and play a regulatory role in cells or organisms [15]. Assuming that miRNA is significantly deregulated in CAAD and CAD samples, and may be transported into epithelial, smooth muscle, or myocyte cells, we undertook in silico functional enrichment analysis. To identify biochemical signaling pathways potentially affected by miRNA modulated in CAAD/CAD patients, we analyzed the predicted target genes of the five significantly affected miRNAs (hsa-miR-451a, hsa-miR-23a-3p, hsa-miR-210-3p, hsa-miR-328-3p, hsa-miR-425-3p, hsa-miR-483-5p) (Table 6). The target genes of the five identified miRNAs were extracted from the miRWalk database (Table S1), and the intersection of identified target genes from GSEA tool programs was chosen [15,16]. We correlated predicted target genes with the KEGG (Kyoto Encyclopedia of Genes and Genomes) to identify potential biochemical pathways as the functional output of five deregulated miRNAs. The enrichment analysis represents pathways significantly enriched with target genes for miRNAs in CAAD and CAD patients in relation to the control group. Significantly (FDR < 0.05) enriched KEGG pathways are presented in Table 6. Annotated notes about predicted target genes for the five dysregulated miRNAs in CAA and CAD groups are presented in supplementary tables (Tables S1–S5).

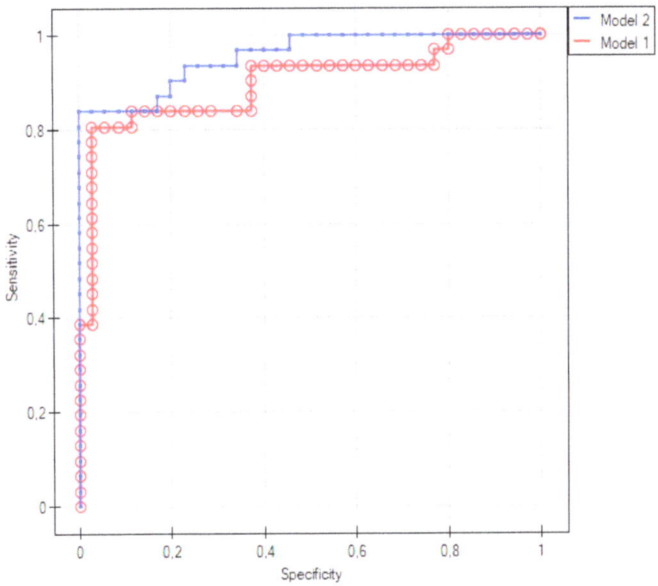

Figure 2. Evaluation of logistic regression models to compare Group 1 and Group 3.

Table 6. Pathways regulated by miRNAs hsa-miR-451a, hsa-miR-23a-3p, hsa-miR-210-3p, hsa-miR-328-3p, hsa-miR-425-3p. The target genes of the five identified miRNAs were extracted from the miRWalk database, and the intersection of identified target genes from GSEA tool programs was chosen. We correlated identified target genes with the KEGG (Kyoto Encyclopedia of Genes and Genomes) to point to biochemical pathways as potentially functional outputs of five deregulated miRNAs.

Pathways	Genes	*p*-Value, Adjusted	*p*-Value (BH)
Non-alcoholic fatty liver disease (NAFLD)	IL6R NDUFV3 PIK3R1 SDHD UQCRFS RAC1	<0.01	0.01
AMPK signaling pathway	TSC1, PIK3R1 PPARGC1A FOXO3 SCD	<0.01	0.01

Table 6. *Cont.*

Pathways	Genes	p-Value, Adjusted	p-Value (BH)
Autophagy	TSC1 ATG7 PIK3R1 HMGB1 VMP1	<0.01	<0.01
Regulation of actin cytoskeleton	VAV3 EZR PIK3R1 MYH10 RAC1	0.02	0.06
Thermogenesis	TSC1 NDUFV3 SDHD PPARGC1A UQCRFS1	0.03	0.06
Huntington disease	NDUFV SDHD PPARGC1A ACTR1A UQCRFS1	0.05	0.08
PI3K-Akt signaling pathway	TSC1 IL6R PIK3R1 FOXO3 RAC1	0.12	0.18
Pathways in cancer	MITF IL6R TPM3 RUNX1T1 PIK3R1 RAC1	0.20	0.25
Herpes simplex virus 1 infection	TSC1 ZNF275 ZNF268 ZNF701 PIK3R1	0.29	0.32
Metabolic pathways	NDUFV3 HMGCS1 GLS SDHD STT3A MGAT5 FUT4 SCD UQCRFS1 STT3B	0.62	0.62

The KEGG enrichment analysis indicated that the five miRNAs targeted genes are involved in ten pathways. The AMPK signaling pathway is involved in vast processes in the heart [17]: e.g., modulation of vascular inflammation and vascular smooth muscle cells phenotype [18]. Several miR-23-3p potential targets, in the AMPK pathway, were selected: TSC1, PIK3R1, PPARGC1A, FOXO3, and one miR-328-3p target SCD.

Endothelial cells, vascular smooth muscle cells, and macrophages play important roles in the development of CAD. In the etiology of atherosclerosis, these cells undergo two types of programmed cell death apoptosis and autophagy [19,20]. Numerous studies have revealed recently that autophagy-related genes (ATGs) are involved in endothelial and smooth-muscle inappropriate autophagy in the processes of atherogenesis [21]. In our in silico analysis, several genes were selected as potential targets for miRNAs. The following miRNAs were selected as regulators of autophagy-related genes: miR-328-3p HMGB1, miR-23-3p TSC1, PIK3R1, miR-210-3p ATG7, and VMP1. An example is HMGB1, the intranuclear non-histone DNA binding protein, which contributes to the stabilization of nucleosomes and the development of atherosclerosis [22].

3. Discussion

Coronary artery aneurysmal disease (CAAD) is a progressive, gradual coronary artery abnormality that mainly results in recurrent angina or, less commonly, severe complications such as distal embolization and myocardial infarction. The critical factor that initiates and deteriorates the course of CAAD is atherosclerosis. Asymptomatic course and progression make an early diagnosis of CAAD difficult. miRNAs play a role in the mechanism of atherosclerosis, and atherosclerotic coronary artery disease is characterized by specific signatures of miRNA in tissue and plasma assays [23–26]. However, scarce data concerning miRNAs involved in CAAD pathology prompted us to study five circulating miRNAs: MiR-125b-5p, miR-210-3p, miR-328-3p, miR-425-3p, and miR-483-5p, which we preselected and described in our former study [14] as potentially associated with CAAD.

This study evaluated the predictive role of circulating miRNAs for CAAD. Through a sampling process, we found miR-451a and miR-328 to be the most effective for CAAD prediction in a general Greater Poland population. Adding miR-451a and miR-328-3p to traditional risk factors significantly improved the detection of CAAD. We found that a higher circulating level of miR-328-3p was significantly associated with a higher risk of CAAD than the control group. On the other hand, the higher plasma level of miR-451a was predictive of the higher risk of CAAD compared to CAD.

As described in the methodology, miRNAs appeared as stable molecules and were recognized as an advantage of good potential markers for cardiovascular issues [27]. Selecting miRNA markers from dozens of circulating miRNA requires a multistep procedure. We overcame this difficulty and applied RT-qPCR to select miRNA markers for CAAD.

Since the presence of miRNAs in plasma was first mentioned in the literature, circulating miRNAs have been reported as a potential biomarker of heart failure and CAD [23,28]. In contrast to the numerous studies in CAD cohorts, the predictive value of miRNAs in CAAD is yet to be discovered. CAAD is usually detected during coronary angiography. Although data suggest a few CAD biomarkers, the need for reliable CAAD biomarkers, discriminating it from CAD, remains a limitation in the clinical practice. Our study suggested miRNA as a minimally invasive biomarker to improve pre-procedure diagnosis of CAAD. We applied the RT-qPCR method using a PCR array system. The custom plates for the RT-qPCR array were designed as in our former study [14].

We confirmed that the concentration of five preselected plasma miRNAs was different in patients with CAAD (Group 1) compared to patients with ischemic heart disease (Group 2) and patients without significant coronary artery disease (Group 3).

By applying logistic regression, we compared two models without and with miRNA. The model without miRNA to compare CAAD (Group 1) and CAD (Group 2) included two independent variables: previous percutaneous coronary intervention (PCI) and dyslipidemia. By adding miR-451a to the model, the prediction increased. miR-451a appeared to be a significant marker of CAAD compared to the group of patients with CAD.

The model without miRNA to compare CAAD (Group 1) and the control (Group 3) included six independent variables, previous MI, P2Y12 inhibitor or aspirin taking, WBC count, RBC count, and serum creatinine level. By adding miR-328-3p to the model, the prediction increased. miR-328-3p appeared to be a significant marker of CAAD compared to patients without significant CAD.

The available papers concerning miRNA as a marker of CAAD are scarce. RT-qPCR evaluated let-7i-3p as a biomarker to identify CAAD patients [29]. Serum exosomal let-7i-3p was proposed as a diagnostic marker for CAAD in patients with Kawasaki disease [29]. miRNAs miR-451a and miR-328, which we elaborated on in the present study, have already been proposed as potential markers of heart failure (HF) and CAD.

miR-451a expression is elevated in the plasma of patients with acute myocardial infarction (AMI) compared with unstable angina pectoris and healthy control groups [30]. Data suggest that miR-451a regulates ischemic heart injury [31]. Overexpression of miR-451a in mice significantly attenuated Ang II-induced cardiac fibrosis and inflammation. miR-451a directly targeted transcription factor TBX1 expression, thus decreasing TBX1-dependent TGFB1 gene expression. In our study, we observed a higher plasma level of miR-451a in CAAD patients compared to CAD patients. Therefore, we conclude that miR-451a could be involved in CAAD pathomechanism since CAAD is a severe consequence of coronary vessel inflammation and atherosclerosis. Recently, miR-451a has been also found to inhibit the expression of both MMP-2 and MMP-9 in cardiomyocytes by macrophage migration inhibitory factor (MIF) [32].

In our study, miR-328 was increased in the CAAD group compared to the control group. An increased miR-328 plasma level was reported in patients with AMI, strongly associated with increased risk of mortality or heart failure [33]. Elevated expression of miR-328 is also a marker of cardiac fibrosis during MI. This observation has recently been demonstrated in vitro and explained by the paracrine mechanism. miR-328 is secreted by cardiomyocytes promoting fibrosis by regulating adjacent fibroblasts [34]. Transgenic mice with miR-328 overexpression in the heart induced cardiac hypertrophy [35]. miR-328 reduces the transcript of sarco/endoplasmic reticulum Ca^{2+}-ATPase (SERCA2a, *ATP2A2*), a transporter of Ca^{2+} from the cytosol into the sarcoplasmic reticulum [35–37].

miR-23a, miR-210, and miR-425 were significantly regulated in the CAAD group versus the control group, but did not significantly influence a logistic regression model improving prediction. Our study demonstrated decreased miR-23a in the CAAD group versus the control. Previous studies suggested increased plasma miR-23a in CAD patients [38,39]. It has been suggested to be used to predict the presence and severity of coronary lesions in patients with CAD. In our study, miR-23a has not been significantly modulated in the CAD group.

So far, several studies evaluating the prognostic value of single circulating miRNAs in cardiovascular disease have been reported [40]. miR-210 derived from peripheral blood predict mortality and represent a valuable biomarker for risk estimation in CAD [41,42]. Circulating miR-210 improved the prediction model, including AUC (area under the receiver-operating characteristic curve). In our study, miR-210 was increased in CAAD vs. control but did not improve the model, including clinical parameters.

miRbase miR-425-3p is a minor transcript in mammalian cells [43]. Several studies demonstrated that predominant miR-425-p functions as a negative regulator of cardiac fibrosis by suppressing *TGFB1* (encoding TGFβ1) expression [44].

Our finding shows a new look at the miRNA signatures for plasma samples from CAAD patients. It provides a composed clinical and molecular predictive model, benefiting the diagnosis of CAAD patients. Developing a method using miRNAs as CAAD markers could improve the early diagnosis of CAAD and identify patients at increased risk of progression and complications.

We proposed a new model of CAAD prediction using miR-451a and miR-328 in addition to independent clinical variables. Our approach with logistic regression models revealed a significant increase in the predictability of discrimination between the CAAD group and the CAD group, as well as the CAAD group against the control group, by adding the status of miR-451a and miR-328, respectively.. This preclinical assessment of potential miRNA markers could be developed in the future in reliable tests for early diagnosis of CAAD.

Our achievement has limits in the number of patients compared to the study. According to several studies, the condition, time of the day, and several other parameters could influence miRNA levels and should be controlled and standardized in future studies.

In future endeavours is the question of how miR-451a and miR-328 contribute to CAAD development. The important question also concerns target cells for miRNA and the source of abnormal levels of selected markers in plasma.

4. Materials and Methods

4.1. Study Design and Patient Selection

The overall cohort enrolled two hundred fifty patients undergoing coronary angiography for angina symptoms between April 2020 and December 2021. The study cohort included 35 consecutive patients with CAAD (Group 1), and two groups of 35 patients matched Group 1 in terms of sex and age from the overall cohort (Group 2 and Group 3) (Figure 3). Group 2 included patients with angiographically documented CAD, while Group 3 enrolled patients with normal coronary arteries (NCA) assessed during coronary angiography. Due to hemolysis in the two CAAD patient samples, we finally analyzed 33 patients.

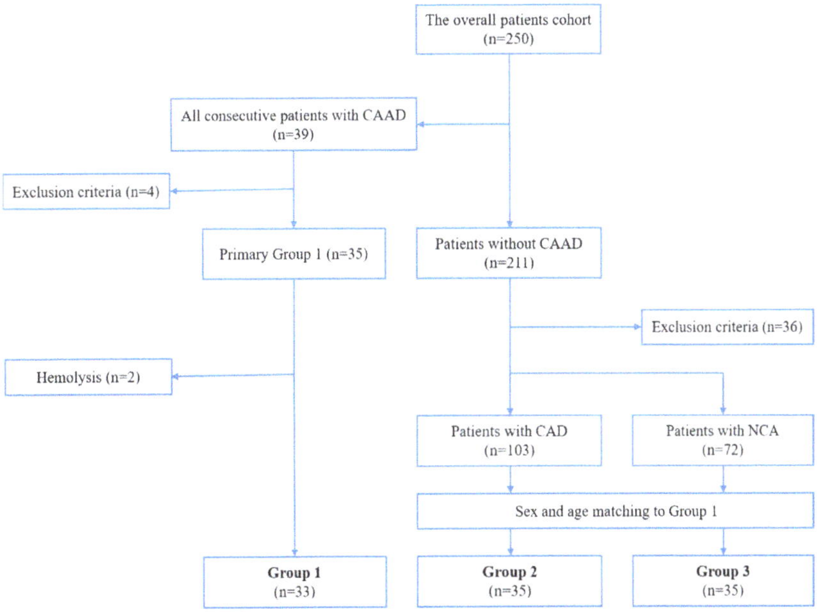

Figure 3. Flowchart of the study population. CAAD, coronary artery aneurysmal disease; CAD, coronary artery disease; NCA, normal coronary arteries.

All patients were qualified for coronary angiography, according to the European Society of Cardiology (ESC) guidelines [45]. According to the available data, CAAD is a diffuse or focal dilatation of the coronary artery with a diameter of 1.5 times the adjacent normal segment. We also enrolled patients with concomitant coronary artery stenosis in Group 1. The angiographic criteria for CAD included: coronary artery stenosis > 90% or intermediate stenosis (50–90%) with documented ischemia or hemodynamically significant, defined as either fractional flow reserve (FFR) $\leq$ 0.80 or an instantaneous wave-free ratio (iFR) $\leq$ 0.89. Patients included in Group 3 had normal ECG and echocardiography apart from no significant lesions in the coronary arteries. In addition, they showed no evidence of ischemia during non-invasive exercise testing.

Patients meeting the following criteria were excluded from the study: (i) acute coronary syndrome; (ii) elevated levels of myocardial injury markers (troponin I or creatine kinase); (iii) severe hepatic and renal dysfunction present or history; (iv) hematological disease; (v) severe inflammatory and malignant diseases; (vi) systemic connective tissue disease; (vii) biological treatment; and (viii) no informed consent.

4.2. Blood Samples Collection and Evaluation of miRNA Profiles

Our previous study precisely described obtaining and preparing blood samples [14]. We treated the new collection of blood samples according to the same procedure described here. We collected blood samples (10 mL) from patients on the first day after the coronary angiography and processed them within 30 min. Samples were centrifuged at 1300× g for 15 min at room temperature. The supernatant was transferred to RNase-free tubes and then stored at −80 °C.

The miRNAs were isolated from individual 200 µL frozen plasma using the miRNeasy Serum/Plasma Advanced Kit (Cat. No 217204, Qiagen, Dusseldorf, Germany) according to the manufacturer's instructions. To correct sample-to-sample variation, a synthetic set of spike-ins, UniSp2, UniSp4, and UniSp5 was applied (RNA Spike-In Kit for reverse transcription (RT) Cat. No 339390, Qiagen, Dusseldorf, Germany). Three of the RNA spike-in templates (UniSp2, UniSp4, and UniSp5) were provided premixed in one vial, each at a different concentration in 100-fold increments (UniSp2 (2 fmol/µL), UniSp4 (0.02 fmol/µL), and UniSp5 (0.0002 fmol/µL)). Before starting the RNA isolation procedure, one microliter of this RNA spike-in mix per RNA prepared with 60 µL Lysis Buffer RPL was aliquoted. Approximate RNA quantity and quality were estimated using a NanoDrop 2000 spectrophotometer (Thermo Scientific, Waltham, MA, USA). Four microliters of each total RNA were reversely transcribed with the miRCURY LNA RT Kit (Cat. No 339340, Qiagen, Dusseldorf, Germany). The UniSp6 RNA spike-in from the miRCURY LNA RT Kit and cel-miR-39-3p RNA from miRCURY LNA RT Kit were applied as controls.

Twenty microliters of cDNA product of the miRCURY LNA RT Kit were applied for each quantitative PCR (qPCR). qPCR was performed using miRCURY LNA SYBR Green PCR Kit (Cat. No 339345, 339346, 339347, Qiagen, Dusseldorf, Germany) with miRCURY LNA Custom PCR Panels (Qiagen) (Table 7).

Table 7. miRCURY LNA Custom PCR Panels (Qiagen) primers on a 96-well plate. Columns 1 and 2 were repeated six times. Six samples were analyzed on a single 96-well plate.

	1	2
A	UniSp2	hsa-miR-125b-5p
B	UniSp4	hsa-miR-210-3p
C	UniSp5	hsa-miR-328-3p
D	UniSp6	hsa-miR-425-3p
E	cel-miR-39-3p	hsa-miR-483-5p
F	UniSp3	hsa-miR-16-5p
G	hsa-miR-451a	hsa-miR-30d-5p
H	hsa-miR-23a-3p	hsa-miR-320d

Table 8 summarizes the genes analyzed in the custom plate, including preselected miRNA markers and spike-in controls applied in the analysis: the control, hemolysis markers, and reference markers. We provided a detailed description of the spike-in controls in the appendix (Appendix A).

Table 8. Genes analyzed on the Custom plate.

Position	Mature ID	Meaning of the Gene
1	UniSp2	Control
2	UniSp4	Control
3	UniSp5	Control
4	UniSp6	Control
5	cel-miR-39-3p	Control
6	UniSp3	Control
7	hsa-miR-451a	Hemolysis marker
8	hsa-miR-23a-3p	Hemolysis marker
9	hsa-miR-125b-5p	Marker
10	hsa-miR-210-3p	Marker
11	hsa-miR-328-3p	Marker
12	hsa-miR-425-3p	Marker
13	hsa-miR-483-5p	Marker
14	hsa-miR-16-5p	Reference marker
15	hsa-miR-30d-5p	Reference marker
16	hsa-miR-320d	Reference marker

4.3. Statistical Analysis

We presented all continuous variables as means with standard deviation for normal distribution or medians (upper and lower quartile) for non-normal distribution. We tested the normality of the distribution of variables using the Kolmogorov–Smirnov test and presented categorical variables as counts and percentages or frequencies. We assessed the significance of differences between the mean values of the continuous data consistent with the normal distribution using one-way ANOVA and Tukey's Test. To compare the continuous data inconsistent with the normal distribution, we used Mann–Whitney and Kruskal–Wallis tests with Benjamini–Hochberg post hoc analysis and compared categorical variables using the χ^2 test. We performed an analysis of array data with GENEGLOBE online software (https://geneglobe.qiagen.com/pl/analyze, accessed on 12 January 2023). We used the global mean as a reference for the PCR-array analysis and calculated the p values based on a Student's t-test of replicate $2^{(-\Delta CT)}$ values for each miRNA in the control and treatment groups. Subsequently, we based the p-value calculation on a parametric, two-sample equal variance, unpaired, two-tailed analysis. In the current analysis, the selection criteria to elaborate potential miRNA markers were the fold-regulation ≥ 2 and $p \leq 0.05$.

We conducted the univariate logistic regression analysis to evaluate confounding factors for each CAAD comparison (Group 1 vs. Group 2 and Group 1 vs. Group 3). The confounding factor was selected based on a p value < 0.1. The multivariate logistic regression analysis was conducted for each comparison with confounding factors that met the criteria in univariate analysis (Model 1). The following step was to create the Model 2 by adding the analyzed miRNA to Model 1, which had an expression that was significantly different between the compared groups (Table 2). We assessed the model fit for Model 1 and Model 2 using the Akaike information criterion (AIC) and pseudo-R-squared values. The model with the lowest AIC value was considered the best. The higher pseudo-R-squared value indicated a better model fit. All statistical analyses were two-sided and were conducted with PQStat Software (PQStat v.1.8.0.476, Poznań, Poland). A p value < 0.05 was considered a statistical significance.

5. Conclusions

The clinical use of the studied miRNAs as markers requires validation on a bigger cohort or advanced sequencing techniques. However, preliminary analysis indicates that miR-451a is a significant marker of CAAD compared to patients without concomitant aneurysms. In turn, miR-328-3p is a significant marker of CAAD compared to patients without significant changes in the coronary arteries.

Supplementary Materials: The following supporting information can be downloaded at: https://www.mdpi.com/article/10.3390/ijms24065817/s1.

Author Contributions: Conceptualization, S.I., T.L. and A.A.; methodology, S.I. and T.L.; software, A.C. and G.K.; validation, A.H. and T.L.; formal analysis, T.L. and S.I.; investigation, S.I. and T.L.; resources, M.G. and M.L.; data curation, A.C., K.M. and G.K.; writing—original draft preparation, S.I. and T.L.; writing—review and editing, P.W., A.C., P.P.J., M.G., M.L. and A.A.; visualization, S.I.; supervision, P.P.J., M.G., M.L. and A.A.; project administration, S.I.; funding acquisition, A.A. All authors have read and agreed to the published version of the manuscript.

Funding: This research was funded by Polish Cardiac Society, grant number 09/2018.

Institutional Review Board Statement: The study was conducted in accordance with the Declaration of Helsinki, and approved by the Institutional Ethics Committee of Poznan University of Medical Sciences (protocol code, 985/18; date of approval, 11 October 2018).

Informed Consent Statement: Informed consent was obtained from all subjects involved in the study.

Data Availability Statement: The data presented in this study are available on request from the corresponding author. The data are not publicly available due to confidential genetic data.

Conflicts of Interest: The authors declare no conflict of interest.

Appendix A

The following spike-in controls were applied: (1) the RNA isolation control spike-in templates UniSp2, UniSp4, and UniSp5 are provided as different concentrations in 100-fold increments, and their primers are present on plates miRCURY LNA miRNA QC PCR Panel, miRNeasy Serum/Plasma Kit (Cat. No 217184) and our custom plates; (2) the cDNA synthesis controls the UniSp6 RNA spike-in template in combination with cel-miR-39-3p, a synthetic version of a C. elegans miRNA; their primers are present on plates miRCURY LNA miRNA QC PCR Panel, miRNeasy Serum/Plasma Kit (Cat. No 217184) and our custom plates; (3) the PCR efficiency control UniSp3, with their primers being present on plates miRCURY LNA miRNA QC PCR Panel, miRNeasy Serum/Plasma Kit (Cat. No 217184) and our custom plates; (4) the hemolysis markers, the red-blood-cell-specific miR-451, and the stable miR-23a were used to monitor hemolysis. A difference of crossing-point Ct values for miR-23a and miR-451 (ΔCq) < 5 in human serum or plasma represents non-hemolyzed samples. If the $\Delta Cq \geq 7$, there is an increased risk that the samples are affected by hemolysis; their primers are present on plates miRCURY LNA miRNA QC PCR Panel, miRNeasy Serum/Plasma Kit (Cat. No 217184) and our custom plates; (5) three reference markers were chosen to normalize variation of RNA extraction efficiency by GeneGlobe using protocols geNorm and Normfinder (Table 2).

References

1. Fichtlscherer, S.; Zeiher, A.M.; Dimmeler, S. Circulating MicroRNAs: Biomarkers or Mediators of Cardiovascular Diseases? *Arterioscler. Thromb. Vasc. Biol.* **2011**, *31*, 2383–2390. [CrossRef] [PubMed]
2. Huntzinger, E.; Izaurralde, E. Gene Silencing by MicroRNAs: Contributions of Translational Repression and MRNA Decay. *Nat. Rev. Genet.* **2011**, *12*, 99–110. [CrossRef]
3. Rocchi, A.; Chiti, E.; Maiese, A.; Turillazzi, E.; Spinetti, I. MicroRNAs: An Update of Applications in Forensic Science. *Diagnostics* **2020**, *11*, 32. [CrossRef]
4. Swaye, P.S.; Fisher, L.D.; Litwin, P.; Vignola, P.A.; Judkins, M.P.; Kemp, H.G.; Mudd, J.G.; Gosselin, A.J. Aneurysmal Coronary Artery Disease. *Circulation* **1983**, *67*, 134–138. [CrossRef]
5. Baman, T.S.; Cole, J.H.; Devireddy, C.M.; Sperling, L.S. Risk Factors and Outcomes in Patients with Coronary Artery Aneurysms. *Am. J. Cardiol.* **2004**, *93*, 1549–1551. [CrossRef] [PubMed]
6. Warisawa, T.; Naganuma, T.; Tomizawa, N.; Fujino, Y.; Ishiguro, H.; Tahara, S.; Kurita, N.; Nojo, T.; Nakamura, S.; Nakamura, S. High Prevalence of Coronary Artery Events and Non-Coronary Events in Patients with Coronary Artery Aneurysm in the Observational Group. *Int. J. Cardiol. Heart Vasc.* **2016**, *10*, 29–31. [CrossRef] [PubMed]
7. Markis, J.E.; Joffe, C.D.; Cohn, P.F.; Feen, D.J.; Herman, M.V.; Gorlin, R. Clinical Significance of Coronary Arterial Ectasia. *Am. J. Cardiol.* **1976**, *37*, 217–222. [CrossRef]
8. Kühl, M.; Varma, C. A Case of Acute Coronary Thrombosis in Diffuse Coronary Artery Ectasia. *J. Invasive Cardiol.* **2008**, *20*, E23–E25.

9. Doi, T.; Kataoka, Y.; Noguchi, T.; Shibata, T.; Nakashima, T.; Kawakami, S.; Nakao, K.; Fujino, M.; Nagai, T.; Kanaya, T.; et al. Coronary Artery Ectasia Predicts Future Cardiac Events in Patients With Acute Myocardial Infarction. *Arterioscler. Thromb. Vasc. Biol.* **2017**, *37*, 2350–2355. [CrossRef]
10. Janus, M.; Iwańczyk, S.; Stanisławska, K.; Mularek-Kubzdela, T.; Lesiak, M. Endovascular Closure of Coronary Artery to Pulmonary Artery Fistula with Giant Aneurysm. *Kardiol. Pol.* **2022**, *80*, 621–622. [CrossRef]
11. Iwańczyk, S.; Borger, M.; Kamiński, M.; Chmara, E.; Cieślewicz, A.; Tykarski, A.; Radziemski, A.; Krasiński, Z.; Lesiak, M.; Araszkiewicz, A. Inflammatory response in patients with coronary artery ectasia and coronary artery disease. *Kardiol. Pol.* **2019**, *77*, 713–715. [CrossRef] [PubMed]
12. Iwańczyk, S.; Lehmann, T.; Cieślewicz, A.; Radziemski, A.; Malesza, K.; Wrotyński, M.; Jagodziński, P.P.; Grygier, M.; Lesiak, M.; Araszkiewicz, A. Involvement of Angiogenesis in the Pathogenesis of Coronary Aneurysms. *Biomedicines* **2021**, *9*, 1269. [CrossRef] [PubMed]
13. Wojciechowska, A.; Braniewska, A.; Kozar-Kamińska, K. MicroRNA in Cardiovascular Biology and Disease. *Adv. Clin. Exp. Med.* **2017**, *26*, 865–874. [CrossRef]
14. Iwańczyk, S.; Lehmann, T.; Cieślewicz, A.; Radziemski, A.; Malesza, K.; Wrotyński, M.; Jagodziński, P.P.; Grygier, M.; Lesiak, M.; Araszkiewicz, A. Circulating MicroRNAs in Patients with Aneurysmal Dilatation of Coronary Arteries. *Exp. Ther. Med.* **2022**, *23*, 404. [CrossRef]
15. Wang, L.; Liu, J.; Xu, B.; Liu, Y.-L.; Liu, Z. Reduced Exosome MiR-425 and MiR-744 in the Plasma Represents the Progression of Fibrosis and Heart Failure. *Kaohsiung J. Med. Sci.* **2018**, *34*, 626–633. [CrossRef] [PubMed]
16. Subramanian, A.; Tamayo, P.; Mootha, V.K.; Mukherjee, S.; Ebert, B.L.; Gillette, M.A.; Paulovich, A.; Pomeroy, S.L.; Golub, T.R.; Lander, E.S.; et al. Gene Set Enrichment Analysis: A Knowledge-Based Approach for Interpreting Genome-Wide Expression Profiles. *Proc. Natl. Acad. Sci. USA* **2005**, *102*, 15545–15550. [CrossRef]
17. Zaha, V.G.; Young, L.H. AMP-Activated Protein Kinase Regulation and Biological Actions in the Heart. *Circ. Res.* **2012**, *111*, 800–814. [CrossRef] [PubMed]
18. Lei, L.; Zhou, Y.; Wang, T.; Zheng, Z.; Chen, L.; Pan, Y. Activation of AMP-Activated Protein Kinase Ablated the Formation of Aortic Dissection by Suppressing Vascular Inflammation and Phenotypic Switching of Vascular Smooth Muscle Cells. *Int. Immunopharmacol.* **2022**, *112*, 109177. [CrossRef]
19. Zhao, F.; Satyanarayana, G.; Zhang, Z.; Zhao, J.; Ma, X.-L.; Wang, Y. Endothelial Autophagy in Coronary Microvascular Dysfunction and Cardiovascular Disease. *Cells* **2022**, *11*, 2081. [CrossRef] [PubMed]
20. Ni, D.; Mo, Z.; Yi, G. Recent Insights into Atherosclerotic Plaque Cell Autophagy. *Exp. Biol. Med.* **2021**, *246*, 2553–2558. [CrossRef] [PubMed]
21. Henderson, J.M.; Weber, C.; Santovito, D. Beyond Self-Recycling: Cell-Specific Role of Autophagy in Atherosclerosis. *Cells* **2021**, *10*, 625. [CrossRef]
22. Naglova, H.; Bucova, M. HMGB1 and Its Physiological and Pathological Roles. *Bratisl. Lekárske Listy* **2012**, *113*, 163–171. [CrossRef] [PubMed]
23. Vavassori, C.; Cipriani, E.; Colombo, G.I. Circulating MicroRNAs as Novel Biomarkers in Risk Assessment and Prognosis of Coronary Artery Disease. *Eur. Cardiol. Rev.* **2022**, *17*, e06. [CrossRef]
24. Samra, M.; Srivastava, K. Non-Coding RNA and Their Potential Role in Cardiovascular Diseases. *Gene* **2023**, *851*, 147011. [CrossRef] [PubMed]
25. Kabłak-Ziembicka, A.; Badacz, R.; Przewłocki, T. Clinical Application of Serum MicroRNAs in Atherosclerotic Coronary Artery Disease. *J. Clin. Med.* **2022**, *11*, 6849. [CrossRef]
26. Bergami, M.; Fabin, N.; Cenko, E.; Bugiardini, R.; Manfrini, O. MicroRNAs as Potential Biomarkers in Coronary Artery Disease. *Curr. Top. Med. Chem.* **2023**, 23. [CrossRef]
27. Shen, N.-N.; Wang, J.-L.; Fu, Y.-P. The MicroRNA Expression Profiling in Heart Failure: A Systematic Review and Meta-Analysis. *Front. Cardiovasc. Med.* **2022**, *9*, 856358. [CrossRef]
28. Sigutova, R.; Evin, L.; Stejskal, D.; Ploticova, V.; Svagera, Z. Specific MicroRNAs and Heart Failure: Time for the next Step toward Application? *Biomed. Pap. Med. Fac. Univ. Palacky Olomouc Czechoslov.* **2022**, *166*, 359–368. [CrossRef]
29. Wang, Y.-F.; Lian, X.-L.; Zhong, J.-Y.; Su, S.-X.; Xu, Y.-F.; Xie, X.-F.; Wang, Z.-P.; Li, W.; Zhang, L.; Che, D.; et al. Serum Exosomal MicroRNA Let-7i-3p as Candidate Diagnostic Biomarker for Kawasaki Disease Patients with Coronary Artery Aneurysm. *IUBMB Life* **2019**, *71*, 891–900. [CrossRef]
30. Xu, L.; Tian, L.; Yan, Z.; Wang, J.; Xue, T.; Sun, Q. Diagnostic and Prognostic Value of MiR-486-5p, MiR-451a, MiR-21-5p and Monocyte to High-Density Lipoprotein Cholesterol Ratio in Patients with Acute Myocardial Infarction. *Heart Vessel.* **2023**, *38*, 318–331. [CrossRef] [PubMed]
31. Deng, H.-Y.; He, Z.-Y.; Dong, Z.-C.; Zhang, Y.-L.; Han, X.; Li, H.-H. MicroRNA-451a Attenuates Angiotensin II-Induced Cardiac Fibrosis and Inflammation by Directly Targeting T-Box1. *J. Physiol. Biochem.* **2022**, *78*, 257–269. [CrossRef]
32. Scrimgeour, N.R.; Wrobel, A.; Pinho, M.J.; Høydal, M.A. MicroRNA-451a Prevents Activation of Matrix Metalloproteinases 2 and 9 in Human Cardiomyocytes during Pathological Stress Stimulation. *Am. J. Physiol. Cell Physiol.* **2020**, *318*, C94–C102. [CrossRef] [PubMed]
33. He, F.; Lv, P.; Zhao, X.; Wang, X.; Ma, X.; Meng, W.; Meng, X.; Dong, S. Predictive Value of Circulating MiR-328 and MiR-134 for Acute Myocardial Infarction. *Mol. Cell. Biochem.* **2014**, *394*, 137–144. [CrossRef] [PubMed]

34. Zhao, D.; Li, C.; Yan, H.; Li, T.; Qian, M.; Zheng, N.; Jiang, H.; Liu, L.; Xu, B.; Wu, Q.; et al. Cardiomyocyte Derived MiR-328 Promotes Cardiac Fibrosis by Paracrinely Regulating Adjacent Fibroblasts. *Cell. Physiol. Biochem.* **2018**, *46*, 1555–1565. [CrossRef]
35. Li, C.; Li, X.; Gao, X.; Zhang, R.; Zhang, Y.; Liang, H.; Xu, C.; Du, W.; Zhang, Y.; Liu, X.; et al. MicroRNA-328 as a Regulator of Cardiac Hypertrophy. *Int. J. Cardiol.* **2014**, *173*, 268–276. [CrossRef] [PubMed]
36. Lu, Y.; Zhang, Y.; Wang, N.; Pan, Z.; Gao, X.; Zhang, F.; Zhang, Y.; Shan, H.; Luo, X.; Bai, Y.; et al. MicroRNA-328 Contributes to Adverse Electrical Remodeling in Atrial Fibrillation. *Circulation* **2010**, *122*, 2378–2387. [CrossRef] [PubMed]
37. Du, W.; Liang, H.; Gao, X.; Li, X.; Zhang, Y.; Pan, Z.; Li, C.; Wang, Y.; Liu, Y.; Yuan, W.; et al. MicroRNA-328, a Potential Anti-Fibrotic Target in Cardiac Interstitial Fibrosis. *Cell. Physiol. Biochem.* **2016**, *39*, 827–836. [CrossRef]
38. Ghafouri-Fard, S.; Gholipour, M.; Taheri, M. Role of MicroRNAs in the Pathogenesis of Coronary Artery Disease. *Front. Cardiovasc. Med.* **2021**, *8*, 632392. [CrossRef]
39. Wang, S.; He, W.; Wang, C. MiR-23a Regulates the Vasculogenesis of Coronary Artery Disease by Targeting Epidermal Growth Factor Receptor. *Cardiovasc. Ther.* **2016**, *34*, 199–208. [CrossRef]
40. Guan, Y.; Song, X.; Sun, W.; Wang, Y.; Liu, B. Effect of Hypoxia-Induced MicroRNA-210 Expression on Cardiovascular Disease and the Underlying Mechanism. *Oxidative Med. Cell. Longev.* **2019**, *2019*, 4727283. [CrossRef]
41. Karakas, M.; Schulte, C.; Appelbaum, S.; Ojeda, F.; Lackner, K.J.; Münzel, T.; Schnabel, R.B.; Blankenberg, S.; Zeller, T. Circulating MicroRNAs Strongly Predict Cardiovascular Death in Patients with Coronary Artery Disease-Results from the Large AtheroGene Study. *Eur. Heart J.* **2017**, *38*, 516–523. [CrossRef] [PubMed]
42. Li, T.; Cao, H.; Zhuang, J.; Wan, J.; Guan, M.; Yu, B.; Li, X.; Zhang, W. Identification of MiR-130a, MiR-27b and MiR-210 as Serum Biomarkers for Atherosclerosis Obliterans. *Clin. Chim. Acta* **2011**, *412*, 66–70. [CrossRef] [PubMed]
43. Landgraf, P.; Rusu, M.; Sheridan, R.; Sewer, A.; Iovino, N.; Aravin, A.; Pfeffer, S.; Rice, A.; Kamphorst, A.O.; Landthaler, M.; et al. A Mammalian MicroRNA Expression Atlas Based on Small RNA Library Sequencing. *Cell* **2007**, *129*, 1401–1414. [CrossRef]
44. Wei, F.; Ren, W.; Zhang, X.; Wu, P.; Fan, J. MiR-425-5p Is Negatively Associated with Atrial Fibrosis and Promotes Atrial Remodeling by Targeting CREB1 in Atrial Fibrillation. *J. Cardiol.* **2022**, *79*, 202–210. [CrossRef] [PubMed]
45. Neumann, F.J.; Sousa-Uva, M.; Ahlsson, A.; Alfonso, F.; Banning, A.P.; Benedetto, U.; Byrne, R.A.; Collet, J.P.; Falk, V.; Head, S.J.; et al. 2018 ESC/EACTS Guidelines on myocardial revascularization. *Eur. Heart J.* **2019**, *40*, 87–165. [CrossRef]

International Journal of
Molecular Sciences

Article

Circulatory miR-411-5p as a Novel Prognostic Biomarker for Major Adverse Cardiovascular Events in Patients with Atrial Fibrillation

Stephan Nopp [1], M. Leontien van der Bent [2], Daniel Kraemmer [1], Oliver Königsbrügge [1], Johann Wojta [3,4], Ingrid Pabinger [1], Cihan Ay [1,*,†] and Anne Yaël Nossent [2,3,5,†]

1 Clinical Division of Haematology and Haemostaseology, Department of Medicine I, Medical University of Vienna, 1090 Vienna, Austria
2 Department of Surgery and Einthoven Laboratory for Experimental Vascular Medicine, Leiden University Medical Center, 2333 Leiden, The Netherlands
3 Department of Internal Medicine II, Medical University of Vienna, 1090 Vienna, Austria
4 Ludwig Boltzmann Institute for Cardiovascular Research, 1090 Vienna, Austria
5 Department of Laboratory Medicine, Medical University of Vienna, 1090 Vienna, Austria
* Correspondence: cihan.ay@meduniwien.ac.at
† These authors contributed equally to this work.

Abstract: The risk stratification of patients with atrial fibrillation (AF) for subsequent cardiovascular events could help in guiding prevention strategies. In this study, we aimed at investigating circulating microRNAs as prognostic biomarkers for major adverse cardiovascular events (MACE) in AF patients. We conducted a three-stage nested case–control study within the framework of a prospective registry, including 347 AF patients. First, total small RNA-sequencing was performed in 26 patients (13 cases with MACE) and the differential expression of microRNAs was analyzed. Seven candidate microRNAs with promising results in a subgroup analysis on cardiovascular death were selected and measured via using RT-qPCR in 97 patients (42 cases with cardiovascular death). To further validate our findings and investigate broader clinical applicability, we analyzed the same microRNAs in a subsequent nested case–control study of 102 patients (37 cases with early MACE) by using Cox regression. In the microRNA discovery cohort ($n = 26$), we detected 184 well-expressed microRNAs in circulation without overt differential expression between the cases and controls. A subgroup analysis on cardiovascular death revealed 26 microRNAs that were differentially expressed at a significance level < 0.05 (three of which with an FDR-adjusted p-value <0.05). We, therefore, proceeded with a nested case–control approach ($n = 97$) focusing on patients with cardiovascular death and selected, in total, seven microRNAs for further RT-qPCR analysis. One microRNA, miR-411-5p, was significantly associated with cardiovascular death (adjusted HR (95% CI): 1.95 (1.04–3.67)). Further validation ($n = 102$) in patients who developed early MACE showed similar results (adjusted HR (95% CI) 2.35 (1.17–4.73)). In conclusion, circulating miR-411-5p could be a valuable prognostic biomarker for MACE in AF patients.

Keywords: circulating microRNA; atrial fibrillation; biomarkers; sequence analysis; RNA; humans

1. Introduction

Atrial fibrillation (AF), the most common clinical cardiac arrhythmia, is associated with high morbidity and mortality. It is largely attributable to an increased risk of stroke, myocardial infarction, heart failure, and bleeding [1]. As the AF population is highly heterogeneous in terms of underlying risk for unfavorable outcomes, risk stratification is of high clinical importance to mitigate the risks for related complications. However, current clinical prediction scores show only moderate predictive performance for the respective outcomes, as they are often solely based on clinical parameters [2]. In contrast, biomarkers can reflect the entire spectrum of disease, improve the biological understanding of underlying pathophysiological mechanisms, and may, therefore, provide a more powerful approach to risk prediction. As such, microRNAs have emerged as a special research interest over the past decade.

MicroRNAs are small non-coding RNAs that consist of 18–25 nucleotides. They regulate gene expression at a post-transcriptional level by binding to the 3′ untranslated region of their target messenger RNA (mRNA) through the so-called "seed sequence" on nucleotides 2 to 8. In addition, they inhibit the translation and/or induce the degradation of its target mRNA. As such, microRNAs function as regulators for hundreds of mRNAs; often, at once within a regulatory network [3]. Thus, they substantially affect the pathophysiological pathways that influence disease progression and hold valuable therapeutic potential [4]. Moreover, microRNAs are present and stable in circulation, making them optimal candidate biomarkers. Previous studies have already linked differentially expressed microRNAs with various cardiovascular diseases [5–11]. Several microRNAs have been identified to be associated with the occurrence or pathological mechanisms of AF [12–15]. However, to our knowledge, no study has yet prospectively investigated the association of differentially expressed microRNAs with the occurrence of adverse cardiovascular outcomes in AF patients. We, therefore, aimed at identifying the microRNAs associated with future major adverse cardiovascular events (MACE) in patients with AF.

2. Results

2.1. Study Population

The total study population consisted of 347 patients with AF (mean age: 71.3 years, 37.3% female) and available plasma samples at baseline (Figure 1). Forty-seven MACE occurred in 42 (12.1%) patients during a median in-person follow-up time of 316 (interquartile range [IQR], 233–497) days. Thirty-seven (10.7%) patients had MACE within the first year after study inclusion. A detailed list of events per outcome is available in the Supplemental Table S1. Follow-up from the national death registry on the survival status and cause of death was available with a median follow-up time of 4.3 years (IQR, 1.8–5.7). In total, 81 patients died (23.3% of the study cohort); in 42 patients (51.9%), death was attributed to cardiovascular causes, while in 4 (4.9%) patients, the cause of death was not reported.

2.2. Selection of Candidate microRNAs in the Discovery Cohort

The results of the differential expression of the microRNAs in the small RNA sequencing are presented in Supplemental Tables S2 and S3. We detected 184 well-expressed microRNAs that were present in circulation of the cases and controls (n = 26). The differences in microRNA expression between the patients who developed three-point (3P)-MACE and their matched controls were not statistically significant when controlling for the false discovery rate (FDR). However, in the subgroup analysis on patients with subsequent cardiovascular death (six cases vs. their matched controls), three microRNAs (miR-150-5p, miR-127-3p, and miR-1908-5p) were significantly associated with cardiovascular mortality (FDR-adjusted $p < 0.05$); an additional 23 microRNAs met the less stringent threshold of $p < 0.05$. We, therefore, decided to proceed with investigating cardiovascular death as the primary endpoint in the subsequent nested case–control study. Of those 26 microRNAs, we selected seven promising microRNAs (miR-150-5p, miR-127-3p, miR-1908-5p, miR-483-3p, miR-411-5p, miR-625-5p, and miR-15a-5p) based on the composite of statistical significance, log fold-change, the visual inspection of the boxplots of the cases vs. controls, and the known biological functions from the cardiovascular literature [16–23] for the purposes of a validation in a nested case–control study with cardiovascular death as the outcome.

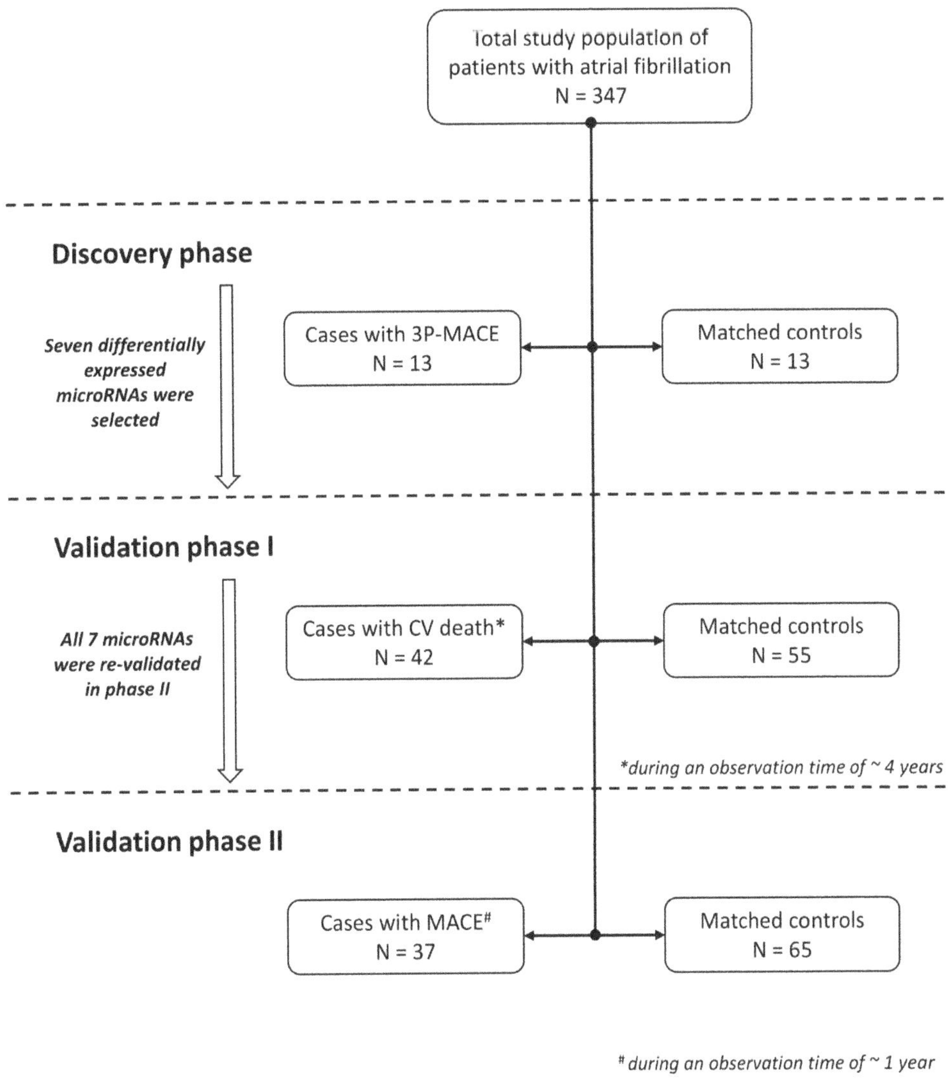

Figure 1. Study flow chart. (Abbreviation: 3P-MACE, three point major adverse cardiovascular event; CV, cardiovascular; and MACE, major adverse cardiovascular event).

2.3. Validation of Candidate microRNAs for Cardiovascular Death

The baseline characteristics of the nested case–control cardiovascular death outcome study are presented in Table 1. Note that this analysis was based on the longer follow-up period available from the national death registry (median follow-up time of 4.3 years). The associations of each microRNA with cardiovascular death, based on RT-qPCR measurements, are shown in Figure 2A, whereby the hazard ratios (HRs) are per one standard deviation increase (1-SD). The levels of miR-411-5p, miR-625-5p, and miR-15a-5p had to be dichotomized due to the low detectability (<20% valid values), and the HRs, therefore, compare detectable with undetectable microRNA in a binary fashion. In a multivariable analysis, adjusted for age and sex (model 1), only the miR-411-5p was significantly associated with the outcome (HR: 1.95; 95%CI 1.04–3.67; $p = 0.038$). After a further adjustment with respect to the history of myocardial infarction, as well as to stroke/TIA/systemic embolism (model 2), the detectable miR-411-5p was associated with cardiovascular death by a HR of 1.89 (95%CI 1.00–3.58; $p = 0.050$).

Table 1. Baseline characteristics of patients with atrial fibrillation: cases with cardiovascular death vs. matched controls without cardiovascular death.

	Cases with Cardiovascular Death (n = 42)	Controls (n = 55)
Demographics		
Age, years	77.0 (71.3–80.8)	76.0 (71.0–80.0)
Male sex	28 (66.7%)	37 (67.3%)
BMI, kg/m^2 [2]	26.6 (24.6–29.0)	26.9 (25.3–30.7)
Information on atrial fibrillation, n (%)		
Type of atrial fibrillation [2]		
First onset	4 (9.5%)	6 (10.9%)
Paroxysmal	18 (42.9%)	26 (47.3%)
Persistent	3 (7.1%)	2 (3.6%)
Permanent	16 (38.1%)	20 (36.4%)
History of electrical cardioversion	10 (23.8%)	20 (36.4%)
History of ablation	6 (14.3%)	4 (7.3%)
Comorbidities		
Arterial hypertension	38 (90.5%)	50 (90.9%)
Diabetes mellitus type 2	14 (33.3%)	18 (32.7%)
Coronary artery disease	15 (35.7%)	17 (30.9%)
Peripheral artery disease	6 (14.3%)	3 (5.5%)
Heart failure	20 (47.6%)	13 (23.6%)
History of myocardial infarction	9 (21.4%)	7 (12.7%)
History of stroke, TIA, and systemic embolism	8 (19.0%)	15 (27.3%)
History of cancer	8 (19.0%)	14 (25.5%)
History of bleeding	9 (21.4%)	10 (18.2%)
CHA_2DS_2-VASc score, median (IQR)	4 (3–6)	4 (3–5)
HAS-BLED score, median (IQR)	2 (1–3)	2 (1–2)
Medication		
Anticoagulant therapy		
DOAC	18 (42.9%)	26 (47.3%)
Vitamin K-Antagonist	23 (54.8%)	26 (47.3%)
Other	0 (0%)	3 (5.5%)
None	1 (2.4%)	0 (0%)
Platelet inhibitor therapy		
Acetylsalicylic acid	10 (23.8%)	6 (10.9%)
P2Y12-Inhibitors	4 (9.5%)	3 (5.5%)

Square brackets indicate the number of missing values. Abbreviations: IQR, interquartile range; TIA, transient ischemic attack.

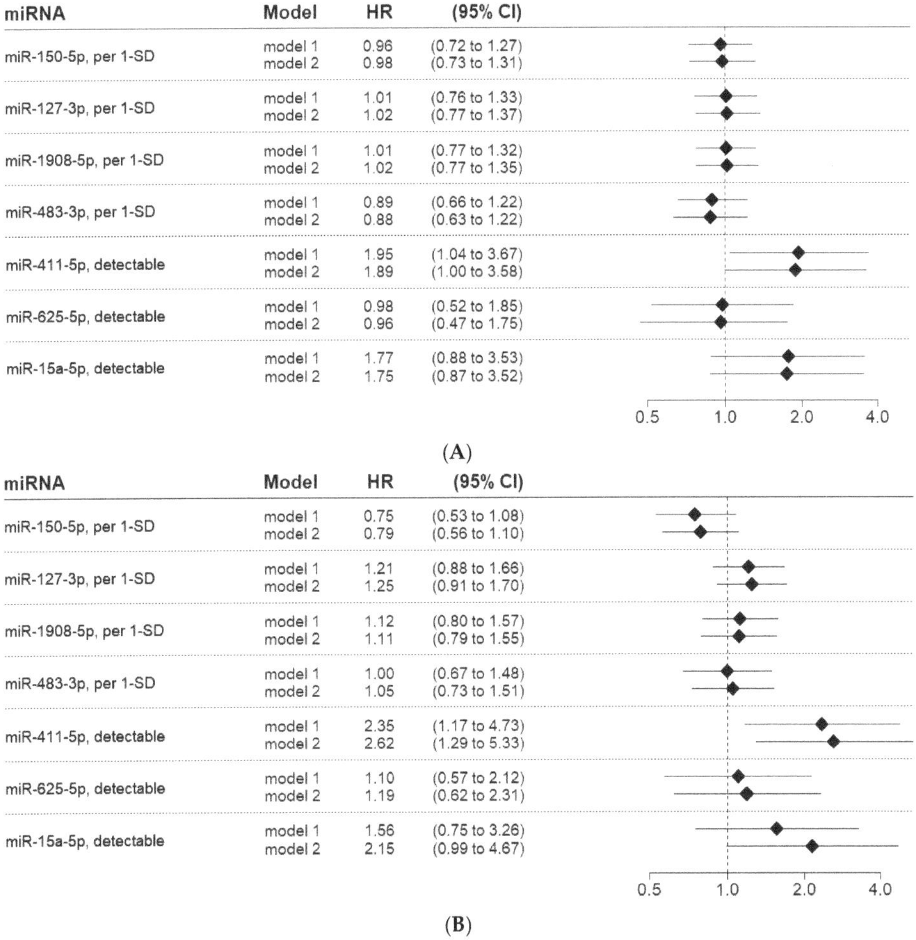

Figure 2. Association of candidate microRNAs with cardiovascular death (**A**) and major adverse cardiovascular events (**B**) in patients with atrial fibrillation. Model 1 includes the adjustment for age and sex. Model 2, in addition to the variables of model 1, further adjusts for the history of myocardial infarction and the history of stroke/TIA/systemic embolism. Due to the low detectability of miR-411-5p, miR-625-5p, and miR-15a-5p in RT-qPCR, the variables were dichotomized based on the detectability of the microRNA. Abbreviations: 1-SD, one standard deviation; CI, confidence interval; HR, hazard ratio; and SD, standard deviation.

We proceeded to test those findings from the first validation phase using the broader and more clinically relevant cardiovascular endpoint MACE and compared patients who developed MACE within one year of follow-up to the matched controls in a subsequent nested case–control study.

2.4. Validation of Candidate microRNAs for Major Adverse Cardiovascular Events

The baseline characteristics of the nested case–control MACE outcome study are presented in Table 2. The associations of each microRNA with MACE are presented in Figure 2B. Notably, this analysis showed similar findings compared to the previous analysis on cardiovascular death. Again, miR-411-5p was significantly associated with the outcome (adjusted HR from model 1: 2.35; 95%CI 1.17–4.73; $p = 0.017$). After adjustment for age, sex, history of myocardial infarction, and stroke/TIA/systemic embolism, the association remained significant (adjusted HR 2.62, 95%CI 1.29–5.33; $p = 0.008$). Eight patients with early cardiovascular death fulfilled the criteria for cases in both validation studies and were, therefore, included in both studies.

Table 2. Baseline characteristics of patients with atrial fibrillation: cases with major adverse cardiovascular events vs. matched controls without major adverse cardiovascular events.

	Cases with MACE at 1 Year (n = 37)	Controls (n = 65)
Demographics		
Age, years	75.0 (69.0–79.0)	74.0 (67.0–79.0)
Male sex	24 (64.9%)	41 (64.6%)
BMI, kg/m^2 [3]	27.1 (23.8–29.2)	26.5 (24.7–31.3)
Information on atrial fibrillation, n (%)		
Type of atrial fibrillation [5]		
First onset	5 (13.5%)	8 (12.3%)
Paroxysmal	15 (40.5%)	26 (40.0%)
Persistent	1 (2.7%)	3 (4.6%)
Permanent	15 (40.5%)	24 (36.9%)
History of electrical cardioversion	14 (37.8%)	18 (27.7%)
History of ablation	2 (5.4%)	8 (12.3%)
Comorbidities		
Arterial hypertension	30 (81.1%)	59 (90.8%)
Diabetes mellitus type 2	14 (37.8%)	21 (32.3%)
Coronary artery disease	11 (29.7%)	21 (32.3%)
Peripheral artery disease	10 (27.0%)	5 (7.7%)
Heart failure	19 (51.4%)	18 (27.7%)
History of myocardial infarction	8 (21.6%)	8 (12.3%)
History of stroke, TIA, systemic embolism	10 (27.0%)	16 (24.6%)
History of cancer	9 (24.3%)	15 (23.1%)
History of bleeding	12 (32.4%)	13 (20.0%)
CHA_2DS_2-VASc score, median (IQR)	4 (3–6)	4 (3–5)
HAS-BLED score, median (IQR)	2 (1–3)	2 (1–2)
Medication		
Anticoagulant therapy		
DOAC	19 (51.4%)	35 (53.8%)
Vitamin K-Antagonist	14 (37.8%)	27 (41.5%)
Other	2 (5.4%)	1 (1.5%)
None	1 (2.7%)	2 (3.1%)
Platelet inhibitor therapy		
Acetylsalicylic acid	7 (18.9%)	10 (15.4%)
P2Y12-Inhibitors	5 (13.5%)	4 (6.2%)

Square brackets indicate the number of missing values. Abbreviations: IQR, interquartile range; MACE, major adverse cardiovascular events; and TIA, transient ischemic attack.

Kaplan–Meier curves for both outcome studies were stratified by the detectability of miR-411-5p, and are shown in Figure 3.

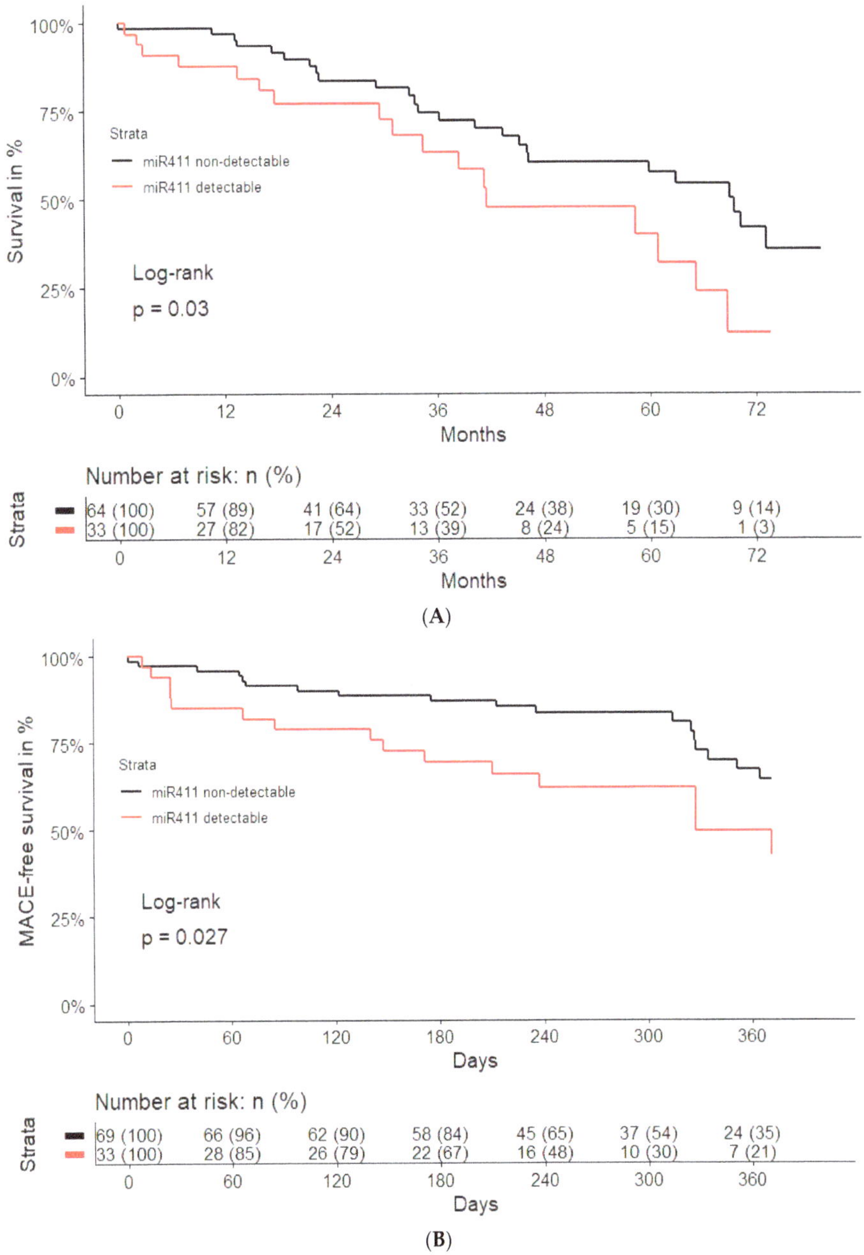

Figure 3. Kaplan–Meier curves for cardiovascular death (**A**) and major adverse cardiovascular events (**B**), as stratified by the detectability of miR-411-5p in patients with atrial fibrillation.

3. Discussion

In this three-stage nested case–control study, miR-411-5p was significantly associated with adverse cardiovascular events in patients with AF. After the initial microRNA screening phase, we selected seven microRNAs and performed two validation phases with different endpoints and follow-up periods. In both validation phases, AF patients with detectable levels of miR-411-5p at study inclusion were at, approximately, a two-fold increased risk for developing the cardiovascular outcome (i.e., cardiovascular death or MACE).

Previous studies mechanistically linked several microRNAs with pathophysiological pathways for AF and proposed microRNAs as a diagnostic tool for AF [12–15]. Furthermore, several circulating microRNAs have already been associated with future cardiovascular events, including myocardial infarction [10,24,25] and stroke [5,26]. The present study is the first to focus on the predictive role of microRNAs with regard to future cardiovascular events in patients with AF. We have chosen seven promising microRNAs that were differentially expressed between cases and controls in the discovery phase of our study for further validation. All of the selected microRNAs have been reported in the literature in the context of cardiovascular biology [16–21], e.g., miR-150-5p appears to be highly involved in platelet physiology and endothelial cell function [17,27]. In addition, it has also been associated with adverse cardiovascular outcomes [17,22]. Further, miR-1908-5p has been found to play an important role in regulating low-density lipoprotein, total cholesterol, fasting glucose, HbA1c, and several lipid-metabolites in blood [18]. Moreover, miR-15a-5p, which belongs to the miR-15 family, affects angiogenesis and might be a promising biomarker in critical limb ischemia [21,23]. However, only miR-411-5p was significantly associated with cardiovascular endpoints in both the validation phases of our study.

Thus far, the role of miR-411-5p has not been explored in prospective cardiovascular outcome studies, but several preclinical studies and one clinical study have suggested a central role of miR-411-5p in cardiovascular diseases [28–33]. MiR-411 is located on chromosome 14 and belongs to the 14q32 microRNA cluster. This genomic region holds one of the largest microRNA clusters, comprising 54 microRNA genes [34] and is known to play a central regulatory role in vascular biology [16,35]. The inhibition of microRNAs from the 14q32 cluster led to increased post-ischemic blood flow recovery and tissue perfusion via neovascularization [29], decreased plasma cholesterol, atherosclerotic lesion formation, and increased plaque stability in mouse models [36,37]. Interestingly, miR-411 also has a 5′ isomiR, which targets a different set of genes than the wildtype miR-411, and is upregulated relative to miR-411 in chronically ischemic human blood vessels [38]. However, in our qPCR analysis, we did not further differentiate between both isoforms. Apart from cardiovascular diseases, miR-411-5p has been attributed a protective role in diabetic retinopathy [39], but most research on miR-411 has been conducted in the field of cancer [40–42]. It has been suggested that miR-411 plays a major role in cancer development [43] and the formation of metastases [44,45]. MiR-411 is abnormally expressed in various cancers but with differing roles between tumor types. Several studies have reported on the downregulation of miR-411 and its role as a tumor suppressor in some cancer types, while also reporting on an overexpression and a role in cancer development in others [46]. In the cardiovascular clinical setting, miR-411 has been suggested as a candidate biomarker in only one study. There, miR-411 was significantly upregulated in the whole blood of patients with peripheral artery disease and of patients with abdominal aortic aneurysm, compared to healthy controls [33]. Our study adds to these findings and suggests a potential role for miR-411-5p in the prediction and risk stratification for cardiovascular death and MACE in patients with AF.

As to why miR-411-5p ends up in circulation, where exactly it originates from, and whether its presence in circulation has a role in cell-to-cell communication, is still unknown. From mouse studies performed on the 14q32 locus, the microRNA cluster from which miR-411 is derived, we could extrapolate that miR-411 may be involved in neovascularization and atherosclerotic lesion formation, as well as, importantly, with lesion (in)stability [29,36,37]. However, further preclinical research is needed to evaluate whether elevated circulatory miR-411-5p is causally linked to cardiovascular events or, on the contrary, upregulated due to cardioprotective effects. More clinical research is needed to externally validate our findings and to further investigate the association of miR-411-5p with future cardiovascular events in patients with AF. Potentially, miR-411-5p plasma levels may, ultimately, be used to guide primary or secondary cardiovascular prevention strategies in patients with AF, preferably in the context of a biomarker panel containing several different small RNA biomarkers.

Our study should be viewed in the context of the following limitations: First, microRNA research, in general, has a pronounced problem with replication due to false positive findings. We tried to account for this by controlling the false discovery rate during the screening phase of our study and then proceeded with a small selection of candidate microRNAs in two validation phases. Second, validation phases were not completely independent as eight cases in validation phase II were also included as cases in validation phase I. Third, each of the three nested case–control studies was limited by small sample sizes. Further, we had to analyze three of the seven microRNAs in the validation phases in a dichotomized fashion due to the low detectability of these microRNAs by qPCR. Both factors limit the power of our analysis. This might be the reason for not detecting significant associations with other microRNAs, e.g., miR-15a-5p, even though we expected otherwise based on previous evidence from the literature [21,23]. Nevertheless, we observed significant associations of miR-411-5p with cardiovascular death and MACE. Given the biological plausibility of this finding, as well as with respect to the same direction and magnitude of effect in the discovery phase and both validation phases, miR-411-5p may indeed prove a valuable biomarker for adverse cardiovascular outcomes in patients with AF.

To conclude, miR-411-5p was associated with adverse cardiovascular outcomes in patients with AF in all phases of our three-stage nested case–control study. If miR-411-5p was detectable in the circulation, patients were at approximately two-fold increased risk for cardiovascular death and MACE. Previous studies have shown that miR-411-5p plays a role in various forms of vascular remodeling, which may help explain its implication in cardiovascular death and MACE. However, future studies are needed to externally validate our findings and to further elucidate the precise biological role of miR-411-5p in cardiovascular events.

4. Materials and Methods

4.1. Study Population and Data Source

Our multistage, nested case–control study was based on a prospective registry that recruited 347 patients with AF, who were referred to the Clinical Division of Hematology and Hemostaseology at the Medical University of Vienna, Austria, between July 2013 and May 2016. Patients with a diagnosis of AF were confirmed by a 12-channel, resting electrocardiogram and the willingness to comply with the study procedures were how the participants were deemed eligible for inclusion; no exclusion criteria were defined. The detailed study procedures have been previously published elsewhere [47]. In brief, a detailed medical history and blood samples were taken at study inclusion. AF patients were then followed for at least 12 months. Follow-up was performed with a particular focus on cardiovascular events, including stroke, transitory ischemic attack (TIA), systemic embolism, myocardial infarction, coronary revascularization, hospitalization for heart failure, and death. Moreover, the national death registry was queried periodically, most recently in May 2020, to identify the deceased patients. Telephone interviews with family members and the review of electronic medical records were used to verify whether the death was due to cardiovascular causes.

This study conforms to the ethical guidelines of the 1975 Declaration of Helsinki. In addition, it was approved by the local ethics committee (number 1665/2012), and participants provided written informed consent.

4.2. Study Design

In the present study, we followed a sequential approach starting with screening for potential candidate microRNAs in a discovery case–control study. Subsequently, the most promising microRNAs of the discovery phase were tested in two nested case–control studies.

4.3. Discovery Phase

In the first stage, RNA sequencing was performed in a discovery case–control study to identify promising candidate microRNAs in order to predict MACE in patients with AF. For this screening phase, we chose the more stringent 3P-MACE endpoint—defined as stroke, myocardial infarction, or cardiovascular death. Patients who developed 3P-MACE during follow-up were eligible as cases. In total, 13 cases were randomly selected from our AF cohort and matched 1:1 to controls by age (±2 years), sex, and CHA_2DS_2-VASc-Score (±1 score point). Notably, the subgroup of patients with cardiovascular death showed the most apparent differential expression pattern compared to their controls. Thus, we first proceeded with investigating the most promising microRNAs in a nested case–control study with cases defined as patients who died from cardiovascular causes during follow-up.

4.4. Validation Phase I: Patients with Cardiovascular Death

Seven candidate microRNAs were selected from the discovery phase, based on statistical significance, log fold change, and known biological functions from the cardiovascular literature. Selected microRNAs were measured using reverse-transcription quantitative polymerase chain reaction (RT-qPCR) in all patients with cardiovascular death (n = 42), labeled as cases, as well as their age- and sex-matched controls (n = 55). Matching was performed in a 1:2 ratio with replacement; therefore, the same control could be matched to several cases.

4.5. Validation Phase II: Patients with MACE at One Year

In a third step, we aimed to further validate our findings from the cardiovascular death outcome study using a more clinically relevant endpoint, i.e., MACE within the first year of follow-up. Patients who experienced MACE within one year after study inclusion were selected as cases (n = 37). Again, cases were matched in a 1:2 ratio to age- and sex-matched controls (n = 65), allowing for a control to be matched to several cases. MACE was defined as the composite endpoint of stroke, TIA, systemic embolism, myocardial infarction, coronary revascularization, hospitalization for heart failure, and all-cause death.

4.6. Laboratory and Statistical Methods

4.6.1. Sample Collection

At study entry, non-fasting venous blood was collected in vacuum tubes containing 3.2% trisodium citrate (Vacuette® Greiner Bio-One, Kremsmünster, Austria). The tubes were centrifuged at 2500 g for 15 min and platelet-poor plasma was isolated. Aliquots with 200 μL each were generated and stored at −80 °C until needed for analysis. All blood samples were processed within the first two hours after blood was drawn.

4.6.2. Discovery Phase: Small RNA Sequencing

For the screening of candidate microRNAs, the total RNA was isolated from 200 μL of citrated plasma using the miRNeasy Serum/Plasma Advanced Kit (Qiagen, Venlo, Netherlands), which was performed according to the manufacturer's instructions. Next, the small RNAs derived from plasma were sent to BGI (Shenzhen, China) for small RNA sequencing. In brief, BGI small RNA sequencing services were executed using unique molecular identifiers (UMIs), small RNA library construction, and RNA sequencing on the DNBseq Platform. Raw data were de-multiplexed and filtered to remove adapter sequences, contamination, and low-quality reads.

4.6.3. Discovery Phase: Statistical Analysis

The small RNA sequencing dataset was analyzed using the virtual machine of the sRNAtoolbox [48], which is a collection of small RNA research tools that can be used for expression profiling and subsequent downstream analysis. The sRNAbench module was used to deduplicate reads based on UMIs and subsequently map them to the human genome. Several libraries were used for annotation, including miRbase, gtRNAdb, RNAcentral, as well as for NCBI ncRNA and cDNA libraries. Finally, an analysis of differential expression was performed using the sRNAde module. With microRNA abundance normalized to the total number of genome-mapped sequencing reads (expressed as reads per million), the Bioconductor package edgeR was used to investigate differential microRNA expression, while also using the standard quantile-adjusted conditional maximum likelihood method. To account for multiple testing during the discovery phase, we adjusted *p*-values using the method of Benjamini and Hochberg for controlling the FDR.

4.6.4. Validation Phases: Quantification Using RT-qPCR

The seven microRNAs selected during the discovery phase were then measured by RT-qPCR in triplets for each sample and each microRNA using Taqman™ microRNA assays (ThermoFisher, Waltham, MA, USA) (Supplemental Table S4). First, the circulating microRNAs were isolated using the miRNeasy Serum/Plasma Kit (Qiagen, Venlo, Netherlands). Second, microRNAs were reverse-transcribed to cDNA with the Taqman™ MicroRNA Reverse Transcription Kit (ThermoFisher, Waltham, MA, USA). Third, samples were measured in triplicate using RT-qPCR on the CFX384 Touch Real-Time PCR Detection System (Bio Rad, Hercules, USA). Cel-miR-39 was used as a spike-in control to normalize the expression levels of microRNAs, using the delta threshold cycle (ΔC_T) method. Three µL of 5 nM cel-miR-39 was added to a total sum of 15 µL of reverse transcription mix [49]. All microRNAs were measured in a blinded fashion, regarding the case vs. control status of the samples.

4.6.5. Validation Phases: Statistical Analysis

Patient characteristics are presented as frequencies (percentage) for the categorical variables, as well as the mean (SD) or median (IQR) for continuous data.

Quantitative PCR results were analyzed according to the data handling pipeline suggested by de Ronde et al. [50]. If the mean quantification cycle (Cq) of the triplets performed for each microRNA was above a threshold of 38, the microRNA was labeled as undetectable and the value set to 39. Values were then normalized to cel-miR-39 using the ΔC_T method. Measures that were labeled as invalid by the data handling pipeline were imputed, through using multiple imputation with chained equations (MICE). Due to the low detectability of some microRNAs, we had to adjust our analysis plan after initial data analysis; the microRNAs with less than 20% of valid values in all participants were dichotomized based on detectability (i.e., miR-625-5p, miR-15a-5p, and miR-411-5p). If the microRNA was measurable in any of the triplets of each sample, it was counted as detectable.

Finally, the associations of microRNAs with the outcomes (i.e., cardiovascular death and MACE) were assessed using Cox proportional hazard models. Multivariable analysis was performed using two models; model 1 was adjusted for age and sex; model 2 included the variables of model 1 and the history of myocardial infarction, as well as the history of stroke/TIA/systemic embolism. For visualization, Kaplan–Maier curves were plotted and stratified by the detectability of the microRNA of interest, as well as were compared using the log-rank test. All statistical tests were two-tailed with an alpha level of 0.05 denoting statistical significance. Data were analyzed in R (Version 4.1.0; R Core Team, 2019) using the MatchIt, mice, and survival package, as well as the R script provided by de Ronde et al. [50].

Supplementary Materials: The following supporting information can be downloaded at: https://www.mdpi.com/article/10.3390/ijms24043861/s1.

Author Contributions: S.N., C.A. and A.Y.N. conceived and planned the experiments. S.N. and A.Y.N. carried out the experiments. S.N., D.K., O.K., I.P. and C.A. contributed to sample collection and preparation. S.N., M.L.v.d.B., J.W., I.P., C.A. and A.Y.N. contributed to the interpretation of the results. S.N. and A.Y.N. took the lead in writing the manuscript. All authors provided critical feedback and helped shape the research, analysis, and manuscript. All authors have read and agreed to the published version of the manuscript.

Funding: This research and the APC was funded by the Austrian Science Fund FWF with the Lise Meitner Grant M2578-B30.

Institutional Review Board Statement: The study was conducted in accordance with the Declaration of Helsinki, and approved by the Ethics Committee of the Medical University of Vienna (number 1665/2012, date of approval: 19 November 2012).

Informed Consent Statement: Informed consent was obtained from all subjects involved in the study.

Data Availability Statement: The data presented in this study are available on request from the corresponding author.

Conflicts of Interest: The authors declare that there are no conflict of interest.

References

1. Gomez-Outes, A.; Suarez-Gea, M.L.; Garcia-Pinilla, J.M. Causes of death in atrial fibrillation: Challenges and opportunities. *Trends Cardiovasc. Med.* **2017**, *27*, 494–503. [CrossRef]
2. Chen, J.Y.; Zhang, A.D.; Lu, H.Y.; Guo, J.; Wang, F.F.; Li, Z.C. CHADS2 versus CHA2DS2-VASc score in assessing the stroke and thromboembolism risk stratification in patients with atrial fibrillation: A systematic review and meta-analysis. *J. Geriatr. Cardiol.* **2013**, *10*, 258–266. [CrossRef]
3. Stojkovic, S.; Nossent, A.Y.; Haller, P.; Jager, B.; Vargas, K.G.; Wojta, J.; Huber, K. MicroRNAs as Regulators and Biomarkers of Platelet Function and Activity in Coronary Artery Disease. *Thromb. Haemost.* **2019**, *119*, 1563–1572. [CrossRef]
4. Hanna, J.; Hossain, G.S.; Kocerha, J. The Potential for microRNA Therapeutics and Clinical Research. *Front. Genet.* **2019**, *10*, 478. [CrossRef]
5. Mens, M.M.J.; Heshmatollah, A.; Fani, L.; Ikram, M.A.; Ikram, M.K.; Ghanbari, M. Circulatory MicroRNAs as Potential Biomarkers for Stroke Risk: The Rotterdam Study. *Stroke* **2021**, *52*, 945–953. [CrossRef]
6. Li, P.; Teng, F.; Gao, F.; Zhang, M.; Wu, J.; Zhang, C. Identification of circulating microRNAs as potential biomarkers for detecting acute ischemic stroke. *Cell. Mol. Neurobiol.* **2015**, *35*, 433–447. [CrossRef]
7. Tiedt, S.; Prestel, M.; Malik, R.; Schieferdecker, N.; Duering, M.; Kautzky, V.; Stoycheva, I.; Bock, J.; Northoff, B.H.; Klein, M.; et al. RNA-Seq Identifies Circulating miR-125a-5p, miR-125b-5p, and miR-143-3p as Potential Biomarkers for Acute Ischemic Stroke. *Circ. Res.* **2017**, *121*, 970–980. [CrossRef]
8. Escate, R.; Padro, T.; Suades, R.; Camino, S.; Muniz, O.; Diaz-Diaz, J.L.; Sionis, A.; Mata, P.; Badimon, L. High miR-133a levels in the circulation anticipates presentation of clinical events in familial hypercholesterolaemia patients. *Cardiovasc. Res.* **2021**, *117*, 109–122. [CrossRef]
9. Karakas, M.; Schulte, C.; Appelbaum, S.; Ojeda, F.; Lackner, K.J.; Munzel, T.; Schnabel, R.B.; Blankenberg, S.; Zeller, T. Circulating microRNAs strongly predict cardiovascular death in patients with coronary artery disease-results from the large AtheroGene study. *Eur. Heart J.* **2017**, *38*, 516–523. [CrossRef]
10. Shen, M.; Xu, X.; Liu, X.; Wang, Q.; Li, W.; You, X.; Peng, R.; Yuan, Y.; Long, P.; Niu, R.; et al. Prospective Study on Plasma MicroRNA-4286 and Incident Acute Coronary Syndrome. *J. Am. Heart Assoc.* **2021**, *10*, e018999. [CrossRef]
11. Su, Y.; Sun, Y.; Tang, Y.; Li, H.; Wang, X.; Pan, X.; Liu, W.; Zhang, X.; Zhang, F.; Xu, Y.; et al. Circulating miR-19b-3p as a Novel Prognostic Biomarker for Acute Heart Failure. *J. Am. Heart Assoc.* **2021**, *10*, e022304. [CrossRef] [PubMed]
12. Lu, Y.; Zhang, Y.; Wang, N.; Pan, Z.; Gao, X.; Zhang, F.; Zhang, Y.; Shan, H.; Luo, X.; Bai, Y.; et al. MicroRNA-328 contributes to adverse electrical remodeling in atrial fibrillation. *Circulation* **2010**, *122*, 2378–2387. [CrossRef]
13. Ling, T.Y.; Wang, X.L.; Chai, Q.; Lau, T.W.; Koestler, C.M.; Park, S.J.; Daly, R.C.; Greason, K.L.; Jen, J.; Wu, L.Q.; et al. Regulation of the SK3 channel by microRNA-499–potential role in atrial fibrillation. *Heart Rhythm* **2013**, *10*, 1001–1009. [CrossRef]
14. Wang, J.; Bai, Y.; Li, N.; Ye, W.; Zhang, M.; Greene, S.B.; Tao, Y.; Chen, Y.; Wehrens, X.H.; Martin, J.F. Pitx2-microRNA pathway that delimits sinoatrial node development and inhibits predisposition to atrial fibrillation. *Proc. Natl. Acad. Sci. USA* **2014**, *111*, 9181–9186. [CrossRef] [PubMed]
15. Chiang, D.Y.; Kongchan, N.; Beavers, D.L.; Alsina, K.M.; Voigt, N.; Neilson, J.R.; Jakob, H.; Martin, J.F.; Dobrev, D.; Wehrens, X.H.; et al. Loss of microRNA-106b-25 cluster promotes atrial fibrillation by enhancing ryanodine receptor type-2 expression and calcium release. *Circ. Arrhythmia Electrophysiol.* **2014**, *7*, 1214–1222. [CrossRef] [PubMed]
16. Welten, S.M.; Goossens, E.A.; Quax, P.H.; Nossent, A.Y. The multifactorial nature of microRNAs in vascular remodelling. *Cardiovasc. Res.* **2016**, *110*, 6–22. [CrossRef] [PubMed]
17. Pordzik, J.; Pisarz, K.; De Rosa, S.; Jones, A.D.; Eyileten, C.; Indolfi, C.; Malek, L.; Postula, M. The Potential Role of Platelet-Related microRNAs in the Development of Cardiovascular Events in High-Risk Populations, Including Diabetic Patients: A Review. *Front. Endocrinol.* **2018**, *9*, 74. [CrossRef] [PubMed]

18. Nikpay, M.; Beehler, K.; Valsesia, A.; Hager, J.; Harper, M.E.; Dent, R.; McPherson, R. Genome-wide identification of circulating-miRNA expression quantitative trait loci reveals the role of several miRNAs in the regulation of cardiometabolic phenotypes. *Cardiovasc. Res.* **2019**, *115*, 1629–1645. [CrossRef]
19. Kemp, J.R.; Unal, H.; Desnoyer, R.; Yue, H.; Bhatnagar, A.; Karnik, S.S. Angiotensin II-regulated microRNA 483-3p directly targets multiple components of the renin-angiotensin system. *J. Mol. Cell. Cardiol.* **2014**, *75*, 25–39. [CrossRef]
20. Cai, K.; Chen, H. MiR-625-5p Inhibits Cardiac Hypertrophy Through Targeting STAT3 and CaMKII. *Hum. Gene Ther. Clin. Dev.* **2019**, *30*, 182–191. [CrossRef]
21. Spinetti, G.; Fortunato, O.; Caporali, A.; Shantikumar, S.; Marchetti, M.; Meloni, M.; Descamps, B.; Floris, I.; Sangalli, E.; Vono, R.; et al. MicroRNA-15a and microRNA-16 impair human circulating proangiogenic cell functions and are increased in the proangiogenic cells and serum of patients with critical limb ischemia. *Circ. Res.* **2013**, *112*, 335–346. [CrossRef] [PubMed]
22. Rhodes, C.J.; Wharton, J.; Boon, R.A.; Roexe, T.; Tsang, H.; Wojciak-Stothard, B.; Chakrabarti, A.; Howard, L.S.; Gibbs, J.S.; Lawrie, A.; et al. Reduced microRNA-150 is associated with poor survival in pulmonary arterial hypertension. *Am. J. Respir. Crit. Care Med.* **2013**, *187*, 294–302. [CrossRef] [PubMed]
23. Yin, K.J.; Olsen, K.; Hamblin, M.; Zhang, J.; Schwendeman, S.P.; Chen, Y.E. Vascular endothelial cell-specific microRNA-15a inhibits angiogenesis in hindlimb ischemia. *J. Biol. Chem.* **2012**, *287*, 27055–27064. [CrossRef] [PubMed]
24. Zampetaki, A.; Willeit, P.; Tilling, L.; Drozdov, I.; Prokopi, M.; Renard, J.M.; Mayr, A.; Weger, S.; Schett, G.; Shah, A.; et al. Prospective study on circulating MicroRNAs and risk of myocardial infarction. *J. Am. Coll. Cardiol.* **2012**, *60*, 290–299. [CrossRef] [PubMed]
25. Bye, A.; Rosjo, H.; Nauman, J.; Silva, G.J.; Follestad, T.; Omland, T.; Wisloff, U. Circulating microRNAs predict future fatal myocardial infarction in healthy individuals—The HUNT study. *J. Mol. Cell. Cardiol.* **2016**, *97*, 162–168. [CrossRef]
26. Mick, E.; Shah, R.; Tanriverdi, K.; Murthy, V.; Gerstein, M.; Rozowsky, J.; Kitchen, R.; Larson, M.G.; Levy, D.; Freedman, J.E. Stroke and Circulating Extracellular RNAs. *Stroke* **2017**, *48*, 828–834. [CrossRef]
27. Yan, B.; Yao, J.; Liu, J.Y.; Li, X.M.; Wang, X.Q.; Li, Y.J.; Tao, Z.F.; Song, Y.C.; Chen, Q.; Jiang, Q. lncRNA-MIAT regulates microvascular dysfunction by functioning as a competing endogenous RNA. *Circ. Res.* **2015**, *116*, 1143–1156. [CrossRef]
28. van der Kwast, R.; Woudenberg, T.; Quax, P.H.A.; Nossent, A.Y. MicroRNA-411 and Its 5′-IsomiR Have Distinct Targets and Functions and Are Differentially Regulated in the Vasculature under Ischemia. *Mol. Ther.* **2020**, *28*, 157–170. [CrossRef] [PubMed]
29. Welten, S.M.; Bastiaansen, A.J.; de Jong, R.C.; de Vries, M.R.; Peters, E.A.; Boonstra, M.C.; Sheikh, S.P.; La Monica, N.; Kandimalla, E.R.; Quax, P.H.; et al. Inhibition of 14q32 MicroRNAs miR-329, miR-487b, miR-494, and miR-495 increases neovascularization and blood flow recovery after ischemia. *Circ. Res.* **2014**, *115*, 696–708. [CrossRef]
30. Goossens, E.A.C.; de Vries, M.R.; Simons, K.H.; Putter, H.; Quax, P.H.A.; Nossent, A.Y. miRMap: Profiling 14q32 microRNA Expression and DNA Methylation Throughout the Human Vasculature. *Front. Cardiovasc. Med.* **2019**, *6*, 113. [CrossRef]
31. Downie Ruiz Velasco, A.; Welten, S.M.J.; Goossens, E.A.C.; Quax, P.H.A.; Rappsilber, J.; Michlewski, G.; Nossent, A.Y. Posttranscriptional Regulation of 14q32 MicroRNAs by the CIRBP and HADHB during Vascular Regeneration after Ischemia. *Mol. Ther. Nucleic Acids* **2019**, *14*, 329–338. [CrossRef]
32. Ai, P.; Shen, B.; Pan, H.; Chen, K.; Zheng, J.; Liu, F. MiR-411 suppressed vein wall fibrosis by downregulating MMP-2 via targeting HIF-1alpha. *J. Thromb. Thrombolysis* **2018**, *45*, 264–273. [CrossRef] [PubMed]
33. Stather, P.W.; Sylvius, N.; Sidloff, D.A.; Dattani, N.; Verissimo, A.; Wild, J.B.; Butt, H.Z.; Choke, E.; Sayers, R.D.; Bown, M.J. Identification of microRNAs associated with abdominal aortic aneurysms and peripheral arterial disease. *Br. J. Surg.* **2015**, *102*, 755–766. [CrossRef]
34. Benetatos, L.; Hatzimichael, E.; Londin, E.; Vartholomatos, G.; Loher, P.; Rigoutsos, I.; Briasoulis, E. The microRNAs within the DLK1-DIO3 genomic region: Involvement in disease pathogenesis. *Cell. Mol. Life Sci.* **2013**, *70*, 795–814. [CrossRef] [PubMed]
35. Dimmeler, S.; Yla-Herttuala, S. 14q32 miRNA cluster takes center stage in neovascularization. *Circ. Res.* **2014**, *115*, 680–682. [CrossRef] [PubMed]
36. Wezel, A.; Welten, S.M.; Razawy, W.; Lagraauw, H.M.; de Vries, M.R.; Goossens, E.A.; Boonstra, M.C.; Hamming, J.F.; Kandimalla, E.R.; Kuiper, J.; et al. Inhibition of MicroRNA-494 Reduces Carotid Artery Atherosclerotic Lesion Development and Increases Plaque Stability. *Ann. Surg.* **2015**, *262*, 841–847; discussion 847–848. [CrossRef] [PubMed]
37. Welten, S.M.J.; de Jong, R.C.M.; Wezel, A.; de Vries, M.R.; Boonstra, M.C.; Parma, L.; Jukema, J.W.; van der Sluis, T.C.; Arens, R.; Bot, I.; et al. Inhibition of 14q32 microRNA miR-495 reduces lesion formation, intimal hyperplasia and plasma cholesterol levels in experimental restenosis. *Atherosclerosis* **2017**, *261*, 26–36. [CrossRef]
38. Muiwo, P.; Pandey, P.; Ahmad, H.M.; Ramachandran, S.S.; Bhattacharya, A. IsomiR processing during differentiation of myelogenous leukemic cell line K562 by phorbol ester PMA. *Gene* **2018**, *641*, 172–179. [CrossRef] [PubMed]
39. Hu, K.; Li, J.L.; Yuan, X.W. MicroRNA-411 plays a protective role in diabetic retinopathy through targeted regulating Robo4. *Eur. Rev. Med. Pharmacol. Sci.* **2019**, *23*, 9171–9179. [CrossRef]
40. Shan, D.; Shang, Y.; Hu, T. MicroRNA-411 Inhibits Cervical Cancer Progression by Directly Targeting STAT3. *Oncol. Res.* **2019**, *27*, 349–358. [CrossRef]
41. Nadal, E.; Zhong, J.; Lin, J.; Reddy, R.M.; Ramnath, N.; Orringer, M.B.; Chang, A.C.; Beer, D.G.; Chen, G. A MicroRNA cluster at 14q32 drives aggressive lung adenocarcinoma. *Clin. Cancer Res.* **2014**, *20*, 3107–3117. [CrossRef] [PubMed]
42. Qi, C.L.; Sheng, J.F.; Huang, M.L.; Zou, Y.; Wang, Y.P.; Wang, F.; Zeng, F.; Hua, Q.Q.; Chen, S.M. Integrated analysis of deregulation microRNA expression in head and neck squamous cell carcinoma. *Medicine* **2021**, *100*, e24618. [CrossRef] [PubMed]

43. Zhang, C.; Wang, H.; Liu, X.; Hu, Y.; Ding, L.; Zhang, X.; Sun, Q.; Li, Y. Oncogenic microRNA-411 promotes lung carcinogenesis by directly targeting suppressor genes SPRY4 and TXNIP. *Oncogene* **2019**, *38*, 1892–1904. [CrossRef] [PubMed]
44. Zhang, C.; Wang, H.; Deng, M.; He, L.; Ping, F.; He, Y.; Fan, Z.; Cheng, B.; Xia, J. Upregulated miR4115p levels promote lymph node metastasis by targeting RYBP in head and neck squamous cell carcinoma. *Int. J. Mol. Med.* **2021**, *47*, 36. [CrossRef] [PubMed]
45. Zhang, Y.; Xu, G.; Liu, G.; Ye, Y.; Zhang, C.; Fan, C.; Wang, H.; Cai, H.; Xiao, R.; Huang, Z.; et al. miR-411-5p inhibits proliferation and metastasis of breast cancer cell via targeting GRB2. *Biochem. Biophys. Res. Commun.* **2016**, *476*, 607–613. [CrossRef] [PubMed]
46. Zou, M.; Shen, J.; Wu, Y.; Zhong, C.; Fang, L.; Zhu, F.; Duan, S. Dysregulation of miR-411 in cancer: Causative factor for pathogenesis, diagnosis and prognosis. *Biomed. Pharmacother.* **2022**, *149*, 112896. [CrossRef]
47. Konigsbrugge, O.; Simon, A.; Domanovits, H.; Pabinger, I.; Ay, C. Thromboembolic events, bleeding, and drug discontinuation in patients with atrial fibrillation on anticoagulation: A prospective hospital-based registry. *BMC Cardiovasc. Disord.* **2016**, *16*, 254. [CrossRef]
48. Aparicio-Puerta, E.; Lebron, R.; Rueda, A.; Gomez-Martin, C.; Giannoukakos, S.; Jaspez, D.; Medina, J.M.; Zubkovic, A.; Jurak, I.; Fromm, B.; et al. sRNAbench and sRNAtoolbox 2019: Intuitive fast small RNA profiling and differential expression. *Nucleic Acids Res.* **2019**, *47*, W530–W535. [CrossRef]
49. Roberts, T.C.; Coenen-Stass, A.M.; Betts, C.A.; Wood, M.J. Detection and quantification of extracellular microRNAs in murine biofluids. *Biol. Proced. Online* **2014**, *16*, 5. [CrossRef]
50. de Ronde, M.W.J.; Ruijter, J.M.; Lanfear, D.; Bayes-Genis, A.; Kok, M.G.M.; Creemers, E.E.; Pinto, Y.M.; Pinto-Sietsma, S.J. Practical data handling pipeline improves performance of qPCR-based circulating miRNA measurements. *RNA* **2017**, *23*, 811–821. [CrossRef]

International Journal of
Molecular Sciences

MDPI

Article

Dysregulation of Plasma miR-146a and miR-155 Expression Profile in Mycosis Fungoides Is Associated with rs2910164 and rs767649 Polymorphisms

Chrysostomos Avgeros [1], Aikaterini Patsatsi [2,3], Dimitrios Dimitriadis [4], Andigoni Malousi [1,3], Triantafyllia Koletsa [5], Despoina Papathemeli [2], Antonia Syrnioti [5], Paraskevi Avgerou [1], Elizabeth Lazaridou [2], Georgios Tzimagiorgis [1,3] and Elisavet Georgiou [1,3,*]

1 Laboratory of Biological Chemistry, School of Medicine, Faculty of Health Sciences, Aristotle University of Thessaloniki, 54124 Thessaloniki, Greece
2 2nd Dermatology Department, School of Medicine, Faculty of Health Sciences, Aristotle University of Thessaloniki, "Papageorgiou" General Hospital, 56403 Thessaloniki, Greece
3 Center for Interdisciplinary Research and Innovation (CIRI-AUTH), 57001 Thessaloniki, Greece
4 School of Economics, Aristotle University of Thessaloniki, 54124 Thessaloniki, Greece
5 Department of Pathology, School of Medicine, Faculty of Health Sciences, Aristotle University of Thessaloniki, 54124 Thessaloniki, Greece
* Correspondence: elgeorgiou@auth.gr; Tel.: +30-2310999171

Abstract: Diagnosis of Mycosis Fungoides (MF) may be challenging, due to its polymorphic nature. The use of miRNAs as biomarkers to assist in diagnosis has been investigated, mainly in skin lesion biopsies. The purpose of this study is to evaluate the plasma levels of miR-146a and miR-155 in MF patients and to investigate their association with SNPs of their genes. Plasma miRNAs were quantified by RT-qPCR. Genomic DNA was used for SNPs' genotyping by Sanger sequencing. Plasma levels of miR-146a and miR-155 were significantly higher in patients vs. controls, in early MF patients vs. controls, and in advanced vs. early MF patients. Both miRNAs' levels were significantly higher in stage IIB vs. early-stage patients. miR-155 plasma levels were significantly higher in patients with skin tumors or erythroderma. CC genotype (rs2910164 C>G) was significantly more frequent in healthy controls and associated with lower MF risk and lower miR-146a levels. The AA genotype (rs767649 T>A) was significantly more frequent in patients and correlated with increased MF risk and increased miR-155 levels. The combination of GG+AA was only detected in patients and was correlated with higher MF susceptibility. Increased mir-146a and mir-155 plasma levels in MF is an important finding to establish putative noninvasive biomarkers. The presence of SNPs is closely associated with miRs' expression, and possibly with disease susceptibility.

Keywords: miRNAs; mycosis fungoides; single-nucleotide polymorphism; CTCL

1. Introduction

Mycosis Fungoides (MF) is the most prevalent primary Cutaneous T Cell Lymphoma subtype, comprising 55–64% of cases [1]. The clinical spectrum of MF is considerably variable. Most patients have early-stage MF, displaying erythematous patches or infiltrated plaques. Nevertheless, a quarter of patients progress to advanced-stage with tumors, erythroderma, and systemic involvement [2–5]. Diagnosis is mostly based on specific histopathological and immunohistochemical findings, while clinicopathological correlation is essential. However, due to its polymorphic nature, especially at early stages, repeated skin biopsies are often required to minimize the possibility of mis- or late diagnosis [6], while molecular analysis and the discovery of novel biomarkers may contribute to the accuracy of diagnosis.

MicroRNAs (miRNAs) are small (~18–25 nucleotides) non-coding RNA molecules that regulate genes' expression [7]. Due to their distinct expression profile, miRNAs have been suggested as putative biomarkers in many malignancies, including MF [8,9]. Two of the most well-studied and highlighted miRNAs in MF are the inflammation-related miR-146a and miR-155 [10–12], which play a key role in the pathogenesis of the disease [13,14]. Circulating miRNAs display high stability and resistance to degradation, featuring as possible non-invasive biomarkers [15]. Nevertheless, there are only two studies evaluating miRNA plasma levels in MF, in a limited number of patients [16,17].

Single-nucleotide polymorphisms (SNPs) are common genetic variations that affect miRNAs' expression. Two SNPs, rs2910164 (C>G) in pre-mir-146a and rs767649 (T>A) in a regulatory region of miR-155, have been associated with the expression of these miRNAs in benign and malignant diseases [18–21]. There are no studies in MF to date that associate miRNAs' expression with SNPs in their genes.

In the present study, we evaluated plasma levels of miR-146a and miR-155 in MF patients to investigate their potential role as non-invasive prognostic and diagnostic biomarkers. Furthermore, we examined whether rs2910164 and rs767649 affect the miRNAs' levels in plasma and susceptibility to MF.

2. Results

2.1. Demographic and Clinical Characteristics of Patients Recruited in This Study

A total of 41 patients were recruited (30 male, 11 female). The median age was 59 years. Among them, 11 had patch-stage disease, 21 plaque-stage, 6 tumor-stage and 3 were erythrodermic. According to EORTC staging, 32 (~78%) patients had early-stage MF (IA–IIA) and 9 (~22%) patients had advanced-stage MF (IIB-IV). Furthermore, 19 patients had not received any treatment prior to sample collection, while 22 had received skin-directed and/or systematic treatment depending on MF status. The demographic characteristics of patients and healthy individuals and the clinical features of patients are listed in Table 1.

Table 1. Demographic and clinical characteristics of the patients and healthy individuals included in the study.

	Patients $n = 41$	**Healthy Individuals** $n = 41$
Demographic characteristics		
Median age (range)	59 (39–85)	56 (34–79)
Male n (%)	30 (73.17%)	28 (68.29%)
Female n (%)	11 (26.83%)	13 (31.71%)
Clinical Features	n **(%)**	
Patch stage	11 (27%)	
Plaque stage	21 (51%)	
Tumor Stage	6 (14%)	
Erythrodermic	3 (8%)	
MF stage	n **(%)**	
Early-stage (IA–IIA)	32 (78.03%)	
IA	22 (53.65%)	
IB	8 (19.51%)	
IIA	2 (4.87%)	
Advanced-stage (IIB–IV)	9 (21.97%)	
IIB	5 (12.19%)	
III	3 (7.31%)	
IV	1 (2.47%)	
Treatment	n **(%)**	
None	19 (46.3%)	
Skin-directed	3 (7.3%)	
Systematic	12 (29.4%)	
Combination of skin-directed and systematic	7 (17%)	

To study the association between rs2910164 (C>G) and rs767649 (T>A) polymorphisms with the expression levels of miR-146a and miR-155, respectively, genotyping was performed in 33/41 MF patients and 26/41 healthy individuals of the initial cohort, where corresponding samples were available. Among these patients, 24/33 were male (~73%) and 9 were female (~27%), with median age equal to 60 years. Furthermore, the genotyping cohort included 24 (~73%) early-stage MF patients and 9 patients with advanced-stage MF (~27%).

2.2. Plasma Levels of miR-146a and miR-155 Differ Significantly between Patients and Healthy Controls, and among MF Stages

The relative quantity (RQ) of both miR-146a and miR-155 was significantly increased in the plasma of MF patients when compared to healthy individuals ($p = 0.001$ and $p = 0.028$, respectively; Figure 1a) and in the plasma of early-stage MF patients vs. healthy controls ($p = 0.001$ for miR-146a and $p < 0.001$ for miR-155; Figure 1b). Moreover, advanced-stage MF showed markedly higher levels of miR-146a and miR-155 in comparison with early-stage MF ($p = 0.009$ and $p = 0.002$, respectively; Figure 1c). Since stage IIB is considered the threshold stage for advanced-stage MF, we also compared plasma levels of miR-146a and miR-155 between patients at clinical stages IA, IB, IIA (comprising early-stage MF) and stage IIB. Interestingly, both miRNAs were significantly upregulated in stage IIB ($p = 0.008$ for miR-146a and $p = 0.001$ for miR-155; Figure 1d). Table S1 summarizes a statistical analysis of miRs expression between groups.

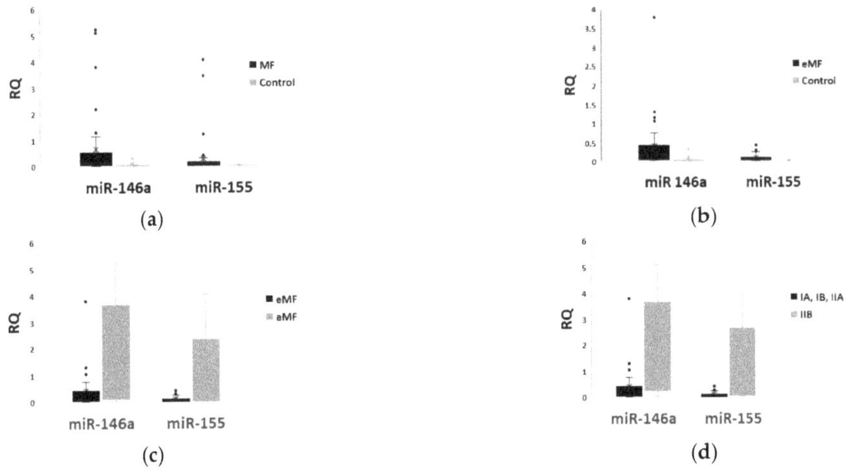

Figure 1. Plasma levels (mean ± SD) of miR-146a and miR-155 in patients vs. control (**a**), in eMF patients vs. control (**b**), in early MF (eMF) vs. advanced MF (aMF) patients (**c**) and in clinical stages IA, IB, IIA vs. IIB (**d**).

Although analysis with ANOVA revealed that both miRNAs' levels were significantly different between clinical stages (Figure S1 and Tables S2 and S3 (see Supplementary Materials)), the small sample size in some stages did not allow for post hoc analysis.

To further investigate the association of miR-146a and miR-155 with skin lesions of MF, we compared plasma levels of both miRNAs between patient groups with different manifestations in the skin. Plasma levels of miR-146a seemed to slightly differ ($p = 0.076$) among groups with different clinical skin lesions (Figure 2a and Table S4) but the difference did not reach statistical significance. However, miR-155 displayed markedly different levels among patients with different lesions ($p = 0.024$) (Figure 2b and Table S4). LSD post hoc analysis showed that miR-155 was upregulated in tumor-stage MF when compared with plaque-stage ($p = 0.018$) or patch-stage MF ($p = 0.039$), as well as in erythrodermic patients when compared with plaque-stage ($p = 0.027$) or patch-stage patients ($p = 0.044$, Table S5). No significant difference was found in miR-155 levels between erythrodermic and tumor-stage MF patients, as well as between plaque- and patch-stage MF patients ($p = 0.701$ and $p = 0.865$, respectively, Table S5).

(a) (b)

Figure 2. Plasma levels (mean ± SD) of miR-146a (**a**) and miR-155 (**b**) among MF patients with different skin lesions.

There was no statistically significant difference in the plasma levels of either miRNA between patients who received treatment and patients with no therapy at all ($p = 0.614$ for miR-146a and $p = 0.243$ for miR-155), or between patients receiving different treatment modalities (skin-directed, systemic, or combination therapy) ($p = 0.446$ for miR-146a and $p = 0.648$ for miR-155) (Figure S2). Similarly, sex ($p = 0.437$ for miR-146a and $p = 0.343$ for miR-155) and age ($p = 0.079$ for miR-146a and $p = 0.336$ for miR-155) of patients had no significant influence on miRNA levels.

Interestingly, a notable positive correlation between plasma levels of mir-146a and miR-155 was observed among patients ($p < 0.001$) (Figure 3a,b).

(a) (b)

Figure 3. Positive correlation between plasma levels of miR-146a and miR-155 in the patients' cohort, scatter plot (**a**) and line of tension (**b**).

2.3. *Allelic and Genotypic Patterns of rs2910164 (C>G) and rs767649 (T>A) Polymorphisms Present a Statistically Significant Difference in the Distribution between MF Patients and Healthy Individuals, Displaying a Strong Association with Susceptibility to MF*

The allelic and genotypic distribution of miR-146a rs2910164 and miR-155 rs767649 polymorphisms among MF patients and healthy individuals is presented in Table 2. Alignment of the DNA sequence coding for pre-mir-155 did not detect any SNPs in any of the tested samples.

In the genotypic model, the CC genotype of rs2910164 was significantly more frequent in healthy individuals and was associated with a reduced risk of MF (OR = 0.011, 95% CI 0.001–0.107, $p < 0.01$). There was no difference in the distribution of GC genotype between the two study groups ($p = 0.066$). In the dominant model, mutant allele-associated genotypes GC+CC displayed an increased prevalence among healthy individuals and were correlated with a decreased risk of MF (OR = 0.079, 95% CI 0.022–0.290, $p < 0.01$). Furthermore, in the recessive and allelic models, both the CC genotype and C allele showed higher distribution in healthy controls when compared with MF cases and were associated with lower risk of MF (OR = 0.020, 95% CI 0.002–0.166, $p < 0.01$ and OR = 0.074, 95% CI 0.030–0.180, $p < 0.01$).

Table 2. Allelic and genotypic distribution of miR-146a rs2910164 (C>G) and miR-155 rs767649 (T>A) between MF patients and controls.

	miR-146a rs2910164 (C>G)		
Genotypes and Alleles	**MF Cases** **n = 33 (%)**	**Controls** **n = 26 (%)**	***p*-Value, Unadjusted OR (95% CI)**
GG	23 (69.7)	4 (15.3)	1 (reference)
GC	9 (27.3)	6 (23)	$p = 0.066$, 0.261 (0.059–1.148)
CC	1 (3)	16 (61.7)	$p < 0.01$, 0.011 (0.001–0.107)
Dominant model			
GG	23 (69.7)	4 (15.3)	1 (reference)
GC+CC	10 (30.3)	22 (84.7)	$p < 0.01$, 0.079 (0.022–0.290)
Recessive model			
GC+GG	32 (97)	10 (38.3)	1 (reference)
CC	1 (3)	16 (61.7)	$p < 0.01$, 0.020 (0.002–0.166)
Allelic model			
G	55 (83.4)	14 (26.9)	1 (reference)
C	11 (16.6)	38 (73.1)	$p < 0.01$, 0.074 (0.030–0.180)
	miR-155 rs767649 (T>A)		
Genotypes and Alleles	**MF Cases** **n = 33 (%)**	**Controls** **n = 26 (%)**	***p*-Value, Unadjusted OR (95% CI)**
TT	18 (54.5)	23 (88.3)	1 (reference)
TA	3 (9)	1 (4)	$p = 0.217$, 4 (0.384–41.701)
AA	12 (36.5)	2 (7.7)	$p < 0.01$, 8 (1.588–40.299)
Dominant model			
TT	18 (54.5)	23 (88.3)	1 (reference)
TA+AA	15 (45.5)	3 (11.7)	$p < 0.01$, 6.667 (1.674–26.554)
Recessive model			
TA+TT	21 (63.5)	24 (92.3)	1 (reference)
AA	12 (36.5)	2 (7.7)	$p = 0.01$, 6.857 (1.374–34.217)
Allelic model			
T	39 (59)	47 (90.4)	1 (reference)
A	27 (41)	5 (9.6)	$p < 0.01$, 6.679 (2.347–19.003)

Regarding rs767649, in the genotypic model, the AA genotype demonstrated higher frequency among MF patients and was associated with increased susceptibility for MF (OR = 8, 95% CI 1.588–40.299, $p < 0.01$). The TA genotype did not display a significant difference in the distribution between the two study groups (OR = 4, 95% CI 0.384–41.701, $p = 0.217$). In the dominant model, an increased prevalence of mutant allele-associated genotypes TA+AA was detected among MF cases and was associated with a high risk of MF (OR = 6.667, 95% CI 1.674–26.554, $p < 0.01$). Moreover, in the recessive and allelic models, the frequencies of the AA genotype and A allele were significantly higher in MF patients and were correlated with an increased risk of MF (OR = 6.857, 95% CI 1.374–34.217, $p = 0.01$ and OR = 4.097, 95% CI 1.386–12.111, $p < 0.01$).

When comparing patients with eMF and aMF (Table S6), there was no significant difference in the distribution of any genotypic and allelic pattern of rs2910164 among these groups. Similarly, for rs767649, there was no difference between aMF and eMF patients in the genotypic, dominant and recessive model. However, in the allelic model, the A allele demonstrated higher distribution among aMF patients and was correlated with increased susceptibility for aMF (OR = 3.143, 95% CI 1.024–9.648, $p = 0.041$).

2.4. rs2910164 (C>G) and rs767649 (T>A) Are Associated with Plasma Levels of miR-146a and miR-155, Respectively, and Their Concomitant Homozygous Presence Is Associated with Susceptibility to MF

Assessment of the possible impact of rs2910164 on the plasma levels of miR-146a showed that the transition of genotypes GG→CG→CC was associated with decreased expression of miR-146a ($p < 0.01$ and $r = -0.551$) Regarding the rs767649 polymorphism, the transition of genotypes TT→AT→AA was associated with increased levels of miR-155 ($p = 0.028$ and $r = 0.286$) (Figure S3).

Considering the positive correlation between mir-146a and miR-155 in patients' plasma and the impact of the SNPs in their expression, we evaluated the distribution of genotypic combinations GG+AA and CC+TT among MF patients and controls (Table 3). The genotypic combination GG+AA was only detected in patients and was correlated with increased susceptibility to MF (OR = 1.49, 95% CI 1.177–1.908, $p < 0.01$). On the contrary, CC+TT genotypic combination presents a higher frequency in healthy individuals and is associated with low risk for MF (OR = 0.025, 95% CI 0.003–0.210, $p < 0.01$).

Table 3. Distribution of genotypic combinations of rs2910164 and rs767649 in MF patients and controls.

Genotypic Combinations	MF $n = 33$ (%)	Controls $n = 26$ (%)	p-Value, Unadjusted OR (95% CI)
Rest cases	22 (67)	26 (0)	1 (reference)
GG+AA	11 (33)	0 (0)	$p < 0.01$, 1.49 (1.177–1.908)
Rest cases	32 (97)	12 (46)	1 (reference)
CC+TT	1 (3)	14 (54)	$p < 0.01$, 0.025 (0.003–0.210)

2.5. rs2910164 and rs767649 Could Affect the Secondary Structure and Minimum Free Energy of miR-146a and Regulatory Region of miR-155, Respectively

No G-quadruplexes or R-loop structures can be formed in the presence of any allele of rs2910164 and rs767649. The mutant allele C of rs2910164 causes a mispairing in the position +60 within the hairpin of pre-mir-146a, increasing the minimum free energy from −41.60 kcal/mol to −39 kcal/mol when compared with the wild-type allele G (Figure 4a). In rs767649, the mutant A allele affects the secondary structure of a regulatory region of miR-155 (Figure 4b), reducing the minimum free energy from −9.60 kcal/mol to −11.10 kcal/mol in comparison with the wild-type allele G (Figure 4c).

Computational analysis of transcription factors (TF) binding prediction revealed TFs that bind in the region around rs767649, in the presence of either the T or A allele (e.g., MEIS1 that presents the highest relative affinity score for both alleles). Other TFs are predicted to bind only in the presence of T allele (Pax2, MEIS3, MEIS2 and MAFG:NFE2L1). Interestingly, T box factors TBX4, TBX3, TBX6, TBX5, TBX1 and TF MGA seem to "recognize" and bind to miR-155 regulatory region only in the presence of the A allele (Table S7).

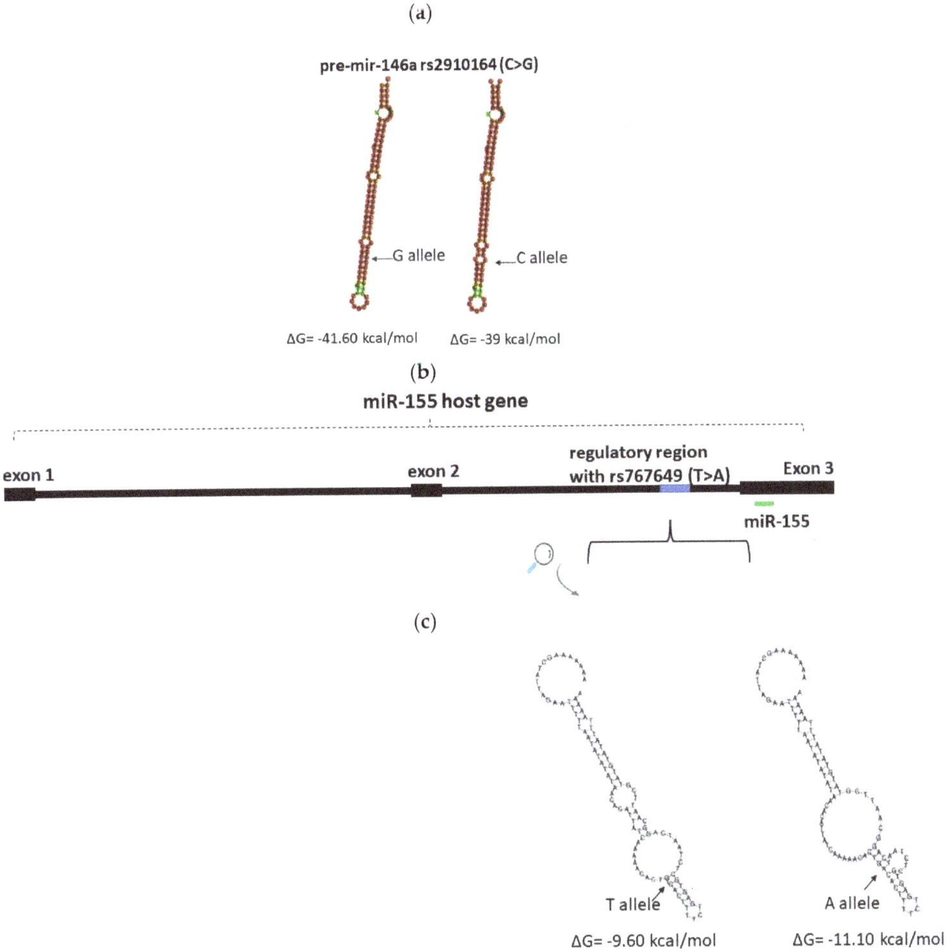

Figure 4. The impact of rs2910164 (C>G) in the secondary structure and minimum free energy of pre-miR-146a, as predicted by RNAfold Web Server (**a**). Genomic location of rs767649 (T>A) in a regulatory region of miR-155 (**b**) and impact of the polymorphism in the secondary structure of single-stranded DNA, as predicted by VectorBuilder (**c**).

3. Discussion

Histopathological examination of skin lesions, immunohistochemical study, and clinicopathological correlation remain the method of choice for diagnosis of MF [22,23]. Despite the advances in diagnostic markers, the polymorphic nature of MF lesions delays definitive diagnosis and has a negative impact on the disease course [24]. Thus, novel, reliable biomarkers are needed to support a timely diagnosis. Among the many proposed biomolecules, miRNAs have been evaluated as diagnostic and prognostic biomarkers for MF, and some have been identified with a possible role in disease pathogenesis [25].

A panel consisting of miRNAs was originally proposed for the diagnosis of MF by Ralfkiaer et al. in 2011 [26], followed by many studies [16,27–31] recognizing their potential role as biomarkers. Nevertheless, the expression of miRNAs was mainly studied in skin biopsies [9,25,27,32–34]. To date, miRNAs plasma levels were only evaluated in two studies [16,17]. In this study, we evaluated the plasma levels of miR-146a and miR-155 in patients with MF and healthy volunteers. We additionally examined how the presence of two SNPs, rs2910164 and rs767649, may affect the levels of miR-146a and miR-155, respectively. Although these SNPs have been associated with many diseases, there was no such correlation study in MF.

The increased plasma levels of miR-146a and miR-155 in MF patients, as compared to healthy individuals, follow a similar expression pattern to skin biopsies [13,14,26,35]. Moreover, our results confirm, in a larger sample, the findings of two previous studies [16,17] detecting increased plasma levels of miR-155 in MF patients. Both miR-146a and miR-155 were found to be significantly upregulated even in early-stage patients, compared to healthy volunteers, rendering them as promising biomarkers for assisting in diagnosis in early stages. Both miRNAs were significantly higher in the plasma of advanced-stage patients when compared to those in early stages. This may be attributed to the increased tumor burden in advanced stages, or to the pathogenetic role of these molecules in disease progression, as, in two studies using cell lines, the activity of STAT5/BIC/miR-155 [13] and JAK/STAT3/STAT4/miR-146a [36] pathways was associated with disease progression. This is further emphasized by the finding that plasma levels of both miRNAs were increased in patients at clinical stage IIB, when compared to patients at lower clinical stages. Similarly, when examining miRNA plasma levels in relation to skin manifestations, patients in erythrodermic and tumor stage had significantly higher miR-155 plasma levels when compared to patients in patch and plaque stages. In a previous study, miR-155 plasma levels were found to be significantly increased in tumor/plaque-stage patients vs. patch-stage patients [16]. To date, miR-146a plasma levels have not been evaluated in MF.

Interestingly, a positive correlation was identified between the plasma levels of the two miRNAs. This could also reflect the tumor burden of each patient or may be the result of a common regulation pathway for the expression of the two miRs. Evidence of a common regulation mechanism is provided by studies in macrophages, where, during the initial stage of activation, NFkBp65 binds in both miR-146a and miR-155 gene loci, increasing their expression, in combination with histone H3 methylation in both genes [37]. Another study suggests that miR-155 and miR-146a form a combined positive and negative regulatory loop controlling NF-κB activity: NF-κB, activated by inflammatory stimuli, rapidly increases miR-155 expression. miR-155 then acts as an amplifier and positive regulator to ensure robust NF-κB activity. miR-146a levels rise more slowly and negatively regulate NF-κB (by targeting IRAK and TRAF6), resulting in a decrease in miR-155 levels [38]. Nevertheless, further clarification of such mechanisms by bioinformatic and functional studies in cutaneous lymphoma cell lines is required.

The single-nucleotide polymorphism rs2910164 (C>G) is located at the +60 nucleotide position of pre-mir-146a, corresponding to the seed area of mature miR-146a. According to the NCBI SNP database, the G allele is more frequent than C, which is also confirmed by our study. Interestingly, the C allele and CC genotype are significantly more frequent among healthy volunteers than among MF patients and are associated with reduced MF risk. Moreover, the C allele and CC genotype are associated with reduced miR-146a plasma levels, in line with previous studies [18,39–42]. This reduction in miR-146a levels in the presence of C allele could be attributed to the altered pri- and/or pre-miR-146a processing [18] associated with an altered secondary conformation and free energy. This conformation contains an additional loop and may affect the protein binding and stability of the molecule. Moreover, the presence of the SNP in the miR's seed area may affect its binding with mRNA, altering its target genes and pathways and contributing to the disease pathogenesis.

Single-nucleotide polymorphism rs767649 (T>A) is located in a regulatory region for miR-155. The dominant allele is T, and it is also detected with a higher frequency in our samples. In the study cohort, the A allele and AA genotype were more frequent among MF patients and were associated with increased MF risk and with increased miR-155 plasma levels. Although in many studies, the A allele is associated with increased miR-155 expression [20,43,44], the results are controversial as there are also studies associating the A allele with reduced expression [21,45]. The position of the SNP (1570 nucleotides upstream of miR-155) is recognized as a regulatory region for miR 155, localized to the second intron of *BIC* (host gene). In the presence of the A allele, there is an alteration in the conformation of the single-stranded DNA around the polymorphism and its free energy. Alternative secondary structures of single-stranded DNA, formed by DNA unwinding during transcription, can affect the binding of RNA polymerase and other enzymes, and may cause RNA polymerase to pause or stall, affecting transcription. Stable structures are more prone to promoter-proximal pausing [46]. Alternatively, the presence of the A allele could alter the binding of transcription factors. The T-Box family of transcription factors, which participate in development and cancer (including lymphomas [47]), are predicted to bind only in the presence of the A allele. These findings of in silico analysis are only suggestive of plausible mechanisms that could explain our results, and should, therefore, be confirmed experimentally.

Finally, the positive correlation between miRNAs levels and the correlation between genotypes and expression suggest a possible synergistic role of the two SNPs as predisposing factors for MF. This cannot be overlooked, since the genotypic combination GG + AA is only detected in patients and is associated with increased disease risk. Combined genotypes of SNPs in separate genes can act synergistically or antagonistically and constitute overlapping risk factors for benign and malignant diseases [48,49].

4. Materials and Methods

4.1. Clinical Samples

Peripheral blood from 41 MF patients and 41 healthy individuals was collected in EDTA. Both newly diagnosed and pre-treated patients with active disease were included in the study cohort. Patients were staged according to WHO-EORTC classification [50] and had no history of autoimmune disease, other malignancies, or recent infection. Healthy volunteers had no history of autoimmune disease, other malignancies, or recent infection and their demographic characteristics were similar to the patient cohort. All patients were followed in the Cutaneous Lymphoma Clinic at "Papageorgiou" Hospital, Thessaloniki, Greece.

Blood samples were centrifuged at 2000× g at 4 °C for 15 min for plasma separation. For genotyping, DNA was extracted from white blood cells of 33/41 MF patients and 26/41 controls.

4.2. microRNAs Isolation

miRNAs were isolated from 200 μL plasma using miRNeasy Serum/Plasma Kit with QIAzol Lysis reagent (Qiagen, Hilden, Germany) according to the protocol. During purification, cel-miR-39 (Caenorhabditis elegans miR-39) was added as spike-in control. The quantity and quality (purity) of the isolated total RNA was assessed spectrophotometrically, with measurements at wavelengths of 230, 260, and 280 nm.

4.3. Reverse Transcription and Quantitative Real-Time Polymerase Chain Reaction

Reverse transcription (RT) and quantitative real-time PCR (qPCR) were carried out using TaqMan miRNA assays (assay ID: 0468 for hsa-miR-146a and 475062_mat for hsa-miR-155, Applied Biosystems, Waltham, MA, USA) using TaqMan microRNA Reverse Transcription kit, and TaqMan Universal PCR master mix with UNG (Applied Biosystems). Amplification was carried out in a StepOnePlus Real Time PCR system (Applied Biosystems). The cycling conditions were 2 min at 50 °C, 10 min at 95 °C, followed by 40 cycles of 15 s at 95 °C and 60 s at 60 °C. All samples were run in duplicate. The relative quantity (RQ) of miRNAs in the plasma was calculated using the $2^{-\Delta CT}$ [$2^{-(CTgene\ of\ interest - CTreference\ gene)}$] method, using cel-miR-39 as exogenous control.

4.4. Genotyping

Genomic DNA was isolated from 100 μL peripheral blood using Monarch Genomic DNA Purification Kit (New England Biolabs, Ipswich, MA, USA). Sequences containing the SNPs were amplified by PCR using KAPA HiFi (Kapa Biosystems, Cape Town, South Africa), followed by Sanger sequencing. Cycling conditions included an initial denaturation at 95 °C for 3 min, followed by 25 cycles of denaturation at 98 °C for 20 s, annealing at 58 °C for 15 s (rs2910164, pre-mir-155) or at 62 °C for 13 s (rs767649), and extension at 72 °C for 15 s. The PCR primers and conditions used in the study are summarized in Table 4.

Table 4. Primers and conditions for PCR, size and genomic location of the products.

Target	Primer Sequence 5′→3′	Product Length	Genomic Location	Annealing Temperature
rs2910164	F: CATTGGATTTACCAGGCTTTT R: CACACTCCTTATACCTTCAGAGC	305 bp	Chr5, GRCh38.p14 (160485260, 160485564)	58 °C
pre-mir-155	F: GCATACACAAACATTTCTTTCTCTCT R: CATCCCAGTGACCAGATTATGA	329 bp	Chr21, GRCh38.p14 (25573862, 25574191)	58 °C
rs767649	F: ACAAAAGGGGACCTGTGTGA R: TTGAAGGTAAATTGCTGGCATACT	281 bp	Chr21, GRCh38.p14 (25572299, 25572579)	62 °C

4.5. Sample Size Calculation

The optimal sample size required to detect a statistically significant difference in the plasma levels of miR-146a and miR-155 was determined by a pilot study, in which an effect size of 0.78 was added to a statistical power of 94% ($1\text{-}b = 0.94$) and a level of significance equal to a = 0.05. Data were analyzed with two tailed Student's *t*-test using G*Power 3.1.9.2. for Windows software [51]. The optimal sample size was calculated at $n = 82$, with a ratio of MF patients/healthy individuals equal to 1:1, i.e., 41 patients and 41 controls.

4.6. Statistical Analysis

RQ values were used to evaluate differences between miRNAs' expression. Normal distribution was examined using the Shapiro–Wilk test and Kolmogorov–Smirnov test. Both tests provided evidence in favor of normal distribution ($p > 0.05$). Homogeneity of variance was examined using Levene's test, which supports equality of variances ($p > 0.05$). A two-tailed Student's *t*-test or one way analysis of variance, followed by Fisher's Least Significant Difference post hoc test, was used to investigate differences in mean expression levels. Associations between clinical characteristics and miRNAs' levels were assessed by Pearson correlation analysis. The genotype and allele frequency differences among study groups were examined using x^2 test. Associations between Mycosis Fungoides risk and genotype/allele patterns were estimated by odds ratios (ORs) and 95% confidence intervals (CIs). The Spearman correlation coefficient was used to evaluate correlations between miRNAs' levels with the presence of polymorphisms. Different genetic models were recruited to examine the allelic/genotypic distribution between MF patients and healthy controls and the association with MF risk. All four models that were used are described below:

(I) The genotypic model, in which the wild-type homozygous genotype was used as a reference (GG for rs2910164 and TT for rs767649) to investigate distribution when compared with the other two genotypes (heterozygous genotype GC for rs2910164 and AT for rs767649, and mutant homozygous genotype CC for rs2910164 and AA for rs767649) in patients and healthy controls.

(II) The dominant model, in which a wild-type homozygous genotype was used as a reference (GG for rs2910164 and TT for rs767649) to investigate the prevalence of mutant-allele-associated genotypes (CC+CG for rs2910164 and AA+AT for rs767649) within MF cases and controls.

(III) The recessive model, in which wild-type-allele-associated genotypes (GC+GG for rs2910164 and AT+TT for rs767649) were used as a reference to investigate the prevalence of the mutant homozygous genotype (CC for rs2910164 and AA for rs767649) within MF and control groups.

(IV) The allelic model, in which the wild-type allele (G for rs2910164 and T for rs767649) was used as a reference to investigate the prevalence of the mutant allele (C for rs2910164 and A for rs767649) within MF and control groups.

IBM SPPS version 25 software was used, and all tests were provided under a 5% level of significance.

4.7. Bioinformatic Analysis

The sequence of pre-mir-146a containing the rs2910164 polymorphism and in the regulatory region of miR-155 containing the rs767649 polymorphism was examined. The prediction of G-quadruplexes (G4) in pre-mir-146a sequence and in the regulatory region of miR-155 polymorphism was performed with G4Hunter online tool [52].

The web tools QmRLFS-finder [53] and Emboss Palindrome [54] were used for the detection of the R-loop-forming structure and inverted repeats (stem loops), respectively.

The online tools RNAfold Web Server [55] and VectorBuilder were utilized to predict the effect of rs2910164 and rs767649 polymorphisms in the secondary structure and minimum free energy (MFE) of pre-mir-146a and the regulatory region of miR-155, respectively.

Jaspar Core database [56] was used to predict, in silico, the impact of rs767649 (T>A) in the binding affinity of transcription factors, setting the following sequence as the input: ACACTG(T/A)CACTTT. The relative score threshold was set at 85%.

5. Conclusions

The evaluation of mir-146a and mir-155 plasma levels in MF patients seems to be of diagnostic and prognostic value and is emerging as a putative, noninvasive biomarker. Moreover, the presence of SNPs is closely associated with miR expression and possibly disease susceptibility, as revealed by the present study. Unveiling the mechanism by which SNPs can affect miR's expression and function may be an important step towards further understanding the pathogenesis of MF and other diseases.

Supplementary Materials: The following supporting information can be downloaded at: https://www.mdpi.com/article/10.3390/ijms24010271/s1.

Author Contributions: Conceptualization, E.G. and A.P.; methodology, E.G. and C.A.; formal analysis, D.D. and C.A.; investigation, C.A.; resources, E.G., A.P., D.P., T.K., A.S., E.L. and G.T.; data curation, C.A. and E.G.; writing—original draft preparation, C.A. and E.G.; writing—review and editing, G.T., T.K., A.P. and A.M.; visualization, C.A., D.D. and P.A.; supervision, E.G. and A.P.; project administration, E.G., A.M. and G.T. All authors have read and agreed to the published version of the manuscript.

Funding: This research received no external funding.

Institutional Review Board Statement: The study was conducted in accordance with the Declaration of Helsinki and approved by the Institutional Review Board of "Papageorgiou" Hospital, Thessaloniki, Greece (protocol code 263/07-03-2017, and date of approval 16 March 2017).

Informed Consent Statement: Informed consent was obtained from all subjects involved in the study.

Data Availability Statement: All data are available from the corresponding author upon reasonable request.

Conflicts of Interest: The authors declare no conflict of interest.

References

1. Dobos, G.; Pohrt, A.; Ram-Wolff, C.; Lebbé, C.; Bouaziz, J.-D.; Battistella, M.; Bagot, M.; De Masson, A. Epidemiology of Cutaneous T-Cell Lymphomas: A Systematic Review and Meta-Analysis of 16,953 Patients. *Cancers* **2020**, *12*, 2921. [CrossRef]
2. Lovgren, M.L.; Scarisbrick, J.J. Update on Skin Directed Therapies in Mycosis Fungoides. *Chin. Clin. Oncol.* **2019**, *8*, 7. [CrossRef]
3. Naraghi, Z.S.; Seirafi, H.; Valikhani, M.; Farnaghi, F.; Kavusi, S.; Dowlati, Y. Assessment of Histologic Criteria in the Diagnosis of Mycosis Fungoides. *Int. J. Dermatol.* **2003**, *42*, 45–52. [CrossRef]
4. Furmanczyk, P.S.; Wolgamot, G.M.; Kussick, S.J.; Sabath, D.E.; Olerud, J.E.; Argenyi, Z.B. Diagnosis of Mycosis Fungoides with Different Algorithmic Approaches. *J. Cutan. Pathol.* **2010**, *37*, 8–14. [CrossRef]

5. Massone, C.; Kodama, K.; Kerl, H.; Cerroni, L. Histopathologic Features of Early (Patch) Lesions of Mycosis Fungoides. *Am. J. Surg. Pathol.* **2005**, *29*, 550–560. [CrossRef]
6. Scarisbrick, J.J.; Quaglino, P.; Prince, H.M.; Papadavid, E.; Hodak, E.; Bagot, M.; Servitje, O.; Berti, E.; Ortiz-Romero, P.; Stadler, R.; et al. The PROCLIPI International Registry of Early-Stage Mycosis Fungoides Identifies Substantial Diagnostic Delay in Most Patients. *Br. J. Dermatol.* **2019**, *181*, 350–357. [CrossRef]
7. O'Brien, J.; Hayder, H.; Zayed, Y.; Peng, C. Overview of MicroRNA Biogenesis, Mechanisms of Actions, and Circulation. *Front. Endocrinol.* **2018**, *9*, 402. [CrossRef]
8. Cui, M.; Wang, H.; Yao, X.; Zhang, D.; Xie, Y.; Cui, R.; Zhang, X. Circulating MicroRNAs in Cancer: Potential and Challenge. *Front. Genet.* **2019**, *10*, 626. [CrossRef]
9. Shen, X.; Wang, B.; Li, K.; Wang, L.; Zhao, X.; Xue, F.; Shi, R.; Zheng, J. MicroRNA Signatures in Diagnosis and Prognosis of Cutaneous T-Cell Lymphoma. *J. Investog. Dermatol.* **2018**, *138*, 2024–2032. [CrossRef]
10. Fava, P.; Bergallo, M.; Astrua, C.; Brizio, M.; Galliano, I.; Montanari, P.; Daprà, V.; Novelli, M.; Savoia, P.; Quaglino, P.; et al. MiR-155 Expression in Primary Cutaneous T-Cell Lymphomas (CTCL). *J. Eur. Acad. Dermatol. Venereol.* **2017**, *31*, e27–e29. [CrossRef]
11. Kopp, K.L.; Ralfkiaer, U.; Nielsen, B.S.; Gniadecki, R.; Woetmann, A.; Ødum, N.; Ralfkiaer, E. Expression of MiR-155 and MiR-126 in Situ in Cutaneous T-Cell Lymphoma. *Apmis* **2013**, *121*, 1020–1024. [CrossRef] [PubMed]
12. Di Raimondo, C.; Han, Z.; Su, C.; Wu, X.; Qin, H.; Sanchez, J.F.; Yuan, Y.C.; Martinez, X.; Abdulla, F.; Zain, J.; et al. Identification of a Distinct Mirna Regulatory Network in the Tumor Microenvironment of Transformed Mycosis Fungoides. *Cancers* **2021**, *13*, 5854. [CrossRef]
13. Kopp, K.L.; Ralfkiaer, U.; Gjerdrum, L.M.R.; Helvad, R.; Pedersen, I.H.; Litman, T.; Jønson, L.; Hagedorn, P.H.; Krejsgaard, T.; Gniadecki, R.; et al. STAT5-Mediated Expression of Oncogenic MiR-155 in Cutaneous T-Cell Lymphoma. *Cell Cycle* **2013**, *12*, 1939–1947. [CrossRef] [PubMed]
14. Manso, R.; Martinez-Magunacelaya, N.; Eraña-Tomás, I.; Monsalvez, V.; Rodriguez-Peralto, J.L.; Ortiz-Romero, P.L.; Santonja, C.; Cristóbal, I.; Piris, M.A.; Rodriguez-Pinilla, S.M. Mycosis Fungoides Progression Could Be Regulated by MicroRNAs. *PLoS ONE* **2018**, *13*, e0198477. [CrossRef] [PubMed]
15. Wang, H.; Peng, R.; Wang, J.; Qin, Z.; Xue, L. Circulating MicroRNAs as Potential Cancer Biomarkers: The Advantage and Disadvantage. *Clin. Epigenetics* **2018**, *10*, 1–10. [CrossRef] [PubMed]
16. Moyal, L.; Arkin, C.; Gorovitz-Haris, B.; Querfeld, C.; Rosen, S.; Knaneh, J.; Amitay-Laish, I.; Prag-Naveh, H.; Jacob-Hirsch, J.; Hodak, E. Mycosis Fungoides-Derived Exosomes Promote Cell Motility and Are Enriched with MicroRNA-155 and MicroRNA-1246, and Their Plasma-Cell-Free Expression May Serve as a Potential Biomarker for Disease Burden. *Br. J. Dermatol.* **2021**, *185*, 999–1012. [CrossRef]
17. Dusílková, N.; Bašová, P.; Polívka, J.; Kodet, O.; Kulvait, V.; Pešta, M.; Trněný, M.; Stopka, T. Plasma MiR-155, Mir-203, and MiR-205 Are Biomarkers for Monitoring of Primary Cutaneous T-Cell Lymphomas. *Int. J. Mol. Sci.* **2017**, *18*, 2136. [CrossRef]
18. Jazdzewski, K.; Murray, E.L.; Franssila, K.; Jarzab, B.; Schoenberg, D.R.; De La Chapelle, A. Common SNP in Pre-MiR-146a Decreases Mature MiR Expression and Predisposes to Papillary Thyroid Carcinoma. *Proc. Natl. Acad. Sci. USA* **2008**, *105*, 7269–7274. [CrossRef]
19. Zhuang, H.; Shen, J.; Zheng, Z.; Luo, X.; Gao, R.; Zhuang, X. MicroRNA-146a Rs2910164 Polymorphism and the Risk of Diffuse Large B Cell Lymphoma in the Chinese Han Population. *Med. Oncol.* **2014**, *31*, 1–5. [CrossRef]
20. Wang, S.; Cao, X.; Ding, B.; Chen, J.; Cui, M.; Xu, Y.; Lu, X.; Zhang, Z.; He, A.; Jin, H. The Rs767649 Polymorphism in the Promoter of MiR-155 Contributes to the Decreased Risk for Cervical Cancer in a Chinese Population. *Gene* **2016**, *595*, 109–114. [CrossRef]
21. Xie, K.; Ma, H.; Liang, C.; Wang, C.; Qin, N.; Shen, W.; Gu, Y.; Yan, C.; Zhang, K.; Dai, N.; et al. A Functional Variant in MiR-155 Regulation Region Contributes to Lung Cancer Risk and Survival. *Oncotarget* **2015**, *6*, 42781–42792. [CrossRef]
22. Miyagaki, T. Diagnosis of Early Mycosis Fungoides. *Diagnostics* **2021**, *11*, 1721. [CrossRef]
23. Cerroni, L. Mycosis Fungoides-Clinical and Histopathologic Features, Differential Diagnosis, and Treatment. *Semin. Cutan. Med. Surg.* **2018**, *37*, 2–10. [CrossRef]
24. Skov, A.G.; Gniadecki, R. Delay in the Histopathologic Diagnosis of Mycosis Fungoides. *Acta Derm. Venereol.* **2015**, *95*, 472–475. [CrossRef]
25. Gluud, M.; Willerslev-Olsen, A.; Gjerdrum, L.M.R.; Lindahl, L.M.; Buus, T.B.; Andersen, M.H.; Bonefeld, C.M.; Krejsgaard, T.; Litvinov, I.V.; Iversen, L.; et al. MicroRNAs in the Pathogenesis, Diagnosis, Prognosis and Targeted Treatment of Cutaneous T-Cell Lymphomas. *Cancers* **2020**, *12*, 1229. [CrossRef]
26. Ralfkiaer, U.; Hagedorn, P.H.; Bangsgaard, N.; Løvendorf, M.B.; Ahler, C.B.; Svensson, L.; Kopp, K.L.; Vennegaard, M.T.; Lauenborg, B.; Zibert, J.R.; et al. Diagnostic MicroRNA Profiling in Cutaneous T-Cell Lymphoma (CTCL). *Blood* **2011**, *118*, 5891–5900. [CrossRef]
27. Marstrand, T.; Ahler, C.B.; Ralfkiaer, U.; Clemmensen, A.; Kopp, K.L.; Sibbesen, N.A.; Krejsgaard, T.; Litman, T.; Wasik, M.A.; Bonefeld, C.M.; et al. Validation of a Diagnostic MicroRNA Classifier in Cutaneous T-Cell Lymphomas. *Leuk. Lymphoma* **2014**, *55*, 957–958. [CrossRef]
28. McGirt, L.Y.; Adams, C.M.; Baerenwald, D.A.; Zwerner, J.P.; Zic, J.A.; Eischen, C.M. MiR-223 Regulates Cell Growth and Targets Proto-Oncogenes in Mycosis Fungoides/Cutaneous T-Cell Lymphoma. *J. Investig. Dermatol.* **2014**, *134*, 1101–1107. [CrossRef]

29. Narducci, M.G.; Arcelli, D.; Picchio, M.C.; Lazzeri, C.; Pagani, E.; Sampogna, F.; Scala, E.; Fadda, P.; Cristofoletti, C.; Facchiano, A.; et al. MicroRNA Profiling Reveals That MiR-21, MiR486 and MiR-214 Are Upregulated and Involved in Cell Survival in Sézary Syndrome. *Cell Death Dis.* **2011**, *2*, e151. [CrossRef]
30. Ito, M.; Teshima, K.; Ikeda, S.; Kitadate, A.; Watanabe, A.; Nara, M.; Yamashita, J.; Ohshima, K.; Sawada, K.; Tagawa, H. MicroRNA-150 Inhibits Tumor Invasion and Metastasis by Targeting the Chemokine Receptor CCR6, in Advanced Cutaneous T-Cell Lymphoma. *Blood* **2014**, *123*, 1499–1511. [CrossRef]
31. Manfè, V.; Biskup, E.; Willumsgaard, A.; Skov, A.G.; Palmieri, D.; Gasparini, P.; Laganá, A.; Woetmann, A.; Ødum, N.; Croce, C.M.; et al. CMyc/MiR-125b-5p Signalling Determines Sensitivity to Bortezomib in Preclinical Model of Cutaneous T-Cell Lymphomas. *PLoS ONE* **2013**, *8*, e59390. [CrossRef]
32. van Kester, M.S.; Ballabio, E.; Benner, M.F.; Chen, X.H.; Saunders, N.J.; van der Fits, L.; van Doorn, R.; Vermeer, M.H.; Willemze, R.; Tensen, C.P.; et al. MiRNA Expression Profiling of Mycosis Fungoides. *Mol. Oncol.* **2011**, *5*, 273–280. [CrossRef]
33. Lindahl, L.M.; Besenbacher, S.; Rittig, A.H.; Celis, P.; Willerslev-Olsen, A.; Gjerdrum, L.M.R.; Krejsgaard, T.; Johansen, C.; Litman, T.; Woetmann, A.; et al. Prognostic MiRNA Classifier in Early-Stage Mycosis Fungoides: Development and Validation in a Danish Nationwide Study. *Blood* **2018**, *131*, 759–770. [CrossRef]
34. Talaat, I.M.; Abdelmaksoud, R.E.S.; Guimei, M.; Agamia, N.F.; Nugud, A.; El-Serafi, A.T. Potential Role for MicroRNA-16 (MiR-16) and MicroRNA-93 (MiR-93) in Diagnosis and Prediction of Disease Progression in Mycosis Fungoides in Egyptian Patients. *PLoS ONE* **2019**, *14*, e0224305. [CrossRef]
35. Rittig, A.H.; Lindahl, L.M.; Johansen, C.; Celis, P.; Ødum, N.; Iversen, L.; Litman, T. The MicroRNA Expression Profile Differs between Erythrodermic Mycosis Fungoides and Sézary Syndrome. *Acta Derm. Venereol.* **2019**, *99*, 1148–1153. [CrossRef]
36. Johnson, V.E.; Vonderheid, E.C.; Hess, A.D.; Eischen, C.M.; McGirt, L.Y. Genetic Markers Associated with Progression in Early Mycosis Fungoides. *J. Eur. Acad. Dermatol. Venereol.* **2014**, *28*, 1431–1435. [CrossRef]
37. Doxaki, C.; Kampranis, S.C.; Eliopoulos, A.G.; Spilianakis, C.; Tsatsanis, C. Coordinated Regulation of MiR-155 and MiR-146a Genes during Induction of Endotoxin Tolerance in Macrophages. *J. Immunol.* **2015**, *195*, 5750–5761. [CrossRef]
38. Mann, M.; Mehta, A.; Zhao, J.L.; Lee, K.; Marinov, G.K.; Garcia-Flores, Y.; Baltimore, D. An NF-KB-MicroRNA Regulatory Network Tunes Macrophage Inflammatory Responses. *Nat. Commun.* **2017**, *8*, 851. [CrossRef]
39. Hu, Z.; Liang, J.; Wang, Z.; Tian, T.; Zhou, X.; Chen, J.; Miao, R.; Wang, Y.; Wang, X.; Shen, H. Common Genetic Variants in Pre-MicroRNAs Were Associated with Increased Risk of Breast Cancer in Chinese Women. *Hum. Mutat.* **2009**, *30*, 79–84. [CrossRef]
40. Kamal, A.; Elgengehy, F.T.; Elawady, Z.; Fawzy, N.A.; El Sisi, O. Role of *miR-146a* rs2910164 and *UTS2* rs228648 Genetic Variants in Behçet's Disease. *Immunol. Investig.* **2021**, *51*, 899–908. [CrossRef]
41. Ahmadi, K.; Soleimani, A.; Soleimani Motlagh, S.; Baharvand Ahmadi, S.; Almasian, M.; Kiani, A.A. Polymorphisms of Pre-Mir-499 Rs3746444 T/C and Pre-Mir-146a Rs2910164 C/G in the Autoimmune Diseases of Rheumatoid Arthritis and Systemic Lupus Erythematosus in the West of Iran. *Iran. J. Public Health* **2020**, *49*, 782–790. [CrossRef]
42. Lv, L.; Gu, H.; Chen, Z.; Tang, W.; Zhang, S.; Lin, Z. *MiRNA-146a* rs2910164 Confers a Susceptibility to Digestive System Cancer: A Meta-Analysis Involving 59,098 Subjects. *Immunol. Investig.* **2020**, *51*, 199–219. [CrossRef]
43. Ayoub, S.E.; Shaker, O.G.; Abdelwahed, M.Y.; Ahmed, N.A.; Abdelhameed, H.G.; Bosilah, A.H.; Mohammed, S.R. Association of MicroRNA-155rs767649 Polymorphism with Susceptibility to Preeclampsia. *Int. J. Mol. Cell. Med.* **2019**, *8*, 247–257. [CrossRef] [PubMed]
44. Polina, E.R.; Oliveira, F.M.; Sbruzzi, R.C.; Crispim, D.; Canani, L.H.; Santos, K.G. Gene Polymorphism and Plasma Levels of MiR-155 in Diabetic Retinopathy. *Endocr. Connect.* **2019**, *8*, 1591–1599. [CrossRef] [PubMed]
45. Shaker, O.G.; Abdelaleem, O.O.; Fouad, N.A.; Ali, A.M.E.A.; Ahmed, T.I.; Ibrahem, E.G.; Abdelghaffar, N.K. Association between MiR-155, Its Polymorphism and Ischemia-Modified Albumin in Patients with Rheumatoid Arthritis. *J. Interf. Cytokine Res.* **2019**, *39*, 428–437. [CrossRef]
46. Szlachta, K.; Thys, R.G.; Atkin, N.D.; Pierce, L.C.T.; Bekiranov, S.; Wang, Y.H. Alternative DNA Secondary Structure Formation Affects RNA Polymerase II Promoter-Proximal Pausing in Human. *Genome Biol.* **2018**, *19*, 1–19. [CrossRef]
47. Timmins, M.A.; Wagner, S.D.; Ahearne, M.J. The New Biology of PTCL-NOS and AITL: Current Status and Future Clinical Impact. *Br. J. Haematol.* **2020**, *189*, 54–66. [CrossRef]
48. Karambataki, M.; Malousi, A.; Tzimagiorgis, G.; Haitoglou, C.; Fragou, A.; Georgiou, E.; Papadopoulou, F.; Krassas, G.E.; Kouidou, S. Association of Two Synonymous Splicing-Associated CpG Single Nucleotide Polymorphisms in Calpain 10 and Solute Carrier Family 2 Member 2 with Type 2 Diabetes. *Biomed. Rep.* **2017**, *6*, 146–158. [CrossRef]
49. Dezfuli, N.K.; Adcock, I.M.; Alipoor, S.D.; Seyfi, S.; Salimi, B.; Mafi Golchin, M.; Dalil Roofchayee, N.; Varhram, M.; Mortaz, E. The MiR-146a SNP Rs2910164 and MiR-155 SNP Rs767649 Are Risk Factors for Non-Small Cell Lung Cancer in the Iranian Population. *Can. Respir. J.* **2020**, *2020*, 1–8. [CrossRef]
50. Willemze, R.; Cerroni, L.; Kempf, W.; Berti, E.; Facchetti, F.; Swerdlow, S.H.; Jaffe, E.S. The 2018 Update of the WHO-EORTC Classification for Primary Cutaneous Lymphomas. *Blood* **2019**, *133*, 1703–1714. [CrossRef]
51. El Maniani, M.; Rechchach, M.; El Mahfoudi, A.; El Moudane, M.; Sabbar, A. G*Power 3: A Flexible Statistical Power Analysis Program for the Social, Behavioral, and Biomedical Sciences. *J. Mater. Environ. Sci.* **2016**, *7*, 3759–3766.
52. Brázda, V.; Kolomazník, J.; Lýsek, J.; Bartas, M.; Fojta, M.; Šťastný, J.; Mergny, J.L. G4Hunter Web Application: A Web Server for G-Quadruplex Prediction. *Bioinformatics* **2019**, *35*, 3493–3495. [CrossRef]

53. Jenjaroenpun, P.; Wongsurawat, T.; Yenamandra, S.P.; Kuznetsov, V.A. QmRLFS-Finder: A Model, Web Server and Stand-Alone Tool for Prediction and Analysis of R-Loop Forming Sequences. *Nucleic Acids Res.* **2015**, *43*, W527–W534. [CrossRef] [PubMed]
54. Rice, P.; Longden, L.; Bleasby, A. EMBOSS: The European Molecular Biology Open Software Suite. *Trends Genet.* **2000**, *16*, 276–277. [CrossRef] [PubMed]
55. Churkin, A.; Retwitzer, M.D.; Reinharz, V.; Ponty, Y.; Waldispühl, J.; Barash, D. Design of RNAs: Comparing Programs for Inverse RNA Folding. *Brief. Bioinform.* **2018**, *19*, 350–358. [CrossRef] [PubMed]
56. Khan, A.; Fornes, O.; Stigliani, A.; Gheorghe, M.; Castro-Mondragon, J.A.; Van Der Lee, R.; Bessy, A.; Chèneby, J.; Kulkarni, S.R.; Tan, G.; et al. JASPAR 2018: Update of the Open-Access Database of Transcription Factor Binding Profiles and Its Web Framework. *Nucleic Acids Res.* **2018**, *46*, D260–D266. [CrossRef] [PubMed]

International Journal of
Molecular Sciences

MDPI

Review

Non-Coding RNAs in Human Cancer and Other Diseases: Overview of the Diagnostic Potential

Roman Beňačka [1], Daniela Szabóová [1], Zuzana Guľašová [2], Zdenka Hertelyová [2] and Jozef Radoňak [3,*]

[1] Department of Pathophysiology, Medical Faculty, Pavol Jozef Šafarik University, 04011 Košice, Slovakia; roman.benacka@upjs.sk (R.B.); daniela.szaboova@student.upjs.sk (D.S.)
[2] Center of Clinical and Preclinical Research MEDIPARK, Pavol Jozef Šafarik University, 04011 Košice, Slovakia; zuzana.gulasova@upjs.sk (Z.G.); zdenka.hertelyova@upjs.sk (Z.H.)
[3] 1st Department of Surgery, Faculty of Medicine, Louis Pasteur University Hospital (UNLP) and Pavol Jozef Šafarik University, 04011 Košice, Slovakia
* Correspondence: jozef.radonak@upjs.sk; Tel.: +421-055-234-3488

Abstract: Non-coding RNAs (ncRNAs) are abundant single-stranded RNA molecules in human cells, involved in various cellular processes ranging from DNA replication and mRNA translation regulation to genome stability defense. MicroRNAs are multifunctional ncRNA molecules of 18–24 nt in length, involved in gene silencing through base-pair complementary binding to target mRNA transcripts. piwi-interacting RNAs are an animal-specific class of small ncRNAs sized 26–31 nt, responsible for the defense of genome stability via the epigenetic and post-transcriptional silencing of transposable elements. Long non-coding RNAs are ncRNA molecules defined as transcripts of more than 200 nucleotides, their function depending on localization, and varying from the regulation of cell differentiation and development to the regulation of telomere-specific heterochromatin modifications. The current review provides recent data on the several forms of small and long non-coding RNA's potential to act as diagnostic, prognostic or therapeutic target for various human diseases.

Keywords: non-coding RNAs; microRNA; small interfering RNA; piwi-interacting RNA; long non-coding RNA; small nuclear RNA; small nucleolar RNA; cancer; biomarkers

1. Introduction

The Encyclopedia of DNA Elements (ENCODE) project indicates that most of the human genome is transcribed and consists of app. 60,000 genes, of which app. 1/3 (app. 20,000 genes) encode proteins and approximately 16,000 are protein non-coding RNAs and 14,000 pseudogenes [1]. It is estimated that at least 80% of the human genome is biologically active, i.e., transcribed. In contrast to this "transcriptional activity", only a small part of human DNA (structural genes) encodes for proteins (2%); the rest (98%) is non-coding DNA and for a long time was considered junk, evolutionarily accumulated reserve copies, supplementary matter, etc. [2]. Recently, the role of this puzzling part of the genome started to be disentangled as the gigantic world of ncRNAs was disclosed. Non-coding RNAs (ncRNAs) are RNA molecules encoded by RNA genes, which are ultimately not translated into proteins [3,4]. Non-coding RNAs can be encoded by their own genes (RNA genes) in areas in between structural genes (intergenic ncRNA), but can also be derived from protein-coding genes, i.e., intronic or even exonic parts of mRNA, which are further spliced or modified [5,6]. Their involvement in cellular processes is very varied, including gene regulation, translation, RNA splicing [7], DNA replication [8], genome defense, and chromosome structure [9]. They participate in virtually all aspects of genetic and epigenetic regulations [5,10] and are of enormous interest due to their use in practical medicine.

According to the size of the chain, ncRNAs are subdivided into (i) *short non-coding RNA (sncRNA)* containing less than 200 nucleotides (nt) and (ii) *long non-coding RNAs (lncRNA)* that contain more than 200 nt. The first group comprises several subgroups, with prominent examples being *transfer RNA (tRNA)* and *ribosomal RNA (rRNA)*, and less known examples include small RNAs such as *micro RNAs (miRNA)*, *small interfering RNA (siRNA)*, *piwi-interacting RNA (piRNA)*, *small nuclear RNA (snRNA)*, *small nucleolar RNAs (snoRNAs)*, *extracellular RNA (exRNA)*, and *small Cajal body-specific RNAs (scaRNAs)*. The second group, *long ncRNAs*, are subdivided based on their roles in the cell—biogenesis (intronic RNA, enhancer RNA, promoter RNA, antisense RNA, sense RNA, intergenic RNA, bidirectional RNA), structure (linear lncRNA, circular lncRNA, long intergenic non-coding RNA, enhancer-derived RNA, transcribed ultra-conserved RNA, natural antisense transcript), and action (cis-acting long non-coding RNA, competing endogenous RNA, trans-acting long non-coding RNA) [5]. With respect to function, ncRNAs can generally be divided into two main classes—*structural ncRNAs* and *regulatory ncRNAs*. The structural ncRNA group includes the abovementioned *rRNAs*, *tRNA*, *snRNA*, and *snoRNA*, as well as *small cytoplasmic RNA*, and *promoter upstream transcripts (PROMPTs)*. Regulatory ncRNAs can be further subdivided into: (i) *small non-coding RNA (miRNA, piRNA, siRNA, centromere repeat-associated small interacting RNA, telomere-specific small RNA)*, (ii) *medium-sized ncRNAs*, and (iii) *long ncRNAs* [11].

In this review, we discuss the role of ncRNAs and their diagnostic and prognostic use in human diseases, including various forms of cancers [12–15]. Some of them, e.g., circRNA and siRNAs, appear to also be promising therapeutical tools and have been reviewed in excellent ways in recent works [16–18]. In the beginning of this work, we highlighted the main features and functions and clinical applications of selected ncRNA species in general, while in the second part we focus on particular body systems and their related disorders.

2. General Overview of ncRNA Biogenesis and Action

2.1. MicroRNAs

MicroRNAs (miRNAs) are a class of small, non-coding, single-stranded RNAs (18–24 nt) involved in the regulation of several cellular processes by repressing the translation of targeted mRNAs. The biogenesis of microRNAs starts in the nucleus via transcription from DNA sequences often located in the non-coding regions of the genome—introns—often near the sequences of mRNA that are subsequently affected by a given miRNA. The products of transcription are long primary transcripts with a characteristic hairpin structure (pri-miRNA), which is recognized and cleaved by RNAase III Drosha. This cleaving generates 70-nucleotide-long pre-microRNA molecules, which are actively exported (Ran-GTP, Ecportin 5) into the cytoplasm. In the cytoplasm, pre-microRNAs are cleaved by the RNAse III Dicer into double-stranded RNA molecules. An enzymatic cleavage by helicase follows, and final single-stranded microRNAs and antisense microRNAs are formed (Figure 1). The final single-stranded microRNA can be then incorporated into three possible complexes—miRISC (miRNA-containing RNA-induced silencing complex), miRNP (miRNA-containing ribonucleoprotein complex), or miRgonaute, and functions as a gene regulator at the post-transcriptional level [19,20].

Function. miRNAs regulate cellular processes via RNA silencing and the post-transcriptional regulation of gene expression through base-pair complementarity to target sequences and, in case of full complementarity, either cleaving the mRNA transcripts into two-strand pieces or destabilizing them by shortening their poly(A) tail. In case of incomplete complementarity, the entire RISC complex binds to the mRNA transcript that is inaccessible for translation [21].

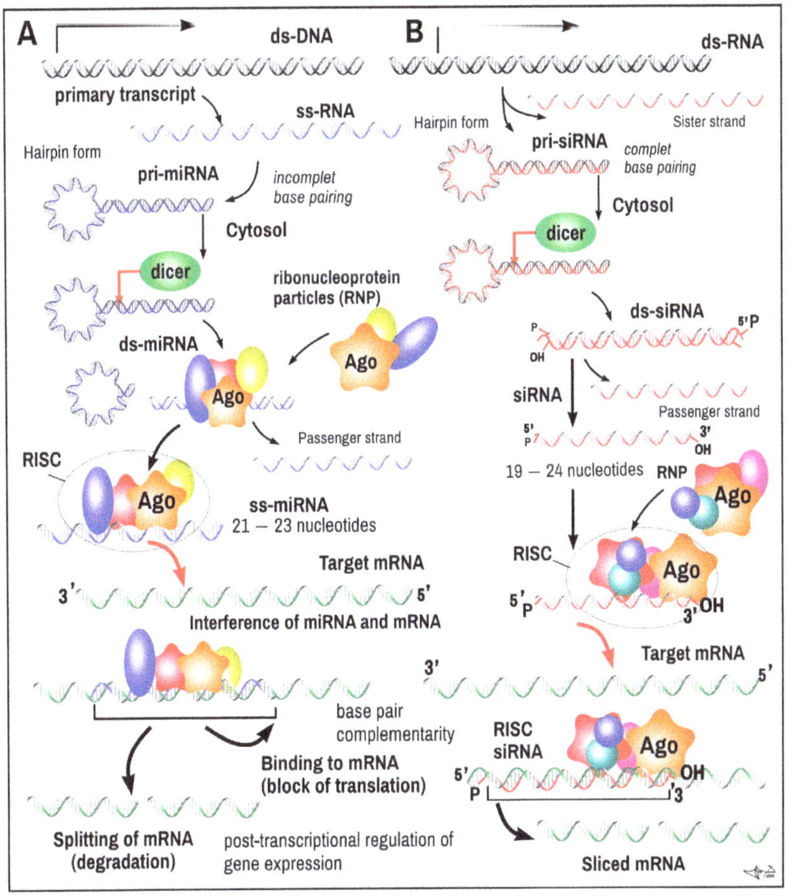

Figure 1. RNA interference pathway in microRNA and silencing RNA action. Primary single-strand transcript of microRNA (known as a pri-miRNA) is processed in the cell nucleus into a 70-nucleotide stem–loop structure called a pre-miRNA by the microprocessor complex (RNAse III). It is folded back to self-assemble into double-stranded form with a small loop (hairpin shape) at one end of the molecule. The complementarity of the bases (C-G; A-U) is not always perfect in miRNA, leading to several mismatched bases. siRNAs derived from long dsRNA precursors differ from miRNAs in that the base pairing is more complete. The hairpin-shaped primary miRNA (pri-miRNA) is exported out of the nucleus in the following step via exportin-5 transport. In cytosol, a hairpin loop is trimmed off by Dicer endo-ribonuclease (RNA helicase), while the rest of the double-stranded remnant of miRNA assembles into a complex with the Argonaute family of nucleoproteins. Precursors of siRNA, long pieces of ds siRNA or hair pin single-stranded siRNA, are called pri-siRNA. Dicer enzyme cuts these molecules to form short double-stranded interfering RNA or siRNA (usually from 20 to 24 bp dsRNA) with phosphorylated 5′ ends and hydroxylated 3′ ends with two overhanging nucleotides. miRNA and siRNA form complexes with nucleoproteins called RNA-Induced Silencing Complexes (RISCs). The RISC of miRNA and siRNA binds to complementary motifs in mRNA. In both cases, the posttranslational mRNA silencing process is started: (1) translation is blocked by the prolonged binding of complex RISC-miRNA, or (2) the mRNA strand is degraded by RISC-miRNA or RISC-siRNA by the splicing to pieces and/or shortening of the poly(A) tail. Legend: red—siRNA sequences; blue—miRNA sequences; green—target mRNA.

Clinical application. Based on their cellular functions, the dysregulation of miRNA expression has been associated with a variety of human diseases. Many miRNAs are capable of directly inhibiting cell cycle genes and controlling cell proliferation, acting either as onco-miRNAs or tumour-suppressor miRNA; therefore, they have been closely connected to various human cancers as diagnostic/prognostic biomarkers or therapeutic agents. Historically, the first human disease associated with miRNA dysregulation was chronic lymphocytic leukemia, but this dysregulation has been since connected to more [22]. In many cancers, miRNAs are the causal factor behind increased cell proliferation, causing DNA repair deficiencies. In colorectal cancers, miR-155 has been observed to epigenetically methylate the *MLH1* gene, causing reduced DNA mismatch repair [23]. In hepatocellular carcinoma, miRNA miR-21 interacts with *MAP2K3*, promoting cell proliferation and giving rise to tumour mass [24]. In 28% of glioblastomas, an increased expression of miR-181b has been connected to a reduced expression of MGMT, a DNA repair enzyme, and a direct target of miR181b [25]. Aberrant levels of miRNAs have been observed in cardiovascular diseases like heart failure, acute myocardial infarction, pulmonary hypertension, or arrhythmias. miRNAs have been linked with atrial remodeling via the regulation of Ca^{2+} channel protein expression, specifically miR-328 and its direct targets: cardiac L-type Ca^{2+} channel α1c- and β1 subunits CACNA1C and CACNB2. Therefore, they affect atrial fibrillations [26]. miRNAs play a vital role in the control of genes significant to addiction. In the nucleus accumbens, a region of the basal forebrain responsible for controlling the sensations of reward that fuel motivational routines, miR-382, was shown to be downregulated. Dopamine receptor D1 (DRD1) is the target of miR-382, whose overexpression upregulates DRD1 and delta fosB (transcription factor), which starts a cascade of transcription events in the nucleus accumbens that eventually leads to addictive behavior [27].

2.2. Small Interfering RNAs

Small interfering RNA (*siRNA*), also called *short interfering RNA* or *silencing RNA*, is a class of *double-stranded RNA*, typically of 20–24 bp in length, with phosphorylated 5′ ends and hydroxylated 3′ ends with two overhanging nucleotides. They operate similarly to miRNA within the *RNA interference* (*RNAi*) *pathway*. However, in contrast to miRNA, they are fully complementary to mRNA, which is targeted. siRNAs are created within the nucleus from long dsRNA (either small hairpin RNAs or complementary RNAs), although siRNAs can also be introduced into cells through transfection. Similar to miRNA, molecules of siRNA bind to specific nucleoproteins (RNPs) from the Argonaut Ago subfamily to build up an *RNA-induced silencing complex* (*RISC*) (Figure 1). Longer ds pre-siRNA precursors (from 30 to over 100 nucleotides) can be cleaved by an endo-ribonuclease type III Dicer (as a part of the RISC complex) to the smaller final ds RNA. Once the double siRNA is part of the RISC complex, the siRNA is unwrapped to form *single-stranded siRNA*. The siRNA, as a part of the RISC complex, can scan around and bind to a complementary mRNA sequence of specific genes. The binding of siRNA-RNP to target mRNA induces mRNA cleavage as a posttranscriptional regulatory mechanism, preventing translation into amino acids and proteins [28].

Function. Similar to miRNA, the major role of siRNA is the *silencing of gene expression using the interfering pathway to inactivate mRNA* and to repress translation. On the contrary, miRNAs have an incomparably bigger repertoire, broader specificity and paralleled actions compared to siRNA. While siRNA very precisely targets its specific mRNA (100% complementarity), miRNA can have multiple targets (possibly over 100 mRNAs at the same time). If the main action of siRNA is the endonucleolytic cleavage of targeted mRNA, in the case of miRNA, endonucleocytic cleavage is rare (in the case of really high complementarity). Rather, miRNA delays and represses translation through mRNA degradation or physical binding to a target mRNA [28].

Clinical application. While miRNAs are more widely used as diagnostic and biomarker tools, siRNAS are promising therapeutical agents. The direct application of *synthetic siRNAs* opens novel possibilities in innovative therapies for human diseases in the post-genomic era. The first siRNA drugs received approval for clinical use by the US Food and Drug Administration and the European Medicines Agency between 2018 and 2022. In addition, research regarding siRNA usage in therapy helps improve the clinical applications of other molecular methodologies. This research includes siRNA dosages in conjugates, exact sequencing, suitable vectors of delivery to intended site to avoid unwanted target-off effects, and gene knockout applications, which include metabolic diseases (hypercholesterolemia, porphyrias), hematology, infectious diseases, oncology, and ocular diseases. [29–31]. A promising means of the intracellular delivery of siRNAs is via nano vectors, nano-scale delivery vehicles, capable of protecting siRNAs against degradation and facilitating the cellular uptake of siRNA. Currently, lipid-based nano vectors are showing great therapeutic potential for central nervous system disorders, as they are able to completely cross the blood–brain barrier [32].

2.3. Circular ncRNAs

Circular ncRNAs (*circRNAs*) are single-stranded regulatory ncRNAs with a closed circular structure (ring shape) formed by covalent bonds, which renders them highly stable against exonuclease-mediated degradation [33,34]. They are ubiquitous across species but not highly conserved among different species ranging from viruses to mammals. Most circRNAs (84%) are created from protein-coding genes by an alternative splicing called back-splicing of pre-mRNAs, connecting the 3′splice end of the downstream exon to the 5′splice site of the upstream exon of genes containing long exons and flanking introns [35,36]. circRNA can be transported from the nucleus to the cytosol by specific mechanisms. In human cells, nuclear RNA helicase UAP56 exports long circRNAs (>1200 nucleotides), while nuclear spliceosome RNA helicase URH49 exports short circRNAs (<400 nucleotides) in size-dependent manner [37,38].

Function. CircRNAs can be implicated in the following important roles: (a) they can regulate protein synthesis and (b) they serve as posttranscriptional gene regulators through acting as so-called "*miRNA sponges*". CircRNAs prevent miRNAs from binding to their target mRNAs, thus serving as competing endogenous miRNAs inhibitors, e.g., *CDR1AS* (*ciRS-7*) contains more than 70 miR-7 target sites and serves as a miRNA sponge in specific tissues [33,39]. *CircHIPK3* also sponges multiple miRNAs; it binds to tumour-suppressive miR-124, and overexpression of this circRNA promotes tumours [37].

Clinical applications. CircRNAs are expressed in human tissues or present with developmental stage-specific expression. Furthermore, they are very stable and resistant to degradation by nucleases, as seen in the nervous tissue [33,34]. Because of the specific expression and long-term stability of circRNAs, they can be used as potential biomarkers in tumour diagnostics (e.g., subgroupings of medulloblastoma based on circRMST) or gastric cancer based on hsa_circRNA_102958 [39]. Recent cumulative data examine the usage of circRNAs in breast cancer, lung cancer, colorectal cancer, kidney or bladder carcinomas, gastric cancer, glioma, hepatocellular carcinoma, pancreatic cancer, and leukemia [33]. Dysregulated circRNAs play important roles in various pathological conditions, including cancer, neurodegenerative diseases, osteoarthritis, diabetes, and cardiovascular diseases. [33,38]. CircRNA can be involved in cancer pathogenesis in several ways: (a) acting through the miRNA sponge, circRNAs may serve as competitive inhibitors of miRNA and behave as *oncogenes* (e.g., circCDDC66 in colon cancer, circCCAC1 in cholangiocarcinoma, WHSC1 in endometrial and ovarian cancer, circANRIL in breast, bladder and gastric cancers, and circRNA_0000285 in cervical cancer [35]). circPCNXL2 and hsa_circ_001895 cause kidney cancer by inhibiting miR-153 and miR-296-5p, respectively. circRNA hsa_circ_001783 in breast cancer acts through the miR-20c-3p sponge. circRNAs such as circHMGB2, circIGF2BP3 and circUSP7 promote the development of non-small cell carcinoma [35,40]. (b) Acting through the miRNA sponge pathway or through other mechanisms, circRNAs may behave as *tumour suppressors* (e.g., circHIPK3 in bladder cancer, circURI1 and hsa_circ_0014717, circORC5 in gastric cancer, circ-AKT3 in kidney cancer through the miR-296-3p sponge, and circPLCE1 in colon cancer) [33,39,40]. circRNA isolated from various body fluids can serve as possible *diagnostic* or *prognostic markers*, determining different subtypes or different staging and grading of the tumours (e.g., circRMST from blood in medulloblastoma, hsa_circ_0014717, hsa_circ_0000190 or hsa_circRNA_102958 from blood or from stomach samples in gastric cancer, hsa_circ_0001874, and hsa_circ_0001971 from local samples in oral carcinoma, circPDLIM5, circSCAF8, circPLXDC2, circSCAMP1, and circCCNT2 collected from urine or blood in prostate cancer) [33,35,38–40].

2.4. Piwi-Interacting ncRNAs

PIWI-interacting RNAs (piRNAs) are single-stranded ncRNAs (26–31 nucleotides in length) with a non-conserved and very diverse set of nucleotide sequences when compared with any other known cellular RNA family. piRNAs were first discovered in germ cells (2006), and now they constitute one of the largest known classes of ncRNA [15]. On the basis of their origin, piRNAs are divided into: (i) transposon-derived piRNA, (ii) lncRNA-derived piRNAs, and (iii) mRNA-derived piRNA. Only the function of the first group is well understood. The vast majority of piRNAs are clustered in relatively short genomic loci on chromosomes 17, 5, 4 and 2. piRNAs received their name due to their interaction with P-element-induced wimpy testis (PIWI) proteins, a germ-line-specific Argonaute family of nucleoproteins. [41,42]. piRNAs are transcribed into long transcripts by RNA polymerase II and then exported to the cytoplasm and cut into smaller pieces (mature piRNAs) in an unknown but most likely Dicer-independent manner [15,43,44]. piRNAs received their name due to their interaction with P-element-induced wimpy testis (PIWI) proteins, a germ-line-specific Argonaute family of nucleoproteins. [41,42]. piRNAs form RNA–protein complexes with the piwi-subfamily of Argonaute proteins. In contrast to the ubiquitous Ago subfamily, the piwi subfamily of Argonaute was found in animals. Two isoforms of PIWI proteins were described—PIWI1 and PIWI2. The overexpression of PIWI1 is associated with cell cycle arrest and the overexpression of PIWI2 is associated with anti-apoptotic signaling and cell proliferation. piRNA plays a role in RNA silencing via the formation of an RNA-induced silencing complex (RISC). Precursors are processed into mature piRNAs by two presumed mechanisms: (i) primary synthesis and (ii) ping pong amplification. For the maturation of piRNA, certain mitochondrial proteins are essential, namely Zuc, Mino, GasZ, and Armi. The piRNA molecule is 5′ monophosphated and 2′-O-methyl modified in the 3′ terminal, processes which have been proposed to increase piRNA stability. Mature piRNAs in cytoplasm form a complex with PIWI proteins and migrate back into the nucleus.

Function. The accumulated data predict the role of piRNA in: (a) *genome rearrangements*, (b) *epigenetic regulation*, (c) *protein regulation*, (d) *transposon silencing*, and (e) *the maintenance of spermatogenesis* [43]. piRNAs are likely involved in the *gene silencing (of transposons)* [41,42] piRNAs have been shown to be implicated in the silencing of retrotransposons, at both the post-transcriptional and epigenetic levels, as well as the silencing of other genetic elements in germ lines, particularly during spermatogenesis. Most piRNAs show antisensing qualities in transposon sequences and are targets of the piRNAs, mainly during embryogenesis. Transposable elements risk damaging the genome through their transposition; therefore, piRNAs are essential for the protection of genome integrity. piRNAs can be transmitted into maternal cells and may be involved in maternally derived epigenetic effects [43,45].

Clinical applications. The research data indicated that the overexpression of specific piRNAs can be associated with particular pathological states and cancers. piRNAs are associated with cancer development [15,45]; in breast cancer, piR-36249, piR-34736, piR-35407, piR-36318, piR-34377, piR-36249piR-36743, piR-36026 and piR-31106 were found to be significantly differentially expressed between tumours and matched non-malignant tissue. piR-651 and piR-823 were found to be overexpressed in gastric cancer paired tissue, colon, lung, and breast cancer tissues, as well as hepatic carcinoma, mesothelioma, cervical, breast, and lung cancer cell lines. piR-32051, piR-39894, and piR-43607 belong to the same cluster on chromosome 17, which is up-regulated in kidney cancer [15,43,45].

2.5. Small Nuclear RNAs

Short nuclear RNA (SnRNA) or *spliceosomal RNA* is group of single-stranded ncRNAs with an average size of 150 nt, which plays pivotal roles in the assembly of spliceosomes (RNA–RNA and RNA–protein interactions). These ribonucleoprotein complexes perform the precise removal of introns and join various exons of polyexonic pre- mRNA transcripts of the protein-coding genes [46–48]. snRNAs are mainly responsible for the correct positioning of the spliceosome on substrate pre-mRNAs. SnRNAs are formed from intronic parts of primary single-stranded pre-mRNA transcripts (catalyzed by RNA polymerase II) (Figure 2). SnRNAs differ from protein-coding gene mRNA transcripts in that the transcript of the snRNA gene is not spliced and the 3′-terminal is not phosphorylated, which may be to prevent translation. Future snRNAs obtain their specific secondary structure and are embedded into complexes with specific nucleoproteins (small nuclear ribonucleoproteins (snRNPs)). A mature snRNA–snRNP complex is released from the pre-RNA molecule and starts its role in spliceosomes. The alternative splicing of the pre-mRNA of protein-coding genes is an evolutionary stem to produce multiple protein variants (many dozens from one gene). Based on shared sequence characteristics, protein cofactors, and functions, snRNAs can be divided into *Sm class* (including U1, U2, U4, U4atac, U5, U7, U11, U12) and *Lsm class* (including U6, U6ata). Based on the frequency of their use, there are major spliceosomal snRNAs (U1, U2, U4, U5 and U6), and two of the four minor spliceosomal snRNAs (U11, U12, U4atac and U6atac) [49]. U4atac ncRNA, often forming a base-paired complex with U6atac ncRNA, is a major component of U12-type spliceosome complex (much lower in abundance: ~10,000 copies per cell) required for the removal of the rarer class of eukaryotic introns (AT-AC, U12-type) [47–49].

Function. The number of various protein isoforms is at least 5–10-fold higher than number of proteins encoded by the human genome. More than 90% of human protein-coding genes produce multiple mRNA isoforms, and this evolutionary move heavily implies alterative splicing. snRNA is a principal *part of the alternative splicing machinery*. SnRNAs are highly expressed in humans during proliferation (cell cycle) and cell development, participating in the regulation of cell fate decisions, and usually exist in clusters in the genome [46].

Clinical application. The dysregulation of spliceosome components including spliceosomal ncRNAs can lead to a variety of pathologies [50,51]. Missplicing, as evidenced by intron retention, occurs in mammals in nearly 75% of multi-exon genes. Defects in *cis-acting splicing* may lead to β+-thalassemia, variants of degenerative muscle dystrophies (limb girdle muscular dystrophy), Duchenne- or Becker-type muscular dystrophies, cystic fibrosis, frontotemporal dementia with parkinsonism, or familial dysautonomia dilated cardiomyopathy (DCM). Defects in *trans-acting splicing* may lead to spinal muscular atrophy or amyotrophic lateral sclerosis (ALS) [49]. Mutations in the U4atac snRNA and defective U12 splicing can cause *microcephalic osteodysplastic primordial dwarfism type I* (*MOPD I*), also called Taybi–Linder syndrome (TALS), with developmental brain and skeletal abnormalities. The dysregulation of alternative splicing is an important factor in several types of cancers, affecting either *tumourignesis* (growth of primary tumour) or progression (local spread) and *metastasis* (e.g., PRPF6, a U5 snRNP protein overexpressed in colorectal carcinoma, and PRPF8 somatic mutations in myelodysplastic syndromes) [51]. Interestingly, 5- Fluorouracil, used in the treatment of a variety of solid tumours such as colorectal, breast, and liver carcinomas for the past six decades, may effect RNA processing and spliceosome mechanism [52].

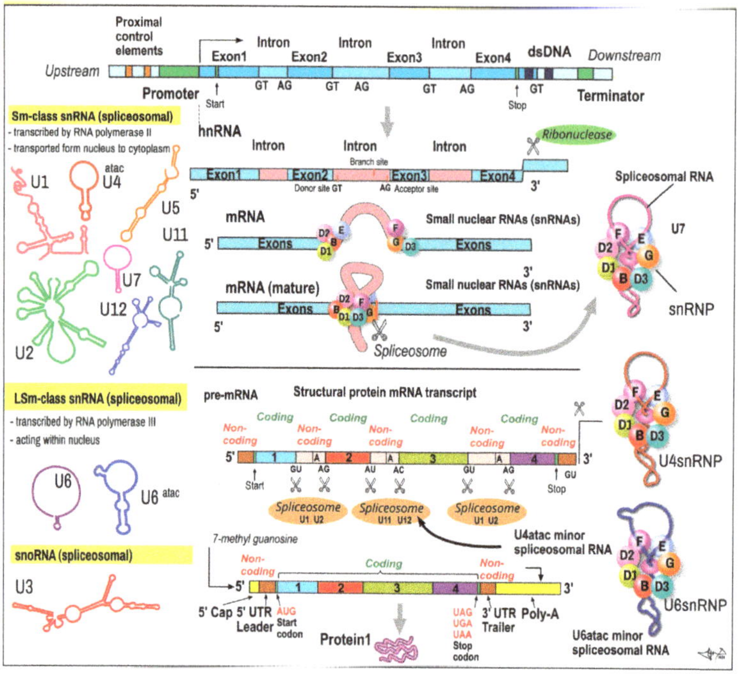

Figure 2. Biogenesis of small nuclear (snRNA). snRNA (average length is 150 nt) is a class of small RNA molecules that are found within the cell nucleus in eukaryotic cells. They are transcribed by either RNA polymerase II or RNA polymerase III [1]. Their primary function is in the processing of pre-messenger RNA (hnRNA) in the nucleus. Examples and classification of spliceosomal snRNA into Sm and LSm class, respectively, is shown on the left. The most common human snRNAs include U1, U2, U4, U5, and U6 spliceosomal RNAs, respectively (for mor detail, see the text). Their nomenclature derives from their high uridine content. U3 belongs to small nucleolar RNA. snRNAs are always associated with a set of specific proteins, and the complexes are referred to as *small nuclear ribonucleoproteins* (*snRNP*). snRNA participates in spliceosomes. U4atac-snRNP and U6atac-snRNP are pivotal components of U11 and U12 spliceosomes typically involved in the posttranscriptional splicing of mRNA in specific locations (AT-AC (AU-AC)).

2.6. Small Nucleolar RNAs

Small nucleolar RNAs (*snoRNAs*) are a class of highly conserved, stable, non-coding RNA molecules (60–300 nucleotides in length) that are located in the nucleolus and very abundant in many organisms [53]. They were among the first identified small ncRNA molecules. snoRNAs are generally transcribed by *RNA polymerase II* from dsDNA. SnoRNAs are either intronic or intergenic; most of the snoRNAs in humans are intronic. In human cells, most snoRNAs are excised from the introns of structural protein-coding genes, which are spliced with debranching and exonucleolytic processing, and the lariat is further processed to a mature snoRNP [53,54].

Each snoRNA associates with at least four core proteins in an RNA/protein complex referred to as a *small nucleolar ribonucleoprotein particle* (*snoRNP*). snoRNA are heavily involved as guide in the sequential chemical modification of various RNAs. Therefore, every snoRNA molecule has to precisely recognize its target and contains an antisense element (a stretch of 10–20 nucleotides), which is complementary to the sequence in targeted RNA [54]. This enables the snoRNP to recognize and bind to exact physical location in the target RNA. There are two main classes of snoRNA, the *C/D box snoRNAs* (genes encoded in the 15q11–q13 region), which are mostly associated with methylation, and the *H/ACA box snoRNAs*, which are associated with the pseudouridylation of target molecules [54–56].

Function. snoRNAs are presumed to play several roles: (a) snoRNAs are involved in synthesis and guide the posttranscriptional chemical modification of other *small ncRNA molecules*, such as ribosomal RNAs (rRNA), transfer RNAs (tRNA), and small nuclear RNAs (snRNA) [54]. For example, nascent pre-rRNA molecules undergo, after transcription, a series of processing steps to produce the mature rRNA molecule. Prior to cleavage by exo- and endonucleases, the pre-rRNA molecule undergoes a series of nucleoside modifications by methylations and pseudouridylations that is guided by snoRNAs [57]. snoRNAs may serve as *miRNAs precursors*, e.g., snoRNA called ACA45 can be processed into a mature miRNA of 21 nt length by the RNAse III endoribonuclease dicer. snoRNAs are also involved in the *regulation of alternative splicing* of the trans-gene transcript, e.g., snoRNA named HBII-52, known as SNORD115), and potentially other functions.

Clinical applications. Historically, it was Gallagher et al., in 2002, who proposed that the deletion of paternal allele SNORD116/PWCR1/HBII-85 could lead to Prader–Willi syndrome (PWS) as a later canonical example of the epigenetic parental imprinting mechanisms [58,59]. snoRNAs are characterized by their stability in body fluids and their clinical relevance, and represent promising tools as diagnostic and prognostic biomarkers. Alterations in snoRNA expression can affect numerous cellular processes, including cell proliferation, angiogenesis, fibrosis, and inflammation, making them a promising target for the diagnostics and treatment of various human pathologies. In recent years, data have suggested that the dysfunction of snoRNAs plays a pivotal role in tumourigenesis and related disorders. snoRNAs have been widely implicated in the pathogenesis of *hepatocellular carcinoma (HCC)* and related liver disorders, such as viral hepatitis B and C and non-alcoholic fatty liver disease (NAFLD) [59]. Growing evidence suggests that snoRNAs act as oncogenes or tumour suppressors in hepatocellular carcinoma (HCC) [59]. SNORD126 and SNORD105 are significantly up-regulated in HCC tissue, whereas SNORA52, SNORD31 and SNORD113 are significantly down-regulated [60,61]. Overexpressed SNORD12B acquired the largest significance as a "driver" role *in colorectal carcinoma*. The high expression of SNORA42 is related to the distant metastasis and poor prognosis of CRC. The overexpression of SNORD44, SNORA15, SNORA41, and SNORD33 may be involved in the progression from chronic intestinal inflammation to colorectal carcinoma. The expression of SNORA71A was significantly correlated with the TNM stage and lymph node metastasis [61,62]. SNORD15A, SNORD15B, SNORD22, SNORD17, and SNORD87 are overexpressed in *human breast cancer* as well as prostate cancer. SNORA71A and 71B are associated with metastatic breast cancer. SNORD3A and SNORD118 regulate tumour occurrence through the inhibition of p53 TSG. In contrast, the expression of SNORD46 and SNORD89 is significantly decreased in breast cancer tissues. SNORD50A delayed the proliferation of breast cancer and improved prognosis [59,63]. SNORD34, SNORD35A, SNORD43, SNORD49A, SNORD55, SNORA74A, SNORD105, SNORD104, SNORD110, and SNORD116-18 show various degrees of association with T-ALL, B-ALL, and CLL leukemias. SNORD42 is a diagnostic marker of AML. SNORD112-114 is mainly associated with leukemias based on stem cell pluripotency [64–67]. The expression of SNORD33, SNORD66, and SNORD76 in the plasma is significantly enhanced in non-small cell lung cancer (NSCLC) compared with that in healthy individuals. SNORA47, SNORA68, and SNORA78 are used to predict the overall survival rate of NSCLC. SNORA42 has been proven to be an oncogene in lung cancer. The expression of SNORA65, SNORA7A, and SNORA7B is also increased in NSCLC [60,68]. The expression of SNORD42, SNORD44, SNORA55, SNORD78, SNORD74, and SNORD81 is significantly increased in *prostate cancer* [69]. *In glioma tumours*, overexpressed SNORD47 may serve as an oncogene negatively correlated with the stage of glioma and positively correlated with survival time, while SNORD44 may be protective and significantly down-regulated in glioma. This is similar to the effect of SNORD 76 [68,70].

2.7. Long Non-Coding RNA

Long non-coding RNA (lncRNA) is a class of single-stranded ncRNAs that are typically longer than 200 nucleotides and do not encode protein. A long-mysterious group of RNAs, lncRNAs are an abundant group of ncRNAs, counting ~60,000 in the human body [71]. Many have been identified in normal tissues and cancer cell lines, yet a significant majority (more than 70%) is still awaiting appropriate analysis [72]. LncRNAs are supposed to play a crucial role in vital cell activities, such as: (a) cell cycle, cell division and proliferation, (b) regulation of cell differentiation (stem cell pluripotency), and (c) cell metabolism [71,73,74].

LncRNAsare coded from *intergenic dsDNA (LincRNAs)*, from *intronic regions* of protein-coding genes, in an antisense manner (asLncRNAs), or in a bidirectional way (bLncRNAs) (Figure 3). Transcripts are often similar to mRNAs as they are transcribed by RNA *polymerase II*, capped and polyadenylated. The molecule contains typical splicing sites and introns/exons, and are also of a similar length to mRNA [73,75]; however, their expression level is usually lower than that of mRNA. They were originally thought to be unstable, but their structure is stabilized by polyadenylation and/or by a secondary structure [12,74]. Enhancer RNAs (eRNAs), first detected in the 2010s, represent a subclass of Lnc RNA (50–2000 nt in lenght) transcribed from the DNA sequences of enhancer regions. eRNAs can be subdivided into two main classes, 1D eRNAs and 2D eRNAs, which differ in their size, transcriptional directionality and polyadenylation place. eRNAs play active roles in transcriptional regulation (positioning of transcription factors) [71].

Function. LncRNAs operate in both the nucleus and cytoplasm, being able to interact with DNA, RNA, and proteins. They are involved in the regulation of gene expression at both the transcriptional and epigenetic levels. Additionally, gene-regulatory elements are embedded in lncRNA genes [71]. LncRNAs can also act post-transcriptionally by acting as sponges for different miRNAs, such as let-7 [76]. Finally, lncRNAs can regulate the translation and posttranslational modification of proteins in the cytoplasm [74].

Clinical application. Many lncRNAs exhibit tissue-specific expressions, including cancers, making them potential diagnostic and/or therapeutic targets. LncRNAs were shown to enhance cancer development and survival by promoting proliferation, growth, migration and drug resistance. LncRNAs can act as tumour suppressors (TSG) or enhance other TSGs; their activity can be downregulated in cancer (e.g., GAS5 in breast cancer and glioblastoma) [12]. Telomeric Repeat-containing RNAs (TERRAs) negatively regulate telomerase [72]. Maternally Expressed 3 (MEG3) is a TSG that can be corroborated with p53 and is downregulated in several cancers [12,77,78]. NAMA and PTCSC3 are dysregulated in thyroid cancer; lncRNA-p21 is dysregulated in rectal cancer [12]. The lncRNA ARHGAP27P1 inhibits gastric cancer cell proliferation and cell cycle progression through the epigenetic regulation of p15 and p16 [75].

LncRNAs work also as inhibitors of TSG, e.g., HOTAIR and retinoblastoma (RB) [12]. The overexpression of lncRNA HOTAIR can promote the development of gastric cancer through the inhibition of TSGs, such as the progesterone receptor, protocadherin 10, protocadherin β5 and junctional adhesion molecule 2 [79]. LncRNA MALAT1 can promote gastric cancer metastasis by inhibiting PCDH10 [12]. LncRNA FAM230B promotes gastric cancer growth and metastasis by regulating the miR-27a-5p/TOP2A axis [12,76].

LncRNAs serve as oncogenes (ONC), e.g., LUCAT1 is highly expressed in various malignant tumours (breast cancer, ovarian cancer, thyroid cancer, and kidney cancer) [80]. LUNAR1 increases the action of IGF1 in T-cell acute lymphoblastic leukemia [76]. NEAT1 is a target of p53 and is overexpressed in liver and ovarian cancer and melanoma [76]. LncRNAs may also enhance the effects of ONC products, e.g., PVT1 prolongs the viability of MYC oncogene [12,76].

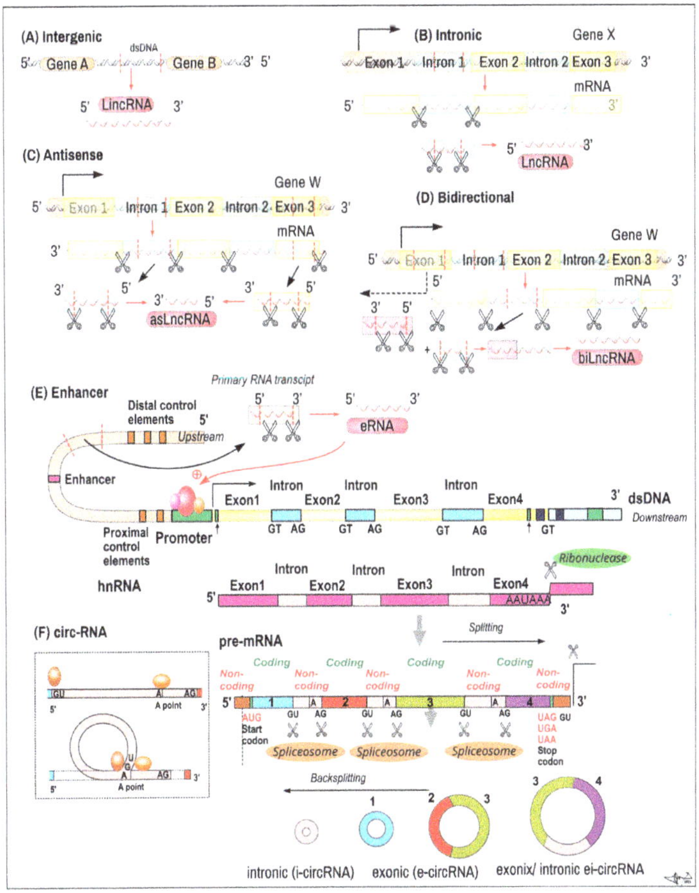

Figure 3. Biogenesis of long non-coding RNA (LncRNA) and circular RNA (circRNA). *LncRNA* classification depends on the genomic position: (**A**) intergenic RNAs (LincRNAs) are transcripts of dsDNA in between two protein-coding genes; (**B**) *intronic LncRNAs* are transcripts from within an intronic region of a protein-coding genes; (**C**) *antisense LncRNAs* (asLncRNAs) are transcribed from complementary strands—either from within the intronic or exonic region of protein-coding genes; (**D**) *bidirectional lncRNAs* (biLncRNA) originate from the bidirectional transcription of protein-coding genes; (**E**) *enhancer LncRNAs (eLncRNAs)* are LncRNAs (50–2000 nt) transcribed from dsDNA of the enhancer regions of genes. These LncRNAs mediate transcription factor positioning in promoters of protein-coding genes. *CircRNAs* are LncRNAs and can originate from primary intronic (i-circ-RNA), exonic (e-circ-RNA) or both exonic and intronic fragments (ei-circRNAs) transcripts of protein coding genes that undergo back splicing arrangement. (**F**) Formation of the circRNA loop by cutting at the 5′ end of RNA—transcript and attachment to the fusion point close to the 3′ end of the molecule. At the bottom of the figure: variants of circRNA. Numbers indicate exons. Legend: Scissors–splicing sites.

LncRNAs may promote cancer by interfering with miRNA, e.g., the increase in miR-675 caused by H19 in hepatocellular carcinoma and rectal cancer. The dysregulation of lncRNA can affect the treatment of cancers by drugs, e.g., through the silencing of PTEN, lncRNA PCAT-1 promotes the cisplatin-resistance of gastric cancer [12]. Through the inhibition of miR-217, lncRNA HOTAIR increases the resistance of gastric cancer cells to doxorubicin and paclitaxe [79].

3. ncRNA Applications in Particular Clinical Areas

3.1. Oncological Diseases

Non-coding RNAs have been extensively studied in the context of various cancers (Figure 4) based on their involvement in tissue development [81], cell proliferation, differentiation [82], and apoptosis [83]. The expression of miRNAs, their amount, and their type, when found in healthy cells and tissues, differs significantly from those found in tumours. Different types of tumours are characterized by an altered representation of miRNAs, which can be studied as tumour biomarkers [84].

Figure 4. Examples of ncRNAs in cancer diagnostics. Types of ncRNAs discussed in the text are depicted. Diagnostic (or prognostic, if denoted) markers from peripheral blood, tissue (endoscopic

biopsy), or other body liquids. In certain markers, their expression levels are indicated by color: red—upregulated markers associated with promoted and invasive growth (protoncogenic) or metastases; blue—downregulated markers associated with antioncogenic, tumour supressor, antimetastatic tumour activites or tumour regression. Certain pictures shown in the background are credited to Servier Medical Art, France (used and modified according to CC-BY-4.0 licence).

Current research in this topic is always exploring new microRNAs as potential biomarkers. A combination of four microRNAs—miR-1246, miR-206, miR-24, and miR-373,—has been reported to detect early-stage breast cancer with 98% sensitivity, 96% specificity, and 97% accuracy [85]. An increased expression of miR-27a, miR-206, and miR-214 has been associated with enhanced tumour chemoresistance and shortened patient survival in patients with triple-negative breast cancer [86]. Non-coding RNAs may also serve as monitoring tools in the treatment and management of oncological diseases, surveilling early responses to neoadjuvant chemotherapy from blood and urine samples. piR-36743, miR-17, miR-19b, and miR-30b have the potential to act as prognostic biomarkers in patients with triple-negative breast cancer [87]. Piwi-interacting RNAs can potentially act as therapeutic agents for breast cancer. piR651 has been shown to be dysregulated in various cancer tissues, including gastric and lung cancer. It appears that, in breast cancer, piR-651 has a dual character. Overexpression leads to tumour growth and metastasis, caused by the upregulation of MDM2, CDK4 and Cyclin D1. However, piR-651 appears to promote PTEN methylation and its downregulation by recruiting DNMT1. This induces the apoptosis of breast cancer cells, acting as a potential therapeutic agent for breast cancer [88]. piRNA-36712 is a tumour-suppressor that is down-regulated in breast cancer cells, targeting SEPW1. The higher expression of SEPW1 due to the down-regulation of piRNA-36712 promotes cell proliferation, invasion, and migration via upregulated Slug, decreased p21 and p53, and E-cadherin levels [89]. In recent years, long non-coding RNAs have been studied as possible markers for breast cancer metastasis, including HOTAIR [90], NR2F1-AS1 [91], ZEB2-AS1 [92], CCAT2 [93], H19 [94], and BLACAT1 [95].

Lung cancer is the most frequent malignancy and a leading cause of cancer death in the male population worldwide. Early stages of lung cancer are often asymptomatic and can only be detected by imaging methods. A microRNA diagnostic assay consisting of miR-1268b and miR-6075 is showing great potential in an early non-invasive screening of lung cancer, with a sensitivity and specificity of 99% regardless of histological type or stage of the tumour [96]. Piwi-interacting RNAs play a role in lung carcinogenesis; however, their involvement has not been fully defined. Significantly increased levels of piR651 have been observed in non-small cell lung cancer [97]; its inhibition has resulted in a decrease in the proliferation and migration of tumour cells [98], making it a potential biomarker and therapeutic agent of lung cancer [99]. piRNA-14633 is a promising prognostic marker in patients with non-small cell lung cancer, as its expression is significantly lowered in tumour cells compared to normal tissues. Down-regulation has correlated with the clinical stage and histological grade of tumour, along with lymph node invasion, increasing the severity of the disease [100].

The expression of small nuclear RNAs U1, U2 and U5 is down-regulated in patients with lung cancer. Expression levels of snRNAs isolated from patients' plasma and healthy controls were quantitated and compared in a study by Dong et al. This dysregulation is correlated with lung cancer progression, serving as a promising biomarker [101]. Fragments of snRNA U2 can serve as lung cancer biomarkers, considering their high relative expression serum levels in lung cancer patients, with a sensitivity of 86% and a specificity of 81% [102].

Colorectal cancer is the third most frequent cancer subtype and fourth most common cause of cancer death in both sexes worldwide. Over 50% of patients with colorectal cancer die due to late diagnosis; diagnosis can be made either by obtaining a sample of the colon, screening stool samples, or using imaging methods to determine the spread of the disease.

Non-coding RNAs may serve as diagnostic or prognostic tools in colorectal cancer. Levels of miR-92a and miR-144 are significantly dysregulated in stool samples of patients with colorectal carcinoma compared to healthy controls, with a diagnostic sensitivity of 89.7% and 78.6%, and a specificity of 51.7% and 66.7%, respectively [103]. Aberrantly expressed piRNAs have been connected to the hallmarks of cancer phenotype and carcinogenic process. PiRNA-823 has been previously reported to be dysregulated in hepatocellular carcinoma and multiple myeloma, promoting tumour growth and metastasis. A recent study has shown that the inhibition of piRNA-823 leads to the inhibition of proliferation, invasion and apoptosis in colorectal cancer. On the other hand, its overexpression correlated with poor overall survival and predicted poor response to adjuvant chemotherapy in patients with colorectal cancer [104]. Based on their aberrant overexpression, several piRNAs have been dubbed as potential early diagnostic markers for colorectal cancer from serum or tissues, piR-24000 [105], piR-020619, piR-020450 [106], piR-54265 [107]. A piwi-interacting RNA expression assay from serum samples consisting of piR-020619 and piR-020450 appears to be capable of detecting early-stage, small-sized tumours, making them another candidate for an early diagnostic tool [106]. Low levels of certain long non-coding RNAs have been associated with poor patient outcome [108], with their involvement in tumour progression making them not only a diagnostic biomarker, but also a therapeutic target [109,110].

3.2. *Neurodegenerative Diseases*

A major fraction of human miRNAs are specifically expressed in the nervous tissue; more than 70% of all miRNAs are expressed in the brain, making these molecules potential biomarkers for the differential diagnosis and prognosis of several neurodegenerative diseases and brain/spine injuries [111]. The differences in the localization-based expression and dynamic regulation of miRNAs during brain development indicates their essential roles in neuronal development and various functions, such as synaptic plasticity, learning and memory [112]. miR-425-5p and miR-502 have been shown to be significantly downregulated in serums of patients with traumatic brain injury during early stages, making them ideal potential markers for mild traumatic brain injury. miR-21 and miR-335 have been shown to be significantly upregulated in serums of patients with traumatic brain injury, making them potential markers of severe traumatic brain injury [113]. Significant upregulation has also been observed in miR-103a-3p, miR-219a-5p, miR-302d-3p, miR-422a, miR-518f-3p, miR-520d-3p, and miR-627 in serums of patients with traumatic brain injury compared with their expression in controls. Particularly significant among these seven microRNAs is miR-219a-5p, which could discriminate patients from controls and also distinguish the severity of the injury (severe or mild) [114].

Neurodegenerative diseases are caused by progressive neuronal damage, resulting in loss of the structure or function of neurons, and ultimately leading to cell death [115]. Neurodegeneration is a complex process, involving several components, including different classes of non-coding RNAs. Mutations of small nuclear RNA U2 cause neuron degeneration as a result of flawed pre-mRNA splicing, which is normally regulated by U2 [116]. Spinal muscular atrophy is caused by the neurodegeneration of spinal motor neurons, a result of SMN1 gene mutation.s The SMN protein is responsible for the assembly of sm-class snRNAs, causing splicing dysfunctions. The SMN1 gene and its product, the SMN protein, are potential therapeutic targets for spinal muscular atrophy, and the dysregulation of snRNAs may serve as a potential biomarker [117].

Alzheimer's disease is the most common neurodegenerative disease in older adults, characterized by a loss of neurons and synapses in the cerebral cortex. Pathological alterations take place years before the appearance of clinical symptoms; early diagnosis and treatment have not been proven successful to date [118]. Studies of microRNA dysregulation in patients with Alzheimer's disease showed a significant deregulation of miR-30b-5p, miR-22-3p, and miR-378a-3p [119], decreased expression of miR-103 in Alzheimer's disease patients in comparison to patients with Parkinson's disease and healthy controls, decreased expression of miR-107 in patients with Alzheimer's disease as well as Parkinson's disease [120], and decreased expression of miR-223-3p in comparison with its upregulation in patients with Parkinson's disease [121]. Jain et al. succeeded in defining a combined microRNA and piwiRNA expression signature from human cerebrospinal fluid that could be used to diagnose patients with Alzheimer's disease. This signature includes miR-27a-3p, miR-30a-5p, miR-34c and piR_019324, piR_019949, and piR_020364 [122]. Possible roles of piwi-interacting RNAs are still emerging; RNA sequencing has revealed the five most upregulated piRNAs (piR-61646, piR-31038, piR-33880, piR-34443, and piR-37213) in patients with Alzheimer's disease, with the potential to act as future diagnostic biomarkers [123]. In a recent study, Cheng et al. report increased levels of long-noncoding RNAs in the serums of patients with Alzheimer's disease, namely ASMTL-AS1, AP001363.1, AC005730.3, and AL133415.1, and consider them potential biomarkers in the Chinese population [124].

Parkinson's disease is a chronic neurodegenerative disorder of the central nervous system, affecting both motor and non-motor systems, and the second most common neurodegenerative disorder. The disease is caused by the selective death of dopaminergic neurons in substantia nigra in the midbrain region [125]. miR-132-3p and miR-146a-5p have been shown to be promising potential biomarkers for the early diagnosis of Parkinson's disease, because of their significant down-regulation in serums of patients with severe Parkinson's disease [126]. Yan et al. has previously reported seven deregulated microRNAs in serums of patients with Parkinson's disease (miR-30c, miR-31, miR-141, miR-146b-5p, miR-181c, miR-214, and miR-193a-3p) and recently, they furthered their research and compared their expression in patients with Parkinson's and multiple system atrophy, revealing a lower expression of miR-31, miR-141, miR-181c, miR-193a-3p, and miR-214 in multiple system atrophy than in Parkinson's disease, accurately discriminating between them [127]. A study by Boros et al. has shown significantly increased levels of long noncoding RNA NEAT1 (core element of nuclear paraspeckles) in the peripheral blood of patients with Parkinson's disease. The aim of the study was to determine a biomarker that is easily detectable by routine laboratory testing [128]. The involvement of long noncoding RNAs is hypothesized to affect neuronal degeneration by promoting apoptosis of dopaminergic neurons [129].

3.3. Cardiovascular Diseases

Cardiovascular diseases (CVDs) are one of the leading causes of death globally. Therefore, it is essential to identify and characterize promising new avenues to better investigate the emergence and progression of CVDs.

The crucial roles of lncRNAs in heart development, pathophysiology, and lipid metabolism reveal their promise as novel diagnostic biomarkers of disease. Abnormalities in lncRNAs' function and expression are associated with cardiovascular biology and disease [130]. The role of lncRNAs in cardiovascular disease is becoming apparent; there is particular interest in the dysregulation of lncRNA regulatory circuits in cell fate, cardiac hypertrophy, vascular disease, atherosclerosis, and metabolic syndrome [130–132].

lncRNAs have been investigated in the context of cerebrovascular diseases. In one of the first studies of lncRNA expression from whole blood samples in ischemic stroke (compared to controls), a significant number of lncRNAs were detected to be differentially expressed according to post-stroke status, with some differentially expressed lncRNAs situated in close genomic proximity to putative stroke risk genes [133,134]. An altered lncRNA expression was found in brain microvasculature under stroke conditions in an in vitro model of oxygen–glucose deprivation in primary brain microvascular endothelial cells [135]. The most upregulated lncRNAs include small nucleolar RNA host gene 12 (Snhg12), Malat1, and lnc-oxygen–glucose deprivation (lnc-OGD) 1006, while the most downregulated lncRNAs include 281008D09Rik, paternally expressed 13 (Peg13), and lnc-OGD 3916, a fact that was confirmed in cerebral micro vessels isolated from mice after experimental stroke [130].

Mhrt is a cardiac-specific lncRNA expressed in mouse and human adult cardiac tissue. This type of lncRNA participates in regulating genes such as *Myh6*, *Myh7*, and osteopontin

(SPP1 or OPN) through its effects on the chromatin remodeling factor Brg1. *Mhrt* is silenced during pressure-overload-induced hypertrophy, and restoring its expression is cardioprotective in mice. Therefore, *Mhrt* levels may be a potential biomarker to predict heart failure in humans [130,133].

lncRNAs could also play a role in the pathogenesis of ischemic stroke. As an example, the lncRNA Malat1 may play a protective role in ischemic stroke through the inhibition of proapoptotic proteins and proinflammatory cytokines, suggesting that lncRNAs may offer novel targets to reduce stroke-related brain damage [135]. The main molecular mechanisms of lncRNAs in the regulation of ischemic brain injury include the epigenetic regulation of transcription via chromatin, influencing miRNA expression, and the translational repression of certain target mRNAs [136–138].

Recently, piwi-intracting RNAs have been found to be differentially expressed in cardiovascular diseases. In patients with thromboembolic pulmonary hypertension (CTEPH), an increased expression level of DQ593039 has been observed in extracellular vesicles. This observation correlated with clinically significant parameters of CTEPH; therefore, DQ593039 may act as a potential diagnostic marker for pulmonary hypertension and disorders of the right heart [139]. Human piRNAs piR-020009 and piR-006426, isolated from serum exosomes of patients with heart failure, are significantly downregulated compared to healthy controls, possibly acting as therapeutic agents in the suppression of heart failure progression [140]. A study by Rajan et al. examined the expression of piRNAs isolated from serum in both in vitro and in vivo conditions in patients with cardiac hypertrophy. Their results extended their previous research, providing insight into piRNAs' role in myocardial infarction. PiRNAs are associated with AKT and Stat3/bcl-xl signaling pathways and may regulate ischemic heart diseases via acting upon downstream targets of these pathways [141].

3.4. Diabetes Mellitus

Type 2 diabetes mellitus (DM2) is a manifestation of insulin resistance, which demonstrates itself through a faulty response to insulin and subsequent altered glucose metabolism. It is a consequence of obesity, metabolic syndrome and altered lipid homeostasis. The metabolic pathway of insulin resistance leads from insulin to glucose transporter type 4 (GLUT4), which regulates glucose uptake. Each of these steps is regulated by several miRNAs that play an important role in pathogenesis and are a prerequisite for their use in therapy [142].

Patients with DM2 exhibit abnormalities in tissue sensitivity to insulin, as well as insulin secretion, by the pancreas, as pancreatic beta cells adapt to changes in insulin action (decreased insulin action means increased insulin secretion and vice versa). Any alterations in this dynamic interaction lead to dysfunctional glucose metabolism [143]. Studies investigating glucose metabolism in DM2 have examined the expression profiles of several miRNAs implicated in the pathogenesis of DM2, comparing newly diagnosed DM2 patients, pre-DM2 subjects, and DM2-prone subjects with normal glucose tolerance. The authors noted a significantly higher expression of miR-9, miR-29a, miR-30d, miR-34a, miR-124a, miR146a, and miR-375 in DM2 patients compared to DM2-prone subjects. In addition, some miRNAs were significantly down-regulated in pre-DM2 compared to DM2 patients. They also found that the expression patterns of circulating miRNAs in patients in the pre-DM2 group were similar to those observed in patients with normal glucose tolerance. On the other hand, the expression levels of the mentioned miRNAs do not change dramatically in the pre-DM2 stage, which reduces their usefulness as a DM2-specific biomarker [144–146]. In another study, the authors aimed to identify patients with metabolic syndrome compared to patients with DM2. miR-150, miR-192, miR-27a, miR-320a, and miR-375 were up-regulated in DM2; thus, they have the potential to participate in the regulation of hyperglycemia [146]. Another study looked at the miRNA profile in normoglycemic, DM2-prone, and DM2 patients. Although the expression of several circulating miRNAs was investigated, only three miRNAs, miR-15a, miR-223, and miR-126, were detected in blood plasma. Levels of miR-126 were analogous between the pre-DM2 and DM2 groups, and significantly lower than in the normoglycemic group. At the same time, the expression pattern of plasma miR-126 was associated with fasting glucose level. The authors concluded that miR-126 could be a potential non-invasive diagnostic tool for the prediction and prevention of DM2 [147,148].

In their study, Liu et al. (2014), in addition to monitoring the differential expression of miR-126 as a potential biomarker for pre-DM2 and DM2, also evaluated the possible therapeutic value of miR-126. The conclusions of their study confirm the diagnostic and therapeutic potential of miR-126 at the early stages of DM2 development [149].

According to studies, miR-375 and miR-9, which affect the regulation of insulin secretion, also show potential biomarker properties. Although, on the one hand, miR-375 is more suitable for distinguishing patients with DM2 from healthy probands, as well as patients with pre-DM2 and DM-2, it is the combination of miR-375 and miR-9 that correlates with the susceptibility to developing DM2 [150].

The presented studies confirm the diagnostic potential of miRNAs (and the therapeutic effect of mi-R126) as they are differentially expressed between persons prone to DM2, persons with DM2, and healthy individuals. Dysfunctions in the glucose metabolism have a significant impact on other cells, tissues, and processes, leading to DM2. miR-126 and miR-375 are among the most frequently monitored miRNAs; therefore, it is necessary to focus on clarifying their role in the pathogenesis of DM2.

Other non-coding RNAs are suspected to play a role in metabolism-related functions, insulin resistance and beta cell function. In a study by Barbosa et al. using a rat model (on a high-fat diet), a transcriptomic analysis of F0 fathers and their F1 female offsprings revealed changes in the expression profile of piRNAs and provided insight into transgenerational epigenetic changes affecting the metabolism of two generations via nutrition [151]. Other evidence for piRNAs' involvement in beta cells' function was presented via the silencing of PIWIL2 and PIWIL4 in rats, causing decreased levels of piRNAs and deficiencies in insulin secretion [152]. piRNA-48383 has been identified as a specific piRNA associated with insulin resistance in a study exploring the relationship between metabolic phenotype and circulating extracellular RNAs [153].

3.5. Gastrointestinal Diseases

miRNAs regulate several biological processes (cell cycle, cell proliferation, cell movement activity, biochemical processes, immunity, inflammatory processes, and apoptosis [154–156]). miRNAs are involved in the regulation of the gastrointestinal system (GIT) and also in its functionality; therefore, dysregulation is associated with many diseases: celiac disease, ulcerative colitis, Crohn's disease, stomach cancer and others [157,158]. miR-143/145 and miR199a/214 clusters are involved in the regulation and proliferation of smooth muscle cells by miRNAs' switching smooth muscle cells between proliferation and differentiation states [159].

The most frequently studied group of miRNAs in GIT diseases is the miR-29 family. miR-29a was upregulated in ulcerative colitis tissues compared with healthy tissue [160]. In a clinical study by Fasseu et al. (2010), the authors found that miRNA-29a and miRNA-29c were significantly increased in GIT tissue from patients with Crohn's disease compared to controls [161]. IFN-γ production was also increased in patients with Crohn's disease, and also plays an important role in ulcerative colitis via miR-29, which is upregulated in these diseases and targets IFN-γ mRNA [154]. In addition, Crohn's disease is associated with the down/up-regulation of miR-19a, miR-1273d, miR886-5p, miR3194, miR192, and miR-200a [162]. The SOCS3 suppressor plays an important role in the inflammatory response in Crohn's disease. miR-19b suppresses this suppressor, and thus prevents the onset and development of Crohn's disease [163]. Schaefer et al. (2015), in their study of colon, blood, and saliva samples, suggested that miRNA may contribute to the pathogenesis of IBD or at least reflect ongoing inflammation, and that miR-19a, miR-21, miR-31, miR101, miR146a, and miR-375 could be reliable markers to identify and distinguish between Crohn's disease and ulcerative colitis [164]. For the monitoring and diagnosis of the disease in patients with IBD or Crohn's disease, a non-invasive examination, namely, the examination of miRNA expression in saliva and blood samples, would be sufficient. Cheng et al. (2015) conducted a clinical study on 60 IBD patients with 120 samples, where they found that miRNA-21 stimulates invasion, intravasation and metastasis in IBD-related carcinogenesis through the post-transcriptional downregulation of the tumour suppressor PDCD4 [163]. Other studies have also confirmed the role of microRNAs in colorectal cancer [165,166]. Duan et al. (2016) identified that miRNA-130 promotes cell proliferation and migration in gastric cancer [167]. Another study by Wu et al. (2015) identified that miR-421 in serum may also be a potential biomarker of gastric cancer [168].

MiR-122 is the most abundant miRNA (~70%) in hepatocytes and in the liver in general, and plays a significant role in normal metabolic functions [169–171]. The downregulation of miR-122 directly controls cholesterol and lipid metabolism, and miR-192, and miR-194 in hepatocytes lead to hepatocellular damage, steatosis, and liver failure [170,172]. Cirrhosis of the liver represents the common end-point of chronic liver injury, and hepatic stellate cells (HSC) play an essential role in this fibrotic process. HSC activation is associated with the upregulation of miR-29 (therapeutic inhibition) and also of miR-501, miR-349, miR-325-5p, miR-328, miR-143, and miR-193. However, others showed significant downregulation (e.g., miR-341, miR-15b, miR-16, miR-375, miR-122, miR-146a, and miR-126 [170–172]. MiR-103 and miR-107 are involved in hepatic insulin effects on glucose/glycogen homeostasis [170,171]. Anti-miR-103/107 (RG-125) were involved in tests for the treatment of nonalcoholic steatohepatitis (NASH) patients who suffered from type 2 diabetes [170,173].

3.6. *Respiratory Disorders*

From non-cancerous pulmonary diagnoses, the main interest in the ncRNA profiling of sputum, epithelial or immune cells, and blood specimens was expectedly recorded in common obstructive disorders such as asthma and chronic obstructive pulmonary disease (COPD) or in their overlap, as shown in Figure 5. The effect of industrial pollution and environmental irritants and allergens was also the focus of ncRNA analysis [174]. These chronic life-time disorders belong to those with high morbidity in young and adult population and in the case of COPD, which also shows the highest mortality worldwide.

In asthma, Th2, an allergic disease, routinely increased serum levels of miR-21 (switch of Th1 vs. Th2), and miR-155 miR-155 (Th2 cell maturation, type 2 innate lymphoid cells ILC2s, eosinophiles) were found—important regulators of the expression of many immunologically relevant genes. Upregulated miRNAs also include the miR-223 family (inflammation machinery), miR-146 family (genetic endotype of asthma), miR-142-5p, miR-142-3p, miR-629-3p (severe asthma), miR-221, and miR-143-3p (inhibit airway remodeling) [174–176]. Downregulated miRNAs included the let-7 family, miR-193b, miR-375, or miR-192-5p [174]. lncRNA CASC2 and BAZ2B are mostly increased in the serum of patients with childhood asthma [177,178]. COPD patients display upregulated miR-497-5p, miR-130b-5p, miR-126-5p miR-221-3p, miR-92a-3p, miR-145, miR-320c, miR-200c-3p, and miR-449c-5p. However, miR-15b-5p appeared to be most consistently downregulated [174,179]. Wang et al. reported that, from 148 significantly identified miRNAs in the blood of COPD patients, 104 miRNAs were upregulated, and 44 miRNAs were downregulated [180]. Asthma and COPD overlap syndrome often reveals upregulated miRNA-338 in the sputum, while the expression of some other miRNAs is lowered (miR-148a-3p, miR-15b-5p, miR-223-3p, and miR-23a-3p) [174].

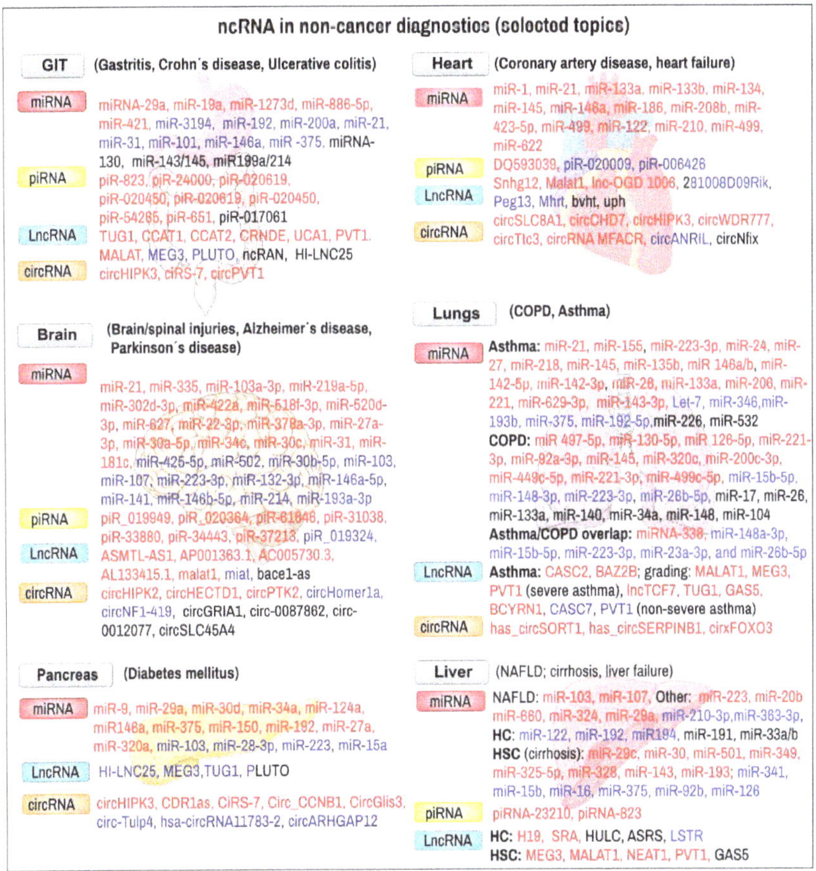

Figure 5. Examples of ncRNAs in non-cancer diagnostic indications. Markers from peripheral blood, tissue (endoscopic biopsy) or other body liquids. In certain markers, their expression levels are indicated by color: red—upregulated markers; blue—downregulated markers. COPD—chronic obstructive pulmonary disease, HC—hepatic cells, HSC—hepatic stellate cells, NAFLD—nonalcoholic fatty liver disease. Certain pictures in the background are credited to Servier Medical Art, France, and used according to the CC-BY-4.0 licence.

4. Methodologies in ncRNAs' Detection

Currently, it is unknown how many non-coding RNAs are present in the human genome; however, new transcriptomic and bioinformatic research indicates that thousands of non-coding transcripts may exist. A large number of non-coding RNAs serve a purpose as epigenetic regulators of several cellular processes [181]. Alterations in non-coding RNAs' expression have been shown to contribute to the dysregulation of gene expression, resulting in various human disorders [182]. Biomarkers are a measurable indicator of a normal biological or pathogenic state that provide information on responses to treatment. A suitable molecular biomarker should be stable, sufficiently specific, easily detectable, and, in particular, its use should be universally applicable to all patients [183]. Dysregulated non-coding RNAs detected in body fluids meet these criteria, and are considered potential diagnostic and prognostic biomarkers for several human diseases. The detection and quantification of ncRNAs is currently becoming easier and more efficient using methods like next-generation sequencing (NGS), RNA microarray, real-time RT-PCR, digital PCR, and Northern blot [184].

Next-generation sequencing and RNA microarray are capable of detecting and quantifying a series of transcripts at the same time; however, a microarray requires larger amounts of RNA and only measures known transcripts, while NGS could be used to identify novel ncRNA transcripts [185]. NGS is a sequencing technology used to determine the order of nucleotides in targeted regions of ncRNAs, using sequencing through synthesis technology, tracking the addition of labeled nucleotides as the RNA chain is copied [186]. The microarray method is used to detect the expression of genes involved in biochemical pathways. Microarrays utilize a solid substrate, which assays a certain amount of target transcripts using multiplex lab-on-a-chip method [187].

Real-time RT-PCR and digital PCR provide fast results and, based on their high sensitivity, require only small amounts of isolated RNAs, but can only detect and quantify a single ncRNA at a time. PCR methods amplify the target DNA obtained after the reverse transcription of sample RNA. The detection of PCR products is conducted via specific primers that are complementary to a specific sequence of DNA [188]. Shorter non-coding RNAs, with an average length of from 21 to 31 nt, have to be extended during reverse transcription by a stem loop primer or by adding a 3′poly A tail, followed by reverse transcription with a poly T primer, with a universal sequence added at the 3′end. Later, qPCR is performed with a forward primer specific to the ncRNA and reverse primer that is complementary to the stem-loop or the poly T primer [189].

Northern blot is currently considered a gold standard, and is mainly used for novel microRNA confirmation and quantification because of its ability to identify the size of the target RNA. However, it is a time-consuming method, requiring large amounts of RNA, and is able to detect only a single transcript at a time [190]. During Northern blot analysis, RNA samples are separated by electrophoresis according to size, transferred onto a nitrocellulose membrane, and detected with a labelled hybridization probe that is complementary to the target RNA. Upregulation or downregulation can be observed based on the abundance of RNA on the Northern blot [191].

5. Conclusions

Research on non-coding RNAs' potential significance in numerous diseases has been based on their abundance in multicellular organisms, varying involvement in cellular processes, and impact on the expression of protein-coding genes. Many diseases can be brought on by changes or dysregulation in the body's non-coding RNA reserve.

The employment of methods such as high-throughput RNA sequencing allows for the expression of mRNAs, lncRNAs, piRNAs, miRNAs, and circRNAs from the blood or other liquids to be analyzed under specified conditions to *build a map of RNA biomarkers* of specific disorders and their variants, and to distinguish in between them. Upregulated or downregulated ncRNAs increasingly appear to be reliable biomarkers for diagnostics, disease stage, therapeutical success, or prognosis. Combining them further enhances their scope, and provides more detailed disease fingerprints for differential diagnostics. As has emerged from recent developments in the field, in addition to being markers, the *multi nc-RNA* signatures help to disclose the *co-expression of circRNA–miRNA–mRNA regulatory networks* enrolled in pathogenesis-specific disease features, or as general mechanisms shared by various disorders (e.g., inflammatory machinery, immune suppression or activation, tissue reconstruction, regeneration, apoptosis, DNA repair, cell migration, proliferation, and cell-to-cell transition). In the future, such a molecular pathogenetic approach will inevitably combine the data from genomics, transcriptomics, and proteomics in a unified framework. Interpretation, however, will always require rational pathophysiological insight and clinical feedback. The biological systems, in either normal or disease conditions, are naturally variable and are permanently evolving. Distinct markers obtained from specific cells or the blood may indicate different stagings and gradings of disease, progress and/or spontaneous remissions, therapeutical regression, known or hidden comorbidities, and other epigenetic modifiers, such as environmental, lifestyle, and nutritional influences. All these provide challenging topics for further research.

Author Contributions: R.B.—Conception and design of the work, Data collection and analysis, Drafting the article and figures. D.S.—Conception and design of the work, Data collection and analysis, Drafting the article Z.G.— Drafting the article, Critical revision of the article Z.H.— Drafting the article, Critical revision of the article, J.R.—Supervision of the final manuscript. All authors have read and agreed to the published version of the manuscript.

Funding: This review article was funded by VEGA grant no. 1/0622/20 of the Scientific Grant Agency of the Ministry of Education, Research and Sport of the Slovak Republic.

Conflicts of Interest: The authors declare no conflict of interest. The funders had no role in the design of the study; in the collection, analyses, or interpretation of data; in the writing of the manuscript; or in the decision to publish the results.

References

1. The ENCODE Project Consortium. An Integrated Encyclopedia of DNA Elements in the Human Genome. *Nature* **2012**, *489*, 57–74. [CrossRef]
2. Kim, M.-S.; Pinto, S.M.; Getnet, D.; Nirujogi, R.S.; Manda, S.S.; Chaerkady, R.; Madugundu, A.K.; Kelkar, D.S.; Isserlin, R.; Jain, S.; et al. A draft map of the human proteome. *Nature* **2014**, *509*, 575–581. [CrossRef]
3. Bartel, D.P. Metazoan MicroRNAs. *Cell* **2018**, *173*, 20–51. [CrossRef]
4. Zhang, B.; Pan, X.; Cobb, G.; Anderson, T. microRNAs as oncogenes and tumor suppressors. *Dev. Biol.* **2007**, *302*, 1–12. [CrossRef] [PubMed]
5. Nozawa, M.; Kinjo, S. Noncoding RNAs, origin and evolution of. In *Encyclopedia of Evolutionary Biology*; Kliman, R.M., Ed.; Academic Press: Cambridge, MA, USA, 2016; pp. 130–135.
6. Mattick, J.S.; Makunin, I.V. Small regulatory RNAs in mammals. *Hum. Mol. Genet.* **2005**, *14*, R121–R132. [CrossRef] [PubMed]
7. Kishore, S.; Stamm, S. The snoRNA HBII-52 Regulates Alternative Splicing of the Serotonin Receptor 2C. *Science* **2006**, *311*, 230–232. [CrossRef] [PubMed]
8. Christov, C.P.; Gardiner, T.J.; Szüts, D.; Krude, T. Functional Requirement of Noncoding Y RNAs for Human Chromosomal DNA Replication. *Mol. Cell. Biol.* **2006**, *26*, 6993–7004. [CrossRef] [PubMed]
9. Park, Y.; Kelley, R.L.; Oh, H.; Kuroda, M.I.; Meller, V.H. Extent of Chromatin Spreading Determined by *roX* RNA Recruitment of MSL Proteins. *Science* **2002**, *298*, 1620–1623. [CrossRef]
10. Esteller, M. Non-coding RNAs in human disease. *Nat. Rev. Genet.* **2011**, *12*, 861–874. [CrossRef]
11. Aliperti, V.; Skonieczna, J.; Cerase, A. Long Non-Coding RNA (lncRNA) Roles in Cell Biology, Neurodevelopment and Neurological Disorders. *Non-Coding RNA* **2021**, *7*, 36. [CrossRef]
12. Wu, S.; Wu, Y.; Deng, S.; Lei, X.; Yang, X. Emerging roles of noncoding RNAs in human cancers. *Discov. Oncol.* **2023**, *14*, 1–16. [CrossRef] [PubMed]
13. Anastasiadou, E.; Jacob, L.S.; Slack, F.J. Non-coding RNA networks in cancer. *Nat. Rev. Cancer* **2018**, *18*, 5–18. [CrossRef]
14. Calore, F.; Lovat, F.; Garofalo, M. Non-coding RNAs and Cancer. *Int. J. Mol. Sci.* **2013**, *14*, 17085–17110. [CrossRef] [PubMed]
15. Romano, G.; Veneziano, D.; Acunzo, M.; Croce, C.M. Small non-coding RNA and cancer. *Carcinogenesis* **2017**, *38*, 485–491. [CrossRef]
16. Ling, H.; Fabbri, M.; Calin, G.A. MicroRNAs and other non-coding RNAs as targets for anticancer drug development. *Nat. Rev. Drug Discov.* **2013**, *12*, 847–865. [CrossRef]
17. Toden, S.; Zumwalt, T.J.; Goel, A. Non-coding RNAs and potential therapeutic targeting in cancer. *Biochim. Biophys. Acta (BBA)-Rev. Cancer* **2020**, *1875*, 188491. [CrossRef] [PubMed]
18. Adams, B.D.; Parsons, C.; Walker, L.; Zhang, W.C.; Slack, F.J. Targeting noncoding RNAs in disease. *J. Clin. Investig.* **2017**, *127*, 761–771. [CrossRef] [PubMed]
19. Iorio, M.V.; Casalini, P.; Piovan, C.; Braccioli, L.; Tagliabue, E. Breast cancer and microRNAs: Therapeutic impact. *Breast* **2011**, *20*, S63–S70. [CrossRef]
20. Khalife, H.; Skafi, N.; Fayyad-Kazan, M.; Badran, B. MicroRNAs in breast cancer: New maestros defining the melody. *Cancer Genet.* **2020**, *246–247*, 18–40. [CrossRef]
21. Shang, R.; Lee, S.; Senavirathne, G.; Lai, E.C. microRNAs in action: Biogenesis, function and regulation. *Nat. Rev. Genet.* **2023**, *24*, 816–833. [CrossRef]
22. Calin, G.A.; Liu, C.-G.; Sevignani, C.; Ferracin, M.; Felli, N.; Dumitru, C.D.; Shimizu, M.; Cimmino, A.; Zupo, S.; Dono, M.; et al. MicroRNA profiling reveals distinct signatures in B cell chronic lymphocytic leukemias. *Proc. Natl. Acad. Sci. USA* **2004**, *101*, 11755–11760. [CrossRef] [PubMed]
23. Valeri, N.; Gasparini, P.; Fabbri, M.; Braconi, C.; Veronese, A.; Lovat, F.; Adair, B.; Vannini, I.; Fanini, F.; Bottoni, A.; et al. Modulation of mismatch repair and genomic stability by miR-155. *Proc. Natl. Acad. Sci. USA* **2010**, *107*, 6982–6987. [CrossRef] [PubMed]
24. Xu, G.; Zhang, Y.; Wei, J.; Jia, W.; Ge, Z.; Zhang, Z.; Liu, X. MicroRNA-21 promotes hepatocellular carcinoma HepG2 cell proliferation through repression of mitogen-activated protein kinase-kinase 3. *BMC Cancer* **2013**, *13*, 469. [CrossRef] [PubMed]
25. Zhang, W.; Zhang, J.; Hoadley, K.; Kushwaha, D.; Ramakrishnan, V.; Li, S.; Kang, C.; You, Y.; Jiang, C.; Song, S.W.; et al. miR-181d: A predictive glioblastoma biomarker that downregulates MGMT expression. *Neuro-Oncology* **2012**, *14*, 712–719. [CrossRef] [PubMed]
26. Zhou, S.-S.; Jin, J.-P.; Wang, J.-Q.; Zhang, Z.-G.; Freedman, J.H.; Zheng, Y.; Cai, L. miRNAS in cardiovascular diseases: Potential biomarkers, therapeutic targets and challenges. *Acta Pharmacol. Sin.* **2018**, *39*, 1073–1084. [CrossRef] [PubMed]

27. Li, J.; Li, J.; Liu, X.; Qin, S.; Guan, Y.; Liu, Y.; Cheng, Y.; Chen, X.; Li, W.; Wang, S.; et al. Micro RNA expression profile and functional analysis reveal that miR-382 is a critical novel gene of alcohol addiction. *EMBO Mol. Med.* **2013**, *5*, 1402–1414. [CrossRef]
28. Carthew, R.W.; Sontheimer, E.J. Origins and Mechanisms of miRNAs and siRNAs. *Cell* **2009**, *136*, 642–655. [CrossRef]
29. Friedrich, M.; Aigner, A. Therapeutic siRNA: State-of-the-Art and Future Perspectives. *BioDrugs* **2022**, *36*, 549–571. [CrossRef]
30. Ahn, I.; Kang, C.S.; Han, J. Where should siRNAs go: Applicable organs for siRNA drugs. *Exp. Mol. Med.* **2023**, *55*, 1283–1292. [CrossRef]
31. Dong, Y.; Siegwart, D.J.; Anderson, D.G. Strategies, design, and chemistry in siRNA delivery systems. *Adv. Drug Deliv. Rev.* **2019**, *144*, 133–147. [CrossRef]
32. Andresen, T.L.; Bruun, J.; Larsen, T.B.; Jølck, R.I.; Eliasen, R.; Holm, R.; Gjetting, T. Investigation of enzyme-sensitive lipid nanoparticles for delivery of siRNA to blood–brain barrier and glioma cells. *Int. J. Nanomed.* **2015**, *10*, 5995–6008. [CrossRef] [PubMed]
33. Zhou, W.Y.; Cai, Z.R.; Liu, J.; Wang, D.S.; Ju, H.Q.; Xu, R.H. Circular RNA: Metabolism, functions and interactions with proteins. *Mol. Cancer* **2020**, *19*, 172. [CrossRef] [PubMed]
34. Jeck, W.R.; Sorrentino, J.A.; Wang, K.; Slevin, M.K.; Burd, C.E.; Liu, J.; Marzluff, W.F.; Sharpless, N.E. Circular RNAs are abundant, conserved, and associated with ALU repeats. *RNA* **2013**, *19*, 141–157. [CrossRef]
35. Cui, J.; Chen, M.; Zhang, L.; Huang, S.; Xiao, F.; Zou, L. Circular RNAs: Biomarkers of cancer. *Cancer Innov.* **2022**, *10*, 5995–6008. [CrossRef]
36. Hansen, T.B.; Jensen, T.I.; Clausen, B.H.; Bramsen, J.B.; Finsen, B.; Damgaard, C.K.; Kjems, J. Natural RNA circles function as efficient microRNA sponges. *Nature* **2013**, *95*, 384–388. [CrossRef]
37. Kristensen, L.S.; Andersen, M.S.; Stagsted, L.V.W.; Ebbesen, K.K.; Hansen, T.B.; Kjems, J. The biogenesis, biology and characterization of circular RNAs. *Nat. Rev. Genet.* **2019**, *20*, 675–691. [CrossRef] [PubMed]
38. Memczak, S.; Jens, M.; Elefsinioti, A.; Torti, F.; Krueger, J.; Rybak, A.; Maier, L.; Mackowiak, S.D.; Gregersen, L.H.; Munschauer, M.; et al. Circular RNAs are a large class of animal RNAs with regulatory potency. *Nature* **2013**, *495*, 333–338. [CrossRef]
39. Wang, F.; Nazarali, A.J.; Ji, S. Circular RNAs as potential biomarkers for cancer diagnosis and therapy. *Am. J. Cancer Res.* **2016**, *6*, 1167–1176.
40. Rybak-Wolf, A.; Stottmeister, C.; Glažar, P.; Jens, M.; Pino, N.; Giusti, S.; Hanan, M.; Behm, M.; Bartok, O.; Ashwal-Fluss, R.; et al. Circular RNAs in the Mammalian Brain Are Highly Abundant, Conserved, and Dynamically Expressed. *Mol. Cell* **2015**, *58*, 870–885. [CrossRef]
41. Anzelon, T.A.; Chowdhury, S.; Hughes, S.M.; Xiao, Y.; Lander, G.C.; MacRae, I.J. Structural basis for piRNA targeting. *Nature* **2021**, *597*, 285–289. [CrossRef]
42. Ozata, D.M.; Gainetdinov, I.; Zoch, A.; O'carroll, D.; Zamore, P.D. PIWI-interacting RNAs: Small RNAs with big functions. *Nat. Rev. Genet.* **2019**, *20*, 89–108. [CrossRef]
43. Liu, Y.; Dou, M.; Song, X.; Dong, Y.; Liu, S.; Liu, H.; Tao, J.; Li, W.; Yin, X.; Xu, W. The emerging role of the piRNA/piwi complex in cancer. *Mol. Cancer* **2019**, *18*, 123. [CrossRef]
44. Czech, B.; Munafò, M.; Ciabrelli, F.; Eastwood, E.L.; Fabry, M.H.; Kneuss, E.; Hannon, G.J. piRNA-Guided Genome Defense: From Biogenesis to Silencing. *Annu. Rev. Genet.* **2018**, *52*, 131–157. [CrossRef]
45. Guo, B.; Li, D.; Du, L.; Zhu, X. piRNAs: Biogenesis and their potential roles in cancer. *Cancer Metastasis Rev.* **2020**, *39*, 567–575. [CrossRef] [PubMed]
46. Chen, J.; Wagner, E.J. snRNA 3′ end formation: The dawn of the Integrator complex. *Biochem. Soc. Trans.* **2010**, *38*, 1082–1087. [CrossRef] [PubMed]
47. Matera, A.G.; Wang, Z. A day in the life of the spliceosome. *Nat. Rev. Mol. Cell Biol.* **2014**, *15*, 108–121. [CrossRef] [PubMed]
48. Egloff, S.; O'Reilly, D.; Murphy, S. Expression of human snRNA genes from beginning to end. *Biochem. Soc. Trans.* **2008**, *36 Pt 4*, 590–594. [CrossRef]
49. Morais, P.; Adachi, H.; Yu, Y.-T. Spliceosomal snRNA Epitranscriptomics. *Front. Genet.* **2021**, *12*, 652129. [CrossRef]
50. Karijolich, J.; Yu, Y.-T. Spliceosomal snRNA modifications and their function. *RNA Biol.* **2010**, *7*, 192–204. [CrossRef] [PubMed]
51. Scotti, M.M.; Swanson, M.S. RNA mis-splicing in disease. *Nat. Rev. Genet.* **2016**, *17*, 19–32. [CrossRef]
52. Sciarrillo, R.; Wojtuszkiewicz, A.; Assaraf, Y.G.; Jansen, G.; Kaspers, G.J.; Giovannetti, E.; Cloos, J. The role of alternative splicing in cancer: From oncogenesis to drug resistance. *Drug Resist. Updat.* **2020**, *53*, 100728. [CrossRef]
53. Bachellerie, J.P.; Cavaillé, J.; Hüttenhofer, A. The expanding snoRNA world. *Biochimie* **2002**, *84*, 775–790. [CrossRef]
54. Lui, L.; Lowe, T. Small nucleolar RNAs and RNA-guided post-transcriptional modification. *Essays Biochem.* **2013**, *54*, 53–77. [CrossRef]
55. Terns, M.P.; Terns, R.M. Small nucleolar RNAs: Versatile trans-acting molecules of ancient evolutionary origin. *Gene Expr.* **2002**, *10*, 17–39. [PubMed]
56. Rabany, O.; Nachmani, D. Small Nucleolar (Sno)RNA: Therapy Lays in Translation. *Non-Coding RNA* **2023**, *9*, 35. [CrossRef]
57. Meier, U.T. The daunting task of modifying ribosomal RNA. *RNA* **2022**, *28*, 1555–1557. [CrossRef] [PubMed]
58. Gallagher, R.C.; Pils, B.; Albalwi, M.; Francke, U. Evidence for the role of PWCR1/HBII-85 C/D box small nucleolar RNAs in Prader-Willi syndrome. *Am. J. Hum. Genet.* **2002**, *71*, 669–678. [CrossRef]
59. Zhang, X.; Wang, C.; Xia, S.; Xiao, F.; Peng, J.; Gao, Y.; Yu, F.; Wang, C.; Chen, X. The emerging role of snoRNAs in human disease. *Genes Dis.* **2022**, *10*, 2064–2081. [CrossRef]

60. Huldani, H.; Gandla, K.; Asiri, M.; Romero-Parra, R.M.; Alsalamy, A.; Hjazi, A.; Najm, M.A.; Fawaz, A.; Hussien, B.M.; Singh, R. A comprehensive insight into the role of small nucleolar RNAs (snoRNAs) and SNHGs in human cancers. *Pathol.-Res. Pract.* **2023**, *249*, 154679. [CrossRef]
61. Yang, X.; Li, Y.; Li, L.; Liu, J.; Wu, M.; Ye, M. SnoRNAs are involved in the progression of ulcerative colitis and colorectal cancer. *Dig. Liver Dis.* **2017**, *49*, 545–551. [CrossRef] [PubMed]
62. Zhang, Z.; Tao, Y.; Hua, Q.; Cai, J.; Ye, X.; Li, H. SNORA71A Promotes Colorectal Cancer Cell Proliferation, Migration, and Invasion. *BioMed Res. Int.* **2020**, *2020*, 8284576. [CrossRef]
63. Su, H.; Xu, T.; Ganapathy, S.; Shadfan, M.; Long, M.; Huang, T.H.-M.; Thompson, I.; Yuan, Z.-M. Elevated snoRNA biogenesis is essential in breast cancer. *Oncogene* **2014**, *33*, 1348–1358. [CrossRef] [PubMed]
64. Dong, J.; Wang, H.; Zhang, Z.; Yang, L.; Qian, X.; Qian, W.; Han, Y.; Huang, H.; Qian, P. Small but strong: Pivotal roles and potential applications of snoRNAs in hematopoietic malignancies. *Front. Oncol.* **2022**, *12*, 939465. [CrossRef] [PubMed]
65. Challakkara, M.F.; Chhabra, R. snoRNAs in hematopoiesis and blood malignancies: A comprehensive review. *J. Cell. Physiol.* **2023**, *238*, 1207–1225. [CrossRef] [PubMed]
66. Ronchetti, D.; Mosca, L.; Cutrona, G.; Tuana, G.; Gentile, M.; Fabris, S.; Agnelli, L.; Ciceri, G.; Matis, S.; Massucco, C.; et al. Small nucleolar RNAs as new biomarkers in chronic lymphocytic leukemia. *BMC Med. Genom.* **2013**, *6*, 27. [CrossRef] [PubMed]
67. Pauli, C.; Liu, Y.; Rohde, C.; Cui, C.; Fijalkowska, D.; Gerloff, D.; Walter, C.; Krijgsveld, J.; Dugas, M.; Edemir, B.; et al. Site-specific methylation of 18S ribosomal RNA by SNORD42A is required for acute myeloid leukemia cell proliferation. *Blood* **2020**, *135*, 2059–2070. [CrossRef] [PubMed]
68. Liu, X.; Ali, K.; Zhao, L.; Dua, K.; Mao, Y. The emerging diagnostic and therapeutic roles of small nucleolar RNAs in lung diseases. *Biomed. Pharmacother.* **2023**, *161*, 114519. [CrossRef] [PubMed]
69. Yi, C.; Wan, X.; Zhang, Y.; Fu, F.; Zhao, C.; Qin, R.; Wu, H.; Li, Y.; Huang, Y. SNORA42 enhances prostate cancer cell viability, migration and EMT and is correlated with prostate cancer poor prognosis. *Int. J. Biochem. Cell Biol.* **2018**, *102*, 138–150. [CrossRef]
70. Xia, X.-R.; Li, W.-C.; Yu, Z.-T.; Li, J.; Peng, C.-Y.; Jin, L.; Yuan, G.-L. Effects of small nucleolar RNA SNORD44 on the proliferation, apoptosis and invasion of glioma cells. *Histochem. Cell Biol.* **2020**, *153*, 257–269. [CrossRef]
71. Fernandes, J.C.R.; Acuña, S.M.; Aoki, J.I.; Floeter-Winter, L.M.; Muxel, S.M. Long Non-Coding RNAs in the Regulation of Gene Expression: Physiology and Disease. *Non-Coding RNA* **2019**, *5*, 17. [CrossRef]
72. Arora, R.; Brun, C.M.; Azzalin, C.M. TERRA: Long noncoding RNA at eukaryotic telomeres. *Prog. Mol. Subcell Biol.* **2011**, *51*, 65–94. [PubMed]
73. Bridges, M.C.; Daulagala, A.C.; Kourtidis, A. LNCcation: lncRNA localization and function. *J. Cell Biol.* **2021**, *220*, e202009045. [CrossRef] [PubMed]
74. Kitagawa, M.; Kitagawa, K.; Kotake, Y.; Niida, H.; Ohhata, T. Cell cycle regulation by long non-coding RNAs. *Cell. Mol. Life Sci.* **2013**, *70*, 4785–4794. [CrossRef]
75. Slack, F.J.; Chinnaiyan, A.M. The Role of Non-coding RNAs in Oncology. *Cell* **2019**, *179*, 1033–1055. [CrossRef]
76. Arun, G.; Diermeier, S.D.; Spector, D.L. Therapeutic Targeting of Long Non-Coding RNAs in Cancer. *Trends Mol. Med.* **2018**, *24*, 257–277. [CrossRef]
77. Zhang, X.; Gejman, R.; Mahta, A.; Zhong, Y.; Rice, K.A.; Zhou, Y.; Cheunsuchon, P.; Louis, D.N.; Klibanski, A. *Maternally Expressed Gene 3*, an Imprinted Noncoding RNA Gene, Is Associated with Meningioma Pathogenesis and Progression. *Cancer Res* **2010**, *70*, 2350–2358. [CrossRef]
78. Chak, W.; Lung, R.W.; Tong, J.H.; Chan, S.Y.; Lun, S.W.; Tsao, S.; Lo, K.; To, K. Downregulation of long non-coding RNA MEG3 in nasopharyngeal carcinoma. *Mol. Carcinog.* **2016**, *56*, 1041–1054. [CrossRef] [PubMed]
79. Wang, H.; Qin, R.; Guan, A.; Yao, Y.; Huang, Y.; Jia, H.; Huang, W.; Gao, J. HOTAIR enhanced paclitaxel and doxorubicin resistance in gastric cancer cells partly through inhibiting miR-217 expression. *J. Cell. Biochem.* **2018**, *119*, 7226–7234. [CrossRef]
80. Xing, C.; Sun, S.-G.; Yue, Z.-Q.; Bai, F. Role of lncRNA LUCAT1 in cancer. *Biomed. Pharmacother.* **2021**, *134*, 111158. [CrossRef] [PubMed]
81. Zhao, Y.; Srivastava, D. A developmental view of microRNA function. *Trends Biochem. Sci.* **2007**, *32*, 189–197. [CrossRef] [PubMed]
82. Chen, J.-F.; Mandel, E.M.; Thomson, J.M.; Wu, Q.; Callis, T.E.; Hammond, S.M.; Conlon, F.L.; Wang, D.-Z. The Role of MicroRNA-1 and MicroRNA-133 in Skeletal Muscle Proliferation and Differentiation. *Nat. Genet.* **2006**, *38*, 228–233. [CrossRef]
83. Lynam-Lennon, N.; Maher, S.G.; Reynolds, J.V. The roles of microRNA in cancer and apoptosis. *Biol. Rev.* **2009**, *84*, 55–71. [CrossRef]
84. Rupaimoole, R.; Slack, F.J. MicroRNA therapeutics: Towards a new era for the management of cancer and other diseases. *Nat. Rev. Drug Discov.* **2017**, *16*, 203–222. [CrossRef]
85. Jang, J.Y.; Kim, Y.S.; Kang, K.N.; Kim, K.H.; Park, Y.J.; Kim, C.W. Multiple microRNAs as biomarkers for early breast cancer diagnosis. *Mol. Clin. Oncol.* **2021**, *14*, 31. [CrossRef] [PubMed]
86. Liu, Y.; Gong, W.; Panoutsopoulou, K.; Singer-Cornelius, T.; Augustin, K.; Bronger, H.; Kiechle, M.; Dorn, J.; Scorilas, A.; Avgeris, M.; et al. Association of high miR-27a, miR-206, and miR-214 expression with poor patient prognosis and increased chemoresistance in triple-negative breast cancer. *Am. J. Cancer Res.* **2023**, *13*, 2471–2487. [PubMed]
87. Ritter, A.; Hirschfeld, M.; Berner, K.; Weiss, D.; Medl, M.; Gassner, S.; Asberger, J.; Erbes, T.; Rücker, G.; Jäger, M.; et al. Circulating non-coding RNA-biomarker potential in neoadjuvant chemotherapy of triple negative breast cancer? *Int. J. Oncol.* **2020**, *56*, 47–68. [CrossRef] [PubMed]

88. Liu, T.; Wang, J.; Sun, L.; Li, M.; He, X.; Jiang, J.; Zhou, Q. Piwi-interacting RNA-651 promotes cell proliferation and migration and inhibits apoptosis in breast cancer by facilitating DNMT1-mediated PTEN promoter methylation. *Cell Cycle* **2021**, *20*, 1603–1616. [CrossRef]
89. Tan, L.; Mai, D.; Zhang, B.; Jiang, X.; Zhang, J.; Bai, R.; Ye, Y.; Li, M.; Pan, L.; Su, J.; et al. PIWI-interacting RNA-36712 restrains breast cancer progression and chemoresistance by interaction with SEPW1 pseudogene SEPW1P RNA. *Mol. Cancer* **2019**, *18*, 9. [CrossRef] [PubMed]
90. Cantile, M.; Di Bonito, M.; Cerrone, M.; Collina, F.; De Laurentiis, M.; Botti, G. Long Non-Coding RNA HOTAIR in Breast Cancer Therapy. *Cancers* **2020**, *12*, 1197. [CrossRef]
91. Liu, Y.; Zhang, P.; Wu, Q.; Fang, H.; Wang, Y.; Xiao, Y.; Cong, M.; Wang, T.; He, Y.; Ma, C.; et al. Long non-coding RNA NR2F1-AS1 induces breast cancer lung metastatic dormancy by regulating NR2F1 and ΔNp63. *Nat. Commun.* **2021**, *12*, 5232. [CrossRef]
92. Zhang, G.; Li, H.; Sun, R.; Li, P.; Yang, Z.; Liu, Y.; Wang, Z.; Yang, Y.; Yin, C. Long non-coding RNA ZEB2-AS1 promotes the proliferation, metastasis and epithelial mesenchymal transition in triple-negative breast cancer by epigenetically activating ZEB2. *J. Cell. Mol. Med.* **2019**, *23*, 3271–3279. [CrossRef]
93. Xu, Z.; Liu, C.; Zhao, Q.; Lü, J.; Ding, X.; Luo, A.; He, J.; Wang, G.; Li, Y.; Cai, Z.; et al. Long non-coding RNA CCAT2 promotes oncogenesis in triple-negative breast cancer by regulating stemness of cancer cells. *Pharmacol. Res.* **2020**, *152*, 104628. [CrossRef]
94. Si, H.; Chen, P.; Li, H.; Wang, X. Long non-coding RNA H19 regulates cell growth and metastasis via miR-138 in breast cancer. *Am. J. Transl. Res.* **2019**, *11*, 3213–3225.
95. Hu, X.; Liu, Y.; Du, Y.; Cheng, T.; Xia, W. Long non-coding RNA BLACAT1 promotes breast cancer cell proliferation and metastasis by miR-150-5p/CCR2. *Cell Biosci.* **2019**, *9*, 14. [CrossRef]
96. Asakura, K.; Kadota, T.; Matsuzaki, J.; Yoshida, Y.; Yamamoto, Y.; Nakagawa, K.; Takizawa, S.; Aoki, Y.; Nakamura, E.; Miura, J.; et al. A miRNA-based diagnostic model predicts resectable lung cancer in humans with high accuracy. *Commun. Biol.* **2020**, *3*, 134. [CrossRef]
97. Li, D.; Luo, Y.; Gao, Y.; Yang, Y.; Wang, Y.; Xu, Y.; Tan, S.; Zhang, Y.; Duan, J. piR-651 promotes tumor formation in non-small cell lung carcinoma through the upregulation of cyclin D1 and CDK4. *Int. J. Mol. Med.* **2016**, *38*, 927–936. [CrossRef] [PubMed]
98. Zhang, S.; Yao, J.; Shen, B.; Li, G.; Kong, S.; Bi, D.; Pan, S.; Cheng, B. Role of piwi-interacting RNA-651 in the carcinogenesis of non-small cell lung cancer. *Oncol. Lett.* **2018**, *15*, 940–946. [CrossRef] [PubMed]
99. Yao, J.; Wang, Y.W.; Fang, B.B.; Zhang, S.J.; Cheng, B.L. piR-651 and its function in 95 D lung cancer cells. *Biomed. Rep.* **2016**, *4*, 546–550. [CrossRef] [PubMed]
100. Peng, W.; Mei, J.; Feng, J. piRNA-14633 underexpression as an independent prognostic marker in non-small cell lung cancer. *J. Clin. Oncol.* **2023**, *41*, e21223. [CrossRef]
101. Dong, X.; Ding, S.; Yu, M.; Niu, L.; Xue, L.; Zhao, Y.; Xie, L.; Song, X.; Song, X. Small Nuclear RNAs (U1, U2, U5) in Tumor-Educated Platelets Are Downregulated and Act as Promising Biomarkers in Lung Cancer. *Front. Oncol.* **2020**, *10*, 1627. [CrossRef]
102. Köhler, J.; Schuler, M.; Gauler, T.C.; Nöpel-Dünnebacke, S.; Ahrens, M.; Hoffmann, A.-C.; Kasper, S.; Nensa, F.; Gomez, B.; Hahnemann, M.; et al. Circulating U2 small nuclear RNA fragments as a diagnostic and prognostic biomarker in lung cancer patients. *J. Cancer Res. Clin. Oncol.* **2015**, *142*, 795–805. [CrossRef] [PubMed]
103. Choi, H.H.; Cho, Y.-S.; Choi, J.H.; Kim, H.-K.; Kim, S.S.; Chae, H.-S. Stool-Based miR-92a and miR-144* as Noninvasive Biomarkers for Colorectal Cancer Screening. *Oncology* **2019**, *97*, 173–179. [CrossRef] [PubMed]
104. Feng, J.; Yang, M.; Wei, Q.; Song, F.; Zhang, Y.; Wang, X.; Liu, B.; Li, J. Novel evidence for oncogenic piRNA-823 as a promising prognostic biomarker and a potential therapeutic target in colorectal cancer. *J. Cell. Mol. Med.* **2020**, *24*, 9028–9040. [CrossRef]
105. Iyer, D.N.; Wan, T.M.-H.; Man, J.H.-W.; Sin, R.W.-Y.; Li, X.; Lo, O.S.-H.; Foo, D.C.-C.; Pang, R.W.-C.; Law, W.-L.; Ng, L. Small RNA Profiling of piRNAs in Colorectal Cancer Identifies Consistent Overexpression of *piR-24000* That Correlates Clinically with an Aggressive Disease Phenotype. *Cancers* **2020**, *12*, 188. [CrossRef] [PubMed]
106. Wang, Z.; Yang, H.; Ma, D.; Mu, Y.; Tan, X.; Hao, Q.; Feng, L.; Liang, J.; Xin, W.; Chen, Y.; et al. Serum PIWI-Interacting RNAs piR-020619 and piR-020450 Are Promising Novel Biomarkers for Early Detection of Colorectal Cancer. *Cancer Epidemiol. Biomark. Prev.* **2020**, *29*, 990–998. [CrossRef] [PubMed]
107. Mai, D.; Zheng, Y.; Guo, H.; Ding, P.; Bai, R.; Li, M.; Ye, Y.; Zhang, J.; Huang, X.; Liu, D.; et al. Serum piRNA-54265 is a New Biomarker for early detection and clinical surveillance of Human Colorectal Cancer. *Theranostics* **2020**, *10*, 8468–8478. [CrossRef] [PubMed]
108. Oehme, F.; Krahl, S.; Gyorffy, B.; Muessle, B.; Rao, V.; Greif, H.; Ziegler, N.; Lin, K.; Thepkaysone, M.-L.; Polster, H.; et al. Low level of exosomal long non-coding RNA HOTTIP is a prognostic biomarker in colorectal cancer. *RNA Biol.* **2019**, *16*, 1339–1345. [CrossRef]
109. Li, N.; Li, J.; Mi, Q.; Xie, Y.; Li, P.; Wang, L.; Binang, H.; Wang, Q.; Wang, Y.; Chen, Y.; et al. Long non-coding RNA ADAMTS9-AS1 suppresses colorectal cancer by inhibiting the Wnt/β-catenin signalling pathway and is a potential diagnostic biomarker. *J. Cell. Mol. Med.* **2020**, *24*, 11318–11329. [CrossRef]
110. Jiang, P.; Han, X.; Zheng, Y.; Sui, J.; Bi, W. Long non-coding RNA NKILA serves as a biomarker in the early diagnosis and prognosis of patients with colorectal cancer. *Oncol. Lett.* **2019**, *18*, 2109–2117. [CrossRef]
111. Fineberg, S.K.; Kosik, K.S.; Davidson, B.L. MicroRNAs Potentiate Neural Development. *Neuron* **2009**, *64*, 303–309. [CrossRef]
112. Wang, W.; Kwon, E.J.; Tsai, L.-H. MicroRNAs in learning, memory, and neurological diseases. *Learn. Mem.* **2012**, *19*, 359–368. [CrossRef]

113. Di Pietro, V.; Ragusa, M.; Davies, D.; Su, Z.; Hazeldine, J.; Lazzarino, G.; Hill, L.J.; Crombie, N.; Foster, M.; Purrello, M.; et al. MicroRNAs as Novel Biomarkers for the Diagnosis and Prognosis of Mild and Severe Traumatic Brain Injury. *J. Neurotrauma* **2017**, *34*, 1948–1956. [CrossRef] [PubMed]
114. Yan, J.; Bu, X.; Li, Z.; Wu, J.; Wang, C.; Li, D.; Song, J.; Wang, J. Screening the expression of several miRNAs from TaqMan Low Density Array in traumatic brain injury: miR-219a-5p regulates neuronal apoptosis by modulating CCNA2 and CACUL1. *J. Neurochem.* **2019**, *150*, 202–217. [CrossRef]
115. Dugger, B.N.; Dickson, D.W. Pathology of Neurodegenerative Diseases. *Cold Spring Harb. Perspect. Biol.* **2017**, *9*, a028035. [CrossRef]
116. Jia, Y.; Mu, J.C.; Ackerman, S.L. Mutation of a U2 snRNA Gene Causes Global Disruption of Alternative Splicing and Neurodegeneration. *Cell* **2012**, *148*, 296–308. [CrossRef]
117. Nik, S.; Bowman, T.V. Splicing and neurodegeneration: Insights and mechanisms. *Wiley Interdiscip. Rev. RNA* **2019**, *10*, e1532. [CrossRef]
118. Liu, S.; Fan, M.; Zheng, Q.; Hao, S.; Yang, L.; Xia, Q.; Qi, C.; Ge, J. MicroRNAs in Alzheimer's disease: Potential diagnostic markers and therapeutic targets. *Biomed. Pharmacother.* **2022**, *148*, 112681. [CrossRef]
119. Dong, Z.; Gu, H.; Guo, Q.; Liang, S.; Xue, J.; Yao, F.; Liu, X.; Li, F.; Liu, H.; Sun, L.; et al. Profiling of Serum Exosome MiRNA Reveals the Potential of a MiRNA Panel as Diagnostic Biomarker for Alzheimer's Disease. *Mol. Neurobiol.* **2021**, *58*, 3084–3094. [CrossRef] [PubMed]
120. Wang, J.; Chen, C.; Zhang, Y. An investigation of microRNA-103 and microRNA-107 as potential blood-based biomarkers for disease risk and progression of Alzheimer's disease. *J. Clin. Lab. Anal.* **2020**, *34*, e23006. [CrossRef] [PubMed]
121. Mancuso, R.; Agostini, S.; Hernis, A.; Zanzottera, M.; Bianchi, A.; Clerici, M. Circulatory miR-223-3p Discriminates between Parkinson's and Alzheimer's Patients. *Sci. Rep.* **2019**, *9*, 9393. [CrossRef]
122. Jain, G.; Stuendl, A.; Rao, P.; Berulava, T.; Centeno, T.P.; Kaurani, L.; Burkhardt, S.; Delalle, I.; Kornhuber, J.; Hüll, M.; et al. A combined miRNA–piRNA signature to detect Alzheimer's disease. *Transl. Psychiatry* **2019**, *9*, 250. [CrossRef] [PubMed]
123. Roy, J.; Sarkar, A.; Parida, S.; Ghosh, Z.; Mallick, B. Small RNA sequencing revealed dysregulated piRNAs in Alzheimer's disease and their probable role in pathogenesis. *Mol. Biosyst.* **2017**, *13*, 565–576. [CrossRef] [PubMed]
124. Cheng, Y.; Zhou, X.; Zou, T.; Zhang, L.; Li, L.; Yang, C.; Ma, L. Plasma long non-coding RNAs ASMTL-AS1, AP001363.1, AC005730.3 and AL133415.1 as a potential biomarker for Alzheimer's disease. *Neurol. Res.* **2023**, *45*, 804–817. [CrossRef]
125. Poewe, W.; Seppi, K.; Tanner, C.M.; Halliday, G.M.; Brundin, P.; Volkmann, J.; Schrag, A.E.; Lang, A.E. Parkinson disease. *Nat. Rev. Dis. Primers* **2017**, *3*, 17013. [CrossRef] [PubMed]
126. Shu, Y.; Qian, J.; Wang, C. Aberrant expression of microRNA-132-3p and microRNA-146a-5p in Parkinson's disease patients. *Open Life Sci.* **2020**, *15*, 647–653. [CrossRef] [PubMed]
127. Yan, J.-H.; Hua, P.; Chen, Y.; Li, L.-T.; Yu, C.-Y.; Yan, L.; Zhang, H.; He, Y.; Zheng, H.; Chen, H.; et al. Identification of microRNAs for the early diagnosis of Parkinson's disease and multiple system atrophy. *J. Integr. Neurosci.* **2020**, *19*, 429–436. [CrossRef]
128. Boros, F.A.; Maszlag-Török, R.; Vécsei, L.; Klivényi, P. Increased level of NEAT1 long non-coding RNA is detectable in peripheral blood cells of patients with Parkinson's disease. *Brain Res.* **2020**, *1730*, 146672. [CrossRef] [PubMed]
129. Fan, Y.; Li, J.; Yang, Q.; Gong, C.; Gao, H.; Mao, Z.; Yuan, X.; Zhu, S.; Xue, Z. Dysregulated Long Non-coding RNAs in Parkinson's Disease Contribute to the Apoptosis of Human Neuroblastoma Cells. *Front. Neurosci.* **2019**, *13*, 1320. [CrossRef]
130. Gurha, P. Noncoding RNAs in cardiovascular diseases. *Curr. Opin. Cardiol.* **2019**, *34*, 241–245. [CrossRef]
131. Sallam, T.; Sandhu, J.; Tontonoz, P. Long Noncoding RNA Discovery in Cardiovascular Disease: Decoding Form to Function. *Circ. Res.* **2018**, *122*, 155–166. [CrossRef]
132. Thum, T. Translational Opportunities and Challenges of Long Noncoding RNAs in Cardiac Regeneration. *J. Am. Coll. Cardiol.* **2018**, *72*, 551–552. [CrossRef] [PubMed]
133. Xuan, L.; Sun, L.; Zhang, Y.; Huang, Y.; Hou, Y.; Li, Q.; Guo, Y.; Feng, B.; Cui, L.; Wang, X.; et al. Circulating long non-coding RNAs *NRON* and *MHRT* as novel predictive biomarkers of heart failure. *J. Cell. Mol. Med.* **2017**, *21*, 1803–1814. [CrossRef]
134. Dykstra-Aiello, C.; Jickling, G.C.; Ander, B.P.; Shroff, N.; Zhan, X.; Liu, D.; Hull, H.; Orantia, M.; Stamova, B.S.; Sharp, F.R.; et al. Altered Expression of Long Noncoding RNAs in Blood After Ischemic Stroke and Proximity to Putative Stroke Risk Loci. *Stroke* **2016**, *47*, 2896–2903. [CrossRef]
135. Zhang, J.; Yuan, L.; Zhang, X.; Hamblin, M.; Zhu, T.; Meng, F.; Li, Y.; Chen, Y.; Yin, K. Altered long non-coding RNA transcriptomic profiles in brain microvascular endothelium after cerebral ischemia. *Exp. Neurol.* **2016**, *277*, 162–170. [CrossRef] [PubMed]
136. Zhang, X.; Tang, X.; Liu, K.; Hamblin, M.H.; Yin, K.-J. Long Noncoding RNA Malat1 Regulates Cerebrovascular Pathologies in Ischemic Stroke. *J. Neurosci.* **2017**, *37*, 1797–1806. [CrossRef] [PubMed]
137. Li, Z.; Li, J.; Tang, N. Long noncoding RNA Malat1 is a potent autophagy inducer protecting brain microvascular endothelial cells against oxygen-glucose deprivation/reoxygenation-induced injury by sponging miR-26b and upregulating ULK2 expression. *Neuroscience* **2017**, *354*, 1–10. [CrossRef]
138. Zhang, L.; Luo, X.; Chen, F.; Yuan, W.; Xiao, X.; Zhang, X.; Dong, Y.; Zhang, Y.; Liu, Y. LncRNA SNHG1 regulates cerebrovascular pathologies as a competing endogenous RNA through HIF-1α/VEGF signaling in ischemic stroke. *J. Cell. Biochem.* **2018**, *119*, 5460–5472. [CrossRef] [PubMed]
139. Lipps, C.; Northe, P.; Figueiredo, R.; Rohde, M.; Brahmer, A.; Krämer-Albers, E.-M.; Liebetrau, C.; Wiedenroth, C.B.; Mayer, E.; Kriechbaum, S.D.; et al. Non-Invasive Approach for Evaluation of Pulmonary Hypertension Using Extracellular Vesicle-Associated Small Non-Coding RNA. *Biomolecules* **2019**, *9*, 666. [CrossRef] [PubMed]

140. Yang, J.; Xue, F.-T.; Li, Y.-Y.; Liu, W.; Zhang, S. Exosomal piRNA sequencing reveals differences between heart failure and healthy patients. *Eur. Rev. Med. Pharmacol. Sci.* **2018**, *22*, 7952–7961. [PubMed]
141. Rajan, K.S.; Velmurugan, G.; Gopal, P.; Ramprasath, T.; Babu, D.V.; Krithika, S.; Jenifer, Y.C.; Freddy, A.; William, G.; Kalpana, K.; et al. Abundant and Altered Expression of PIWI-Interacting RNAs during Cardiac Hypertrophy. *Heart Lung Circ.* **2016**, *25*, 1013–1020. [CrossRef] [PubMed]
142. Angelescu, M.A.; Andronic, O.; Dima, S.O.; Popescu, I.; Meivar-Levy, I.; Ferber, S.; Lixandru, D. miRNAs as Biomarkers in Diabetes: Moving towards Precision Medicine. *Int. J. Mol. Sci.* **2022**, *23*, 12843. [CrossRef] [PubMed]
143. Pordzik, J.; Jakubik, D.; Jarosz-Popek, J.; Wicik, Z.; Eyileten, C.; De Rosa, S.; Indolfi, C.; Siller-Matula, J.M.; Czajka, P.; Postula, M. Significance of circulating microRNAs in diabetes mellitus type 2 and platelet reactivity: Bioinformatic analysis and review. *Cardiovasc. Diabetol.* **2019**, *18*, 113. [CrossRef] [PubMed]
144. Al-Muhtaresh, H.A.; Al-Kafaji, G. Evaluation of two-diabetes related microR NAs suitability as earlier blood biomarkers for detecting prediabetes and type 2 diabetes mellitus. *J. Clin. Med.* **2018**, *7*, 12. [CrossRef] [PubMed]
145. Kong, L.; Zhu, J.; Han, W.; Jiang, X.; Xu, M.; Zhao, Y.; Dong, Q.; Pang, Z.; Guan, Q.; Gao, L.; et al. Significance of serum microRNAs in pre-diabetes and newly diagnosed type 2 diabetes: A clinical study. *Acta Diabetol.* **2011**, *48*, 61–69. [CrossRef]
146. Karolina, D.S.; Tavintharan, S.; Armugam, A.; Sepramaniam, S.; Pek, S.L.T.; Wong, M.T.K.; Lim, S.C.; Sum, C.F.; Jeyaseelan, K. Circulating miRNA Profiles in Patients with Metabolic Syndrome. *J. Clin. Endocrinol. Metab.* **2012**, *97*, E2271–E2276. [CrossRef]
147. Zhang, T.; Lv, C.; Li, L.; Chen, S.; Liu, S.; Wang, C.; Su, B. Plasma miR-126 Is a Potential Biomarker for Early Prediction of Type 2 Diabetes Mellitus in Susceptible Individuals. *BioMed Res. Int.* **2013**, *2013*, 761617. [CrossRef] [PubMed]
148. Zhang, T.; Li, L.; Shang, Q.; Lv, C.; Wang, C.; Su, B. Circulating miR-126 is a potential biomarker to predict the onset of type 2 diabetes mellitus in susceptible individuals. *Biochem. Biophys. Res. Commun.* **2015**, *463*, 60–63. [CrossRef] [PubMed]
149. Liu, Y.; Gao, G.; Yang, C.; Zhou, K.; Shen, B.; Liang, H.; Jiang, X. The Role of Circulating MicroRNA-126 (miR-126): A Novel Biomarker for Screening Prediabetes and Newly Diagnosed Type 2 Diabetes Mellitus. *Int. J. Mol. Sci.* **2014**, *15*, 10567–10577. [CrossRef]
150. Ghorbani, S.; Mahdavi, R.; Alipoor, B.; Panahi, G.; Esfahani, E.N.; Razi, F.; Taghikhani, M.; Meshkani, R. Decreased serum microRNA-21 level is associated with obesity in healthy and type 2 diabetic subjects. *Arch. Physiol. Biochem.* **2017**, *124*, 300–305. [CrossRef]
151. Barbosa, T.d.C.; Ingerslev, L.R.; Alm, P.S.; Versteyhe, S.; Massart, J.; Rasmussen, M.; Donkin, I.; Sjögren, R.; Mudry, J.M.; Vetterli, L.; et al. High-fat diet reprograms the epigenome of rat spermatozoa and transgenerationally affects metabolism of the offspring. *Mol. Metab.* **2015**, *5*, 184–197. [CrossRef]
152. Henaoui, I.S.; Jacovetti, C.; Mollet, I.G.; Guay, C.; Sobel, J.; Eliasson, L.; Regazzi, R. PIWI-interacting RNAs as novel regulators of pancreatic beta cell function. *Diabetologia* **2017**, *60*, 1977–1986. [CrossRef] [PubMed]
153. Shah, R.; Murthy, V.; Pacold, M.; Danielson, K.; Tanriverdi, K.; Larson, M.G.; Hanspers, K.; Pico, A.; Mick, E.; Reis, J.; et al. Extracellular RNAs Are Associated With Insulin Resistance and Metabolic Phenotypes. *Diabetes Care* **2017**, *40*, 546–553. [CrossRef] [PubMed]
154. Hossian, A.K.M.N.; Sajib, S.; Tullar, P.E.; Mikelis, C.M.; Mattheolabakis, G. Multipronged activity of combinatorial miR-143 and miR-506 inhibits Lung Cancer cell cycle progression and angiogenesis in vitro. *Sci. Rep.* **2018**, *8*, 10495. [CrossRef]
155. Hossian, A.K.M.N.; Mackenzie, G.G.; Mattheolabakis, G. miRNAs in gastrointestinal diseases: Can we effectively deliver RNA-based therapeutics orally? *Nanomedicine* **2019**, *14*, 2873–2889. [CrossRef]
156. Su, Z.; Yang, Z.; Xu, Y.; Chen, Y.; Yu, Q. MicroRNAs in apoptosis, autophagy and necroptosis. *Oncotarget* **2015**, *6*, 8474–8490. [CrossRef]
157. Schaefer, J.S. MicroRNAs: How many in inflammatory bowel disease? *Curr. Opin. Gastroenterol.* **2016**, *32*, 258–266. [CrossRef] [PubMed]
158. Orang, A.V.; Barzegari, A. MicroRNAs in Colorectal Cancer: From Diagnosis to Targeted Therapy. *Asian Pac. J. Cancer Prev.* **2014**, *15*, 6989–6999. [CrossRef]
159. Krishna, C.V.; Singh, J.; Thangavel, C.; Rattan, S. Role of microRNAs in gastrointestinal smooth muscle fibrosis and dysfunction: Novel molecular perspectives on the pathophysiology and therapeutic targeting. *Am. J. Physiol. Liver Physiol.* **2016**, *310*, G449–G459. [CrossRef]
160. Ma, F.; Xu, S.; Liu, X.; Zhang, Q.; Xu, X.; Liu, M.; Hua, M.; Li, N.; Yao, H.; Cao, X. The microRNA miR-29 controls innate and adaptive immune responses to intracellular bacterial infection by targeting interferon-γ. *Nat. Immunol.* **2011**, *12*, 861–869. [CrossRef]
161. Fasseu, M.; Tréton, X.; Guichard, C.; Pedruzzi, E.; Cazals-Hatem, D.; Richard, C.; Aparicio, T.; Daniel, F.; Soulé, J.-C.; Moreau, R.; et al. Identification of Restricted Subsets of Mature microRNA Abnormally Expressed in Inactive Colonic Mucosa of Patients with Inflammatory Bowel Disease. *PLoS ONE* **2010**, *5*, e13160. [CrossRef]
162. Palmieri, O.; Creanza, T.M.; Bossa, F.; Latiano, T.; Corritore, G.; Palumbo, O.; Martino, G.; Biscaglia, G.; Scimeca, D.; Carella, M.; et al. Functional Implications of MicroRNAs in Crohn's Disease Revealed by Integrating MicroRNA and Messenger RNA Expression Profiling. *Int. J. Mol. Sci.* **2017**, *18*, 1580. [CrossRef]
163. Cheng, X.; Zhang, X.; Su, J.; Zhang, Y.; Zhou, W.; Zhou, J.; Wang, C.; Liang, H.; Chen, X.; Shi, R.; et al. miR-19b downregulates intestinal SOCS3 to reduce intestinal inflammation in Crohn's disease. *Sci. Rep.* **2015**, *5*, 10397. [CrossRef]
164. Schaefer, J.S.; Attumi, T.; Opekun, A.R.; Abraham, B.; Hou, J.; Shelby, H.; Graham, D.Y.; Streckfus, C.; Klein, J.R. MicroRNA signatures differentiate Crohn's disease from ulcerative colitis. *BMC Immunol.* **2015**, *16*, 5. [CrossRef]
165. Liu, T.; Liu, D.; Guan, S.; Dong, M. Diagnostic role of circulating MiR-21 in colorectal cancer: A update meta-analysis. *Ann. Med.* **2020**, *53*, 87–102. [CrossRef] [PubMed]

166. Masuda, T.; Hayashi, N.; Kuroda, Y.; Ito, S.; Eguchi, H.; Mimori, K. MicroRNAs as Biomarkers in Colorectal Cancer. *Cancers* **2017**, *9*, 124. [CrossRef] [PubMed]
167. Duan, J.; Zhang, H.; Qu, Y.; Deng, T.; Huang, D.; Liu, R.; Zhang, L.; Bai, M.; Zhou, L.; Ying, G.; et al. Onco-miR-130 promotes cell proliferation and migration by targeting TGFβR2 in gastric cancer. *Oncotarget* **2016**, *7*, 44522–44533. [CrossRef]
168. Wu, J.-H.; Yao, Y.-L.; Gu, T.; Wang, Z.-Y.; Pu, X.-Y.; Sun, W.-W.; Zhang, X.; Jiang, Y.-B.; Wang, J.-J. MiR-421 Regulates Apoptosis of BGC-823 Gastric Cancer Cells by Targeting Caspase-3. *Asian Pac. J. Cancer Prev.* **2014**, *15*, 5463–5468. [CrossRef]
169. Girard, M.; Jacquemin, E.; Munnich, A.; Lyonnet, S.; Henrion-Caude, A. miR-122, a paradigm for the role of microRNAs in the liver. *J. Hepatol.* **2008**, *48*, 648–656. [CrossRef]
170. Roy, S.; Trautwein, C.; Luedde, T.; Roderburg, C. A General Overview on Non-coding RNA-Based Diagnostic and Therapeutic Approaches for Liver Diseases. *Front. Pharmacol.* **2018**, *9*, 805. [CrossRef]
171. Doghish, A.S.; Elballal, M.S.; Elazazy, O.; Elesawy, A.E.; Elrebehy, M.A.; Shahin, R.K.; Midan, H.M.; Sallam, A.-A.M. The role of miRNAs in liver diseases: Potential therapeutic and clinical applications. *Pathol.-Res. Pract.* **2023**, *243*, 154375. [CrossRef]
172. Schueller, F.; Roy, S.; Vucur, M.; Trautwein, C.; Luedde, T.; Roderburg, C. The Role of miRNAs in the Pathophysiology of Liver Diseases and Toxicity. *Int. J. Mol. Sci.* **2018**, *19*, 261. [CrossRef]
173. Zhao, E.; Keller, M.P.; Rabaglia, M.E.; Oler, A.T.; Stapleton, D.S.; Schueler, K.L.; Neto, E.C.; Moon, J.Y.; Wang, P.; Wang, I.-M.; et al. Obesity and genetics regulate microRNAs in islets, liver, and adipose of diabetic mice. *Mamm. Genome* **2009**, *20*, 476–485. [CrossRef]
174. Albano, G.D.; Gagliardo, R.; Montalbano, A.M.; Profita, M. Non-Coding RNAs in Airway Diseases: A Brief Overview of Recent Data. *Cancers* **2022**, *15*, 54. [CrossRef] [PubMed]
175. Maes, T.; Cobos, F.A.; Schleich, F.; Sorbello, V.; Henket, M.; De Preter, K.; Bracke, K.R.; Conickx, G.; Mesnil, C.; Vandesompele, J.; et al. Asthma inflammatory phenotypes show differential microRNA expression in sputum. *J. Allergy Clin. Immunol.* **2016**, *137*, 1433–1446. [CrossRef] [PubMed]
176. Kho, A.T.; McGeachie, M.J.; Moore, K.G.; Sylvia, J.M.; Weiss, S.T.; Tantisira, K.G. Circulating microRNAs and prediction of asthma exacerbation in childhood asthma. *Respir. Res.* **2018**, *19*, 128. [CrossRef] [PubMed]
177. Yang, Y.; Sun, Z.; Ren, T.; Lei, W. Differential Expression of lncRNA CASC2 in the Serum of Childhood Asthma and Its Role in Airway Smooth Muscle Cells Proliferation and Migration. *J. Asthma Allergy* **2022**, *15*, 197–207. [CrossRef]
178. Xia, L.; Wang, X.; Liu, L.; Fu, J.; Xiao, W.; Liang, Q.; Han, X.; Huang, S.; Sun, L.; Gao, Y.; et al. lnc-BAZ2B promotes M2 macrophage activation and inflammation in children with asthma through stabilizing BAZ2B pre-mRNA. *J. Allergy Clin. Immunol.* **2021**, *147*, 921–932.e9. [CrossRef]
179. Zhuang, Y.; Hobbs, B.D.; Hersh, C.P.; Kechris, K. Identifying miRNA-mRNA Networks Associated with COPD Phenotypes. *Front. Genet.* **2021**, *12*, 748356. [CrossRef]
180. Wang, L.; Zhao, H.; Raman, I.; Yan, M.; Chen, Q.; Li, Q.-Z. Peripheral Blood Mononuclear Cell Gene Expression in Chronic Obstructive Pulmonary Disease: miRNA and mRNA Regulation. *J. Inflamm. Res.* **2022**, *15*, 2167–2180. [CrossRef]
181. Thind, A.S.; Thind, A.S.; Monga, I.; Monga, I.; Thakur, P.K.; Thakur, P.K.; Kumari, P.; Kumari, P.; Dindhoria, K.; Dindhoria, K.; et al. Demystifying emerging bulk RNA-Seq applications: The application and utility of bioinformatic methodology. *Briefings Bioinform.* **2021**, *22*, bbab259. [CrossRef]
182. Lekka, E.; Hall, J. Noncoding RNAs in disease. *FEBS Lett.* **2018**, *592*, 2884–2900. [CrossRef] [PubMed]
183. Sauter, E.R. Reliable Biomarkers to Identify New and Recurrent Cancer. *Eur. J. Breast Health* **2017**, *13*, 162–167. [CrossRef] [PubMed]
184. Ho, P.T.B.; Clark, I.M.; Le, L.T.T. MicroRNA-Based Diagnosis and Therapy. *Int. J. Mol. Sci.* **2022**, *23*, 7167. [CrossRef] [PubMed]
185. Sun, Y.-M.; Chen, Y.-Q. Principles and innovative technologies for decrypting noncoding RNAs: From discovery and functional prediction to clinical application. *J. Hematol. Oncol.* **2020**, *13*, 109. [CrossRef] [PubMed]
186. Slatko, B.E.; Gardner, A.F.; Ausubel, F.M. Overview of Next-Generation Sequencing Technologies. *Curr. Protoc. Mol. Biol.* **2018**, *122*, e59. [CrossRef]
187. Li, S.; Teng, S.; Xu, J.; Su, G.; Zhang, Y.; Zhao, J.; Zhang, S.; Wang, H.; Qin, W.; Lu, Z.J.; et al. Microarray is an efficient tool for circRNA profiling. *Briefings Bioinform.* **2018**, *20*, 1420–1433. [CrossRef]
188. Chen, Z.; Halford, N.G.; Liu, C. Real-Time Quantitative PCR: Primer Design, Reference Gene Selection, Calculations and Statistics. *Metabolites* **2023**, *13*, 806. [CrossRef]
189. Lu, T.X.; Rothenberg, M.E. MicroRNA. *J. Allergy Clin. Immunol.* **2018**, *141*, 1202–1207. [CrossRef]
190. Ye, J.; Xu, M.; Tian, X.; Cai, S.; Zeng, S. Research advances in the detection of miRNA. *J. Pharm. Anal.* **2019**, *9*, 217–226. [CrossRef]
191. Ouyang, T.; Liu, Z.; Han, Z.; Ge, Q. MicroRNA Detection Specificity: Recent Advances and Future Perspective. *Anal. Chem.* **2019**, *91*, 3179–3186. [CrossRef]

International Journal of
Molecular Sciences

Review

Exploring the Feasibility of Circulating miRNAs as Diagnostic and Prognostic Biomarkers in Osteoarthritis: Challenges and Opportunities

Kyriacos Felekkis, Myrtani Pieri and Christos Papaneophytou *

Department of Life Sciences, School of Life and Health Sciences, University of Nicosia, 46 Makedonitissas Avenue, Nicosia 2417, Cyprus; felekkis.k@unic.ac.cy (K.F.); pieri.m@unic.ac.cy (M.P.)
* Correspondence: papaneophytou.c@unic.ac.cy; Tel.: +357-22841941

Abstract: Osteoarthritis (OA) is a prevalent degenerative joint disease characterized by progressive cartilage degradation and joint inflammation. As the most common aging-related joint disease, OA is marked by inadequate extracellular matrix synthesis and the breakdown of articular cartilage. However, traditional diagnostic methods for OA, relying on clinical assessments and radiographic imaging, often need to catch up in detecting early-stage disease or i accurately predicting its progression. Consequently, there is a growing interest in identifying reliable biomarkers that can facilitate early diagnosis and prognosis of OA. MicroRNAs (miRNAs) have emerged as potential candidates due to their involvement in various cellular processes, including cartilage homeostasis and inflammation. This review explores the feasibility of circulating miRNAs as diagnostic and prognostic biomarkers in OA, focusing on knee OA while shedding light on the challenges and opportunities associated with their implementation in clinical practice.

Keywords: osteoarthritis; circulating miRNAs; diagnosis; prevention

1. Introduction

Osteoarthritis (OA) is a degenerative joint disease and the leading cause of disability in adults, causing significant pain, functional limitations, and impairments in quality of life [1]. While OA can affect nearly any joint in the body, it commonly affects the hands, knees, hips, and feet [2]. The pathological features of OA involve various joint structures, including cartilage, bone, synovium, ligament, muscle, and periarticular fat, resulting in joint dysfunction, pain, stiffness, functional limitations, and loss of mobility [3]. Primary clinical symptoms include chronic pain, joint instability, stiffness, joint deformities, and radiographic joint space narrowing [4].

The severity of OA is evaluated through clinical examination of symptoms like pain, joint stiffness, and restrictions in normal functioning, as well as through imaging tests such as X-rays or magnetic resonance images (MRIs), using the Kellgren and Lawrence (KL) grading system [5]. The KL scale ranges from 0 to 4, where 0 represents a healthy joint with no OA signs, and grades 1 through 4 show progressively worsening disease severity, with a score above 2 indicating the presence of OA. However, it is essential to note that radiographic imaging mainly captures late-stage changes in bone and cartilage [6].

The activity of chondrocytes regulates the balance between anabolic and catabolic processes in articular cartilage homeostasis. Altered joint loading due to malalignment or trauma can tip this balance towards catabolic activities, leading to joint abnormalities such as synovitis, cartilage degradation, subchondral bone sclerosis, and osteophyte formation, collectively referred to as OA [7,8]. The etiology of OA is complex and multifactorial, involving age, sex, obesity, joint trauma, environmental factors, and genetic factors, with the latter being one of the best-known contributors to OA pathogenesis [9].

Existing OA therapies aim to alleviate symptoms like pain, offering only modest, long-term benefits and leaving many patients dissatisfied with their effectiveness [10]. As mentioned above, the diagnosis of OA relies on radiographic criteria and clinical symptoms [11], with radiography remaining the primary imaging modality for diagnosis and followup of OA [12]. However, limitations of radiography have sparked interest in alternate parameters for monitoring OA, potentially serving as biomarkers in drug development [13].

Biochemical markers in blood, urine, or synovial fluid (SF) samples reflecting dynamic changes in joint remodeling and disease progression have emerged as alternatives [14]. In OA, a biochemical marker may function as a molecule, directly causing joint damage, a byproduct of such damage, or both [15]. For instance, fragments of cartilage extracellular matrix like hyaluronan can act in dual roles: they indicate the disease and instigate the ongoing immune-mediated wound-healing response in joints affected by osteoarthritis [16].

Recent studies also underscore the regulatory roles of non-coding RNAs (ncRNAs) in OA development [17]. Different classes of ncRNAs, including long ncRNAs (lncRNAs), microRNAs (miRNAs or miRs), and circular RNAs (circRNAs), have been implicated in the development of OA [17]. Specifically, these ncRNAs can influence the expression of genes that regulate inflammation, cell death, and matrix degradation in chondrocytes, thereby impacting the progression of OA [17,18].

Furthermore, circulating ncRNAs (in biological fluids/plasma/serum) exhibit high potential as biomarkers for OA due to the easy accessibility of the sample [19]. Interestingly, the primary roles of the lncRNAs and circRNAs are to function as molecular absorbers for miRNAs. Together, these ncRNAs impact the proliferation and apoptosis of chondrocytes, inflammatory reactions, and the breakdown of the extracellular matrix [17]. Nevertheless, miRNAs in the circulation (circulating miRNAs/c-miRNAs) have attracted particular attention due to their unique stability among the different classes of ncRNA [20].

Despite the identification of several genetic (e.g., *COL6A3* and *ACTG1* genes, DNA methylation, etc.) [21,22] and biochemical biomarkers (e.g., proinflammatory cytokines, C-telopeptide fragments of type II collagen, hyaluronan, etc.) [23,24], miRNAs offer a unique advantage as biomarkers for OA. Compared to genetic markers, which are "static", miRNAs provide a dynamic and real-time view of disease states, reflecting the complex and ever-changing regulatory networks within the disease pathology [25]. Unlike protein biomarkers, which often struggle to find accurate detection methods due to low concentrations in several cases [26], miRNAs can be more easily detected and quantified using advanced techniques such as RT-qPCR and next-generation sequencing (NGS) [27]. MiRNAs also offer the advantage of potentially serving as components in multimarker models, enhancing precision in diagnosis, treatment guidance, and assessment of treatment response. In contrast to analyzing multiple protein markers, which can be costly and time-intensive, employing panels that include a variety of miRNAs may furnish a more cost-effective and non-invasive approach for diagnosing and forecasting the progression of the disease [28]. Notably, the stability of miRNAs in various body fluids enables non-invasive monitoring of several diseases, including OA, while their high sensitivity and specificity to particular pathological conditions present opportunities for more targeted diagnosis and therapy [28].

This review explores the current status, future perspectives, and limitations of c- miRNAs as prognostic and diagnostic biomarkers for OA, focusing on knee OA (KOA). We searched Pubmed, Scopus, and Science Direct databases for publications up to May 2023. "Osteoarthritis", "non-coding RNAs", "circulating non-coding RNAs", "micro-RNAs", "circulating miRNAs", "biomarkers", "prognosis", "diagnosis", and their respective medical subject headings (MeSH) terms were used as keywords. We also selected studies on c-miRNAs linked to OA. Only studies (both basic and clinical) published in English were included in this review.

2. MicroRNAs: Basic Concepts

2.1. MicroRNA Biogenesis and Biological Roles

MiRNAs are small non-coding RNA strands, approximately 20–25 nucleotides in length, that act as key post-transcriptional regulators of gene expression. They function by binding to the 3′-untranslated regions of target mRNAs, leading to their degradation or inhibition of translation [29,30]. They originate in the nucleus, transcribed by either RNA polymerase II or III into a lengthy precursor molecule called primary miRNA (pri-miRNA) [31]. This pri-miRNA undergoes processing by the Drosha-DGCR8 complex, an RNase III-type enzyme, leading to the formation of a roughly 70–100 nucleotide precursor known as pre-miRNA [29,32,33]. This pre-miRNA is subsequently exported into the cytoplasm through exportin 5 [34,35]. The ribonuclease Dicer processes pre-miRNA in the cytoplasm, resulting in a miRNA duplex of about 22 nucleotides [36,37]. One strand from this duplex is then integrated into the RNA-induced silencing complex (RISC), which is guided to target mRNA molecules [38–40]. MiRNAs play a crucial role in the post-transcriptional regulation of gene expression, targeting nearly 30% of the protein-coding genes by creating a complementary base pair with the 3′-untranslated region (UTR) of the target mRNA. This interaction results in translational suppression or direct mRNA degradation [41]. So far, over 1600 human miRNAs have been identified (reviewed in [30]), and their significant contributions to a diverse range of developmental processes and pathologies have been widely recognized [42]. For example, miRNA-21 has been implicated in the development of both neoplastic and non-neoplastic diseases. Specifically, the downregulation of miR-21 has been observed to increase the rate of cell death by targeting genes involved in inflammation and cancer, such as HIF-1α, PTEN, and PDCD4 [43,44]. Conversely, miR-21 has also increased cell migration by targeting TPM1 and PDCD4 in neoplastic diseases [45]. These findings highlight the multifaceted roles that miR-21 plays in various pathological conditions, demonstrating its potential as both a diagnostic and therapeutic target. Additionally, miR-206 has been shown to affect muscle cell function significantly. Specifically, the overexpression of miR-206 suppressed muscle cell proliferation and induced cell cycle arrest in the G0/G1 phase. This effect was mediated by inhibition of the glucose-6-phosphate dehydrogenase (G6PD) gene, highlighting the precise regulatory role of miR-206 in cellular growth and development [46]. In addition to skeletal muscle development, miR-206 acts as a tumor suppressor by regulating the expression of different genes, such as *Sox9* [47], *Cyclin D2* [48], *Notch 3* [49], and *FMNL2* [50], in various tissues. miRNA-206 is also downregulated and suppresses cell proliferation by targeting Forkhead box protein 1 (FOXP1) in brain gliomas [51].

Some miRNAs synthesized within cells can be actively secreted into the bloodstream, thereby increasing the concentration of these corresponding miRNAs. As previously mentioned, these miRNAs are known as circulating miRNAs (c-miRNAs). These c-miRNAs can then reach receptor cells and perform distal regulatory functions. Present in virtually all biological fluids, c-miRNAs are gaining recognition as potential non-invasive biomarkers for numerous human pathologies, including cancer, cardiovascular diseases, and infectious diseases [52]. Although most miRNAs are located intracellularly, these c-miRNAs found in plasma, serum, and other body fluids demonstrate stable existence despite the presence of RNases [52]. Remarkably, they withstand even severe conditions, such as repeated boiling, extreme pH values, freezing–thawing cycles, and room temperature storage, ensuring their reliability [53].

Intriguingly, c-miRNAs are primarily found within exosomes, accounting for about 83–99% of the total miRNA population detected in circulation [54]. These exosomal miRNAs include AGO2-bound miRNAs, free miRNAs, and mature miRISCs [55]. Exosomes are extracellular vesicles (Evs) (ranging from 40 to 100 nm in diameter) that play an essential role in the transport and stability of miRNAs in biological fluids, including blood, urine, and saliva [56,57]. These specific EVs are lipid bilayer-enclosed structures actively secreted by cells into the extracellular environment. Packaging various molecules, including proteins lipids, messenger RNAs, and ncRNAs, in exosomes protects them from enzymatic degradation [58]. Furthermore, it facilitates their transport to recipient cells. Due to specific proteins on the surface of exosomes, these EVs engage with target cells through a ligand–receptor interaction, leading to endocytosis. This process enables the delivery of encapsulated materials from one cell to another [59]. Exosomes play a role as mediators of cell-to-cell communication, exhibiting pleiotropic activities and participating in homeostasis regulation. Exosomal miRNAs involve various physiological and pathological processes, including inflammation, tissue repair, immune response, and cancer progression (for a review on the topic, see Ref. [60]). Notably, miRNAs are the cargo of exosomes that have attracted specific attention [61]. Encapsulation of miRNAs within exosomes enables precise targeting and cell-to-cell communication [62]. In musculoskeletal diseases, including osteoarthritis and osteoporosis, exosomes derived from host cells and gut microbiota have been explored as potential nanocarriers for targeted therapy [63].

Another key factor contributing to their stability is that both intravesicular and extravesicular c-miRNAs are frequently protein-bound. Research has indicated that about 90% of c-miRNAs are associated with RNA-binding proteins (RBPs), such as the AGO2 protein [64], high-density lipoprotein (HDL) [65], and low-density lipoprotein (LDL) [66], with most circulating miRNAs being transported as AGO2-bound miRNAs [64].

The dysregulation of miRNAs in various diseases, including OA, signifies their potential as diagnostic and prognostic biomarkers [20,67]. Regarding OA, altered expression of miRNAs can influence key pathological processes like cartilage degradation and inflammation, making them strong candidates for non-invasive biomarkers [68]. In addition to their diverse roles in regulating biological processes, miRNAs have pivotal roles in OA progression, and abnormal expressions of specific miRNAs can reflect pathological changes in OA [69] (discussed further below).

2.2. The Roles of miRNAs in Bone Development

Bone development begins with the establishment of mesenchymal stem cell (MSC) condensations that prefigure the shape, size, and location of mature bone elements [9]. Recent studies have elucidated that miRNAs participate in different stages of bone growth and development [10,11], and reviews have highlighted their role in regulating bone formation [70,71]. Several in vivo studies using animal models have shed light on the role of miRNAs in bone formation. For example, the expression of members of the miR-34 family (miRs-34a, b, and c) was found to increase during osteoblast differentiation of murine calvarial cells [72]. These miRNAs inhibited osteoblast proliferation by suppressing cyclin D1, CDK4, and CDK6 accumulation, as well as terminal differentiation, at least partly through the inhibition of SATB2, a nuclear matrix protein critical for osteoblast differentiation [72]. Deletion of miR-34b and miR-34c in Col1a1-producing cells in mice led to increased bone mass during embryonic development and postnatally, while the opposite was observed when miR-34c was overexpressed [72]. The essential role of miR-206 in bone development, particularly in osteoblast differentiation, was highlighted by Inose and coworkers [73]. They demonstrated that the expression of miR-206 in osteoblasts decreased over the course of differentiation and that overexpression inhibited differentiation while knockdown promoted it.

Additionally, the deletion of the miR-181 family in mice resulted in smaller sizes among survivors [74]. Another study revealed that miR-182 is a crucial osteoclastogenic regulator in bone homeostasis, inhibiting osteoclastogenesis in vitro and affecting trabecular bone mass in vivo [75]. Furthermore, miR-21 appeared to enhance both osteogenesis and osteoclastogenesis [76,77]. Its global knockout in mice did not appear to affect bone development. Still, it did promote trabecular bone mass with age and prevented OVX-induced bone loss during aging, partially due to suppressed osteoclast function [78]. Crosstalk mediated by miRNAs in bone and exosomes containing miRNAs derived from differentiated bone cells can regulate other cell types involved in bone formation and turnover [21,22]. For example, miR-23a-5p from osteoclast-derived exosomes can suppress osteoblast differentiation, in part, by targeting Runx2 [79].

2.3. The Roles of miRNAs in OA Pathogenesis

The vital role of miRNAs in the development of the musculoskeletal system and their contribution to OA pathology was first brought to light through genetically modified animal models [80]. These models underwent either a knockout or overexpression of key genes involved in miRNA biogenesis and processing. Focusing specifically on chondrocyte homeostasis, studies using tissue-specific knockout mice have underscored the significance of miRNA processing enzymes [80]. A particular example is Dicer, a crucial enzyme for miRNA biogenesis; its deficiency in Col2α1-expressing cells leads to early postnatal fatality and aberrant skeletal growth due to decreased proliferating chondrocytes [81]. In detail, these results were obtained using a Dicer-null mouse model, which displayed diminished proliferating chondrocytes in its growth plates and hastened differentiation into a hypertrophic type. These alterations led to significant skeletal growth anomalies and premature death, thereby illuminating the critical role of miRNA in chondrocytes and, by extension, in OA [81]. Moreover, miRNAs have a crucial function in cartilage formation and, consequently, significantly influence the pathophysiology of OA. Specifically, variations in the expression levels of specific miRNAs can disrupt the balance in chondrocyte function by impacting apoptosis, senescence, and autophagy and can enhance the level of inflammation and cartilage matrix breakdown by promoting the production of proinflammatory agents (like TNF-α, IL-1, IL-6, COX-2, NO, and ROS) and matrix metalloproteinases (MMPs) [82].

Furthermore, it has been demonstrated that some miRNAs can regulate the expression of inflammation factors in OA, such as miR-146a [83], miR-142-3p [84], and miR-130a [85]. Notably, the expression of miR-140 is unique to cartilage; however, its function in growth and/or tissue preservation is still uncertain [86]. In contrast, miR-146a displays increased expression levels in OA cartilage compared to healthy cartilage [83]. Interestingly, miR-146a-5p was found to be significantly overexpressed in OA patients both in the articular cartilage tissue and in serum, with a positive correlation observed between levels in both sample types. These data imply that serum levels of miR-146a-5p can mirror the molecular activities in the cartilage, suggesting its potential use as a clinical biomarker for the management of OA [87]. Furthermore, miR-146a has garnered significant interest due to the consistency of reports of its deregulation in OA across multiple studies. For instance, in the case of rheumatoid arthritis (RA), miR-146 was observed to be overexpressed in synovial tissue following stimulation with inflammatory cytokines tumor necrosis factor-alpha (TNF-α) and interleukin-1β (IL-1β) [88]. It was previously demonstrated that miR-146a targets TRAF6 and IRAK1, playing a significant role in inflammation control [89].

Song et al. [90] demonstrated that miR-21 was suppressed in OA patients and that the modulation of miR-21 influenced apoptosis and autophagy of OA chondrocytes. Interestingly, growth-arrest-specific 5 (GAS5), a long ncRNA, plays a pivotal role in controlling miR-21 during the progression of osteoarthritis [90]. It has been suggested that GAS5 contributes to the pathogenesis of OA by acting as a sponge for miR-21, thereby influencing cell survival. Nevertheless, the potential interaction network between miR-21 and GAS5 remains inconclusive. It is worth mentioning that miR-21 has drawn attention due to its significant role in developing and spreading tumors, particularly in cell proliferation and differentiation processes essential for chondrogenesis and cartilage remodeling [91]. Despite this, the specific role and underlying mechanism of miR-21 in OA progression remain uncertain. In addition, GDF-5 functions as a signaling molecule for chondrogenesis, facilitating chondrocyte differentiation [92].

Moreover, several recent reviews have highlighted the contribution of miRNAs in normal conditions of the articular cartilage (AC) (osteoclastogenesis, osteoblastogenesis, and chondrogenesis), as well as in pathological conditions (cartilage degradation, synovial inflammation, and OA progression) [93,94]. Numerous miRNAs, including miR-101 [95], miR-181c [96], miR-675 [97], miR-770 [98], miR-140 [86], miR-146a [99], miR-27a [100], miR-122 [101], miR-130a [85], miR-15a [102], and miR-127-5P [103] have been reported to be engaged in OA pathogenesis through regulation of cartilage homeostasis, chondrocyte metabolism, and inflammatory responses, as well as proteolytic enzyme activity. A recent meta analysis of 191 miRNAs by Liu and coworkers [104] revealed that several miRNAs consistently exhibited altered expression in OA. Remarkably, miR-146a-5p, miR-34a-5p, miR-127-5p, and miR-140-5p emerged as prominent candidates with potential diagnostic and prognostic value. The study also highlighted the significance of downstream effectors, including mesenchymal stem cells and transforming growth factor-β, regulated by miRNAs in OA progression. These findings underscore the importance of miRNA signaling and provide insights into potential biomarkers and therapeutic targets for OA. For example, miR-140 targets ADAMTS-5 and has been implicated in chondrogenesis, providing an essential link between miRNA regulation and cartilage homeostasis [105]. Other recent reviews underscore the significant role of miRNAs in the onset and progression of OA. In one such study, Boehme and Ralauffs [106] emphasized the importance of 27 miRNAs or miRNA families, which exhibit significant upregulation or downregulation in human AC of OA patients compared to normal AC. The authors discussed how these differentially regulated miRNAs interfere with the downstream signaling of inflammatory cytokines, fibroblast growth factor 2 (FGF2), and transforming growth factor β (TGF-β), displaying significant interplay at the protein level. In the context of OA AC, procatabolic target gene expression is facilitated by NF-κB, MAPK, and SMAD signaling, while proinflammatory targets are activated downstream of MAPK and NF-κ B signaling. Interestingly, although cell proliferation is dominant during the early stages of OA, catabolic proteins are simultaneously induced, which degrade the extracellular matrix (ECM).

The essential role of FGF signaling in correctly forming and maintaining articular cartilage is evident, as irregular FGF signaling can lead to joint malformation and OA initiation and advancement [107]. Research involving FGF2 transduction in human knee articular cartilage samples has illustrated how it can emulate the early stages of OA by triggering cell proliferation, influenced by miRNAs such as miR-140, which has been shown to regulate Wnt signaling, often interconnected with FGF signaling [86,108]. Additionally, the TGF-β superfamily, comprising 42 human growth factors and categorized into two distinct subfamilies, has garnered considerable interest due to its multifaceted roles in the physiological conditions of joints and the onset of OA [109]. The subfamilies are the TGF-β/activin/nodal and the BMP/GDF/MIS branches [110]. The impact of TGF-β on numerous cellular functions, including cell growth, is paramount, and any disruptions in its signaling pathways could have substantial repercussions for OA [111]. MiR-146a has been implicated in modulating TGF-β signaling in OA by targeting SMAD4 [112], and miR-21 has been shown to influence MSC osteogenic differentiation through the TGF-β signaling pathways [113]. Furthermore, bone regeneration relies on the activation and transformation of MSCs, which fulfill essential immunomodulatory and regenerative functions by evolving into osteoblastic cells that form and mineralize the bone matrix [114]. The role of miRNAs in guiding the development process, including the function of MSCs in bone formation, has been emphasized in recent research, with miR-21 playing a significant role in MSC differentiation [115].

Table 1 shows some examples of miRNAs that have been linked to the development of OA. These miRNAs can be attributed to tissue damage due to direct cartilage injury, persistent inflammation, apoptosis, or even cells affected during the OA process that could potentially release these miRNAs [20]. These miRNAs have been identified in tissues and/or biological fluids using different methods and protocols.

Table 1. Examples of miRNAs implicated in the development of osteoarthritis.

miRNA	Contribution in OA	Studies on the Role of miRNA in OA [a]			
		Source/ Extraction	**Detection/ Normalization**	**Main Findings/Regulation**	**Ref**
9	Involved in chondrocyte hypertrophy and matrix degradation	Bone and cartilage/ TRIzol	RT-qPCR/ 18S RNA	Upregulated in OA bone and cartilage tissue	[116]
	Involved in the development of KOA through the NF-kB1 pathway in chondrocytes	KOA [b] cartilage/ TRIzol	RT-qPCR/ U6 snRNA	Downregulated in KOA compared to the control group	[117]
16	Involved in cartilage homeostasis and structure	**Plasma**/ miRNeasy kit	RT-qPCR/ MammU6s	Upregulated in KOA patients	[118]
		Articular cartilage from hip or KOA/N.A. [c]	Northern Blot & RT-qPCR/U6 snRNA	Upregulated in chondrocytes of OA (KL [d] grades 3 and 4)	[119]
		SF [e] and plasma/ Phenol chloroform and High Pure miRNA Isolation Kit	RT-qPCR/ cel-miR-39	Downregulated in KOA patients	[120]
16-5p	Controls the development of osteoarthritis by targeting SMAD3 in chondrocytes	Cartilage/ TRIzol	RT-qPCR/ U6 snRNA	Upregulated	[121]
21	• Influences apoptosis and autophagy of OA chondrocytes • Suppresses chondrogenesis by directly targeting growth differentiation factor 5 • Contributes to catabolic NF-κB signaling and MMP activation in response to inflammatory cytokines	Articular chondrocytes/ TRIzol	RT-qPCR/ U6 snRNA	Upregulated in KOA patients	[122]
27	Associated with chondrocyte degradation	**SF and serum** (animal model)/miRNeasy Mini Kit	RT-qPCR/ cel-miR-39	Upregulated in both serum and synovial fluid	[123]
27a-3p 27b-3p 27a-5p		**SF**/miRCURY RNA isolation kit	RT-qPCR/ mean Cq values of all detected miRNAs detected	Increased levels of miR27a-3p and 27b-3p and decreased levels of miR27a-5p in late-stage OA (compared to early-stage OA)	[124]
29a	• miR-29 family regulates collagen synthesis and cartilage formation • Involved in cartilage damage	Articular cartilage from hip or knee OA/N.A.	Northern blot & RT-qPCR/ U6 snRNA	Downregulated	[119]
29c		**Plasma**/ miRNeasy kit	RT-qPCR/ MammU6s	Upregulated in the plasma of KOA patients compared to healthy controls	[118]

Table 1. *Cont.*

miRNA	Contribution in OA	Studies on the Role of miRNA in OA [a]			
		Source/ Extraction	Detection/ Normalization	Main Findings/Regulation	Ref
30a	Members of the miR-30 family: • Participate in bone development • Mediate cartilage degradation	Cartilage/ TRIzol	RT-qPCR/ U6 snRNA	Upregulated in primary AC [f] cells from KOA patients compared to healthy controls	[125]
30b		**Plasma**/ miRNeasy kit	RT-qPCR/ MammU6s	Increased in the plasma of KOA patients compared to healthy controls	[118]
34a and 34b	• Induce chondrocyte apoptosis and ECM degradation • Contribute to the breakdown of cartilage and globally suppress signals related to tissue repair and remodeling	Bone and cartilage/ TRIzol	RT-qPCR/ 18S RNA	Both are overexpressed in the AC of KOA patients	[116]
126	• Participates in anabolic and anticatabolic processes • Regulates angiogenesis and de novo vascularization	**Plasma**/ miRNeasy kit	RT-qPCR/ MammU6s	Upregulated in the plasma of KOA patients compared to healthy controls	[118]
132	Downregulation of miR-132: • Decreases cell proliferation and induces apoptosis in chondrocytes • Promotes the expression of Bax protein and activated caspase-3/9 (inflammation)	**SF** and **Plasma**/ Phenol & High Pure miRNA Isolation Kit (Roche)	RT-qPCR/ cel-miR-39	Downregulated: • Plasma miR-132 differentiated HCs [g] from patients with RA or OA	[120]
		Serum/TRIzol	RT-qRCT/ U6 snRNA	Downregulated	[126]
140	Influences chondrocyte differentiation and cartilage homeostasis and suppresses catabolic gene expression	**SF** and cartilage/microRNA Kit (Norgen)	RT-qRCT/ U6 snRNA	Downregulated: Expression levels correlated with OA severity	[127]
146a	Regulates inflammatory and catabolic gene expression	**Plasma**/ miRNeasy kit	RT-qPCR/ MammU6s	Upregulated	[118]
146a-5p	Regulates the expression of inflammatory cytokines	**Serum and cartilage**/ miRCURY RNA Isolation Kit	RT-qRCR/hsa-miR-103a-3p, -423-5p, and -191-5p	Upregulated	[87]
		Serum/ miRCURY Kit	NGS [g]/N.A.	Upregulated	[128]

Table 1. *Cont.*

miRNA	Contribution in OA	Studies on the Role of miRNA in OA [a]			
		Source/ Extraction	Detection/ Normalization	Main Findings/Regulation	Ref
186	Overexpression of miR-186 inhibits chondrocyte apoptosis in OA (see Ref. [129])	**Plasma**/ miRNeasy kit	RT-qPCR/ MammU6s	Upregulated	[118]
186-5p	Regulates chondrocyte apoptosis	**Serum**/ miRCURY Kit	NGS [h]/ N.A.	Upregulated and significantly associated with incident KOA in women	[128]

[a] OA: osteoarthritis; [b] KOA: knee osteoarthritis; [c] N.A.: not available; [d]: KL: Kellgren–Lawrence score; [e] SF: synovial fluid, [f] AC: articular cartilage; [g] HCs: healthy controls; [h] NGS: next-generation sequencing. Biological fluids are indicated in bold.

3. The Potential of Circulating miRNA as Biomarkers for Osteoarthritis

As previously mentioned, some miRNAs involved in the pathogenesis of OA can be detected in the blood or other biological fluids. These c-miRNAs can originate from various sources, including tissue damage from direct cartilage lesions, chronic inflammation, and apoptosis, or cells affected by the OA process that release these miRNAs. Thus c-miRNAs can be discharged into the extracellular space surrounding the chondrocyte, with the specific type of miRNA potentially being influenced by the stage of the disease [118]. Recent studies have explored the potential of c-miRNAs as diagnostic markers in OA and analyzed their expression levels in the circulation and bone/cartilage to understand the biological process occurring in different pathologies. Several studies aimed at validating c-miRNAs in plasma or serum, as well as whole blood, as prognostic or diagnostic markers for the discrimination of OA patients from healthy individuals [68]. Over the years, additional research has reinforced the significance of miRNA in the onset and progression of OA.

Jones et al. [116] determined a set of 17 miRNAs that contribute to cartilage alteration by profiling 157 miRNAs; the authors suggested that the altered expression of these miRNA could be a consequence of hypomethylated states in the miRNA promoter regions. Some specific miRNAs have also shown essential roles in OA; for example, downregulation of miR-140 inhibits interleukin-1β (IL-1β) by inducing *ADAMTS* expression, miR-27b regulates *MMP-13* expression in human chondrocytes and is downregulated in patients with OA, and miR-146a is overexpressed in patients with low-grade OA, suggesting its involvement in OA pathogenesis [86,130,131].

Furthermore, miRNAs found in SF, such as miR-29b3p and miR-140, strongly correlated with radiographic KOA severity and were therefore suggested as potential OA biomarkers [132]. Castanheira and coworkers [133] reported a significant reduction in miR-223 in SFs obtained from patients with early OA, while miR-23b, let-7a-2, snord96A, and snord13 were significantly increased. Notably, it has been suggested that let-7a controls the IL-6 receptor (IL6R). Its suppression can enhance cell growth, reduce cell death, and inhibit inflammation in an in vitro OA model [134]. The let-7 family is frequently mentioned in OA research; a broad study found that serum let-7e can potentially forecast OA risk independent of age, gender, and body mass index [135]. A recent study also affirmed this, noting diminished let-7e expression in the serum of KOA patients [135]. Although the exact functions of let-7 family miRNAs are still unknown, their potential as OA biomarkers is increasingly recognized. Similarly, OA-specific miRNAs were identified in SF that are differentially regulated in early- and late-phase OA [124,125]. Ntoumou et al. [136] found 279 miRNAs differentially expressed in the serum of osteoarthritic conditions, and three signature miRNAs (140-3p, 33b-3p, and 671-3p) were downregulated. These miRNAs are known to be involved in several molecular pathways, including Wnt, ErbB, and TGF-beta.

However, the utilization of c-miRNAs (found in blood/plasma/serum) as OA biomarkers stems from the simplicity of sample acquisition. While SF samples have been utilized for similar purposes, collecting blood/plasma/serum samples has a clear advantage due to their minimally invasive extraction procedures [137]. In detail, Murata et al. [120] demonstrated that differentially expressed miRNAs, especially miR-132, in the plasma and SF can distinguish patients with OA or RA from healthy controls. Borgonio- Cuarda and coworkers [118] identified ten circulating miRNAs (miR-16, -29c, -93, -126, -146a, -184, -186, -195, -345, and -885-5p) in the plasma of patients with primary OA. Notably, the potential of miR-16 and miR-146a as biomarkers for OA has been highlighted by other authors [120,131]. Similar to Murata et al. [120], in their study, Borgonio-Cuadra et al. [118] also observed the deregulation of miR-16 and miR-146. Interestingly, miRNAs were detected in both plasma and SF samples; however, in plasma, the levels were higher than in SF [120]. It has also been demonstrated that knee and hip AC of OA patients exhibit elevated miR-16 expression compared to healthy AC [119]. SMAD3 is a direct target of miR-16, implicating this miRNA in the switch to catabolic TGF signaling [121]. Notably, FGF2 is a direct transcriptional target of miR-16 in human nasopharyngeal carcinoma cells [138], indicating the additional repression of FGF2 signaling by this miRNA. In addition, Iliopoulos et al. [119] reported increased expression of miR-16 in chondrocytes of OA patients with KL scores of 3 and 4. In the same study, the expression of 365 miRNA in articular cartilage tissues from patients with OA was examined, and 16 miRNAs were altered, representing one of the first miRNA signatures to distinguish osteoarthritic from normal chondrocytes [119]. Notably, it has been demonstrated that miR-146b regulates cell differentiation, proliferation, and migration in various tumor cells [139,140]. MiR-146b has also been implicated in the chondrogenic differentiation of human bone marrow-derived skeletal stem cells (SSCs) through the modulation of SOX5 [141].

In another study, Wan et al. [142] reported a significant reduction in plasma miR-136 in KOA patients, correlating negatively with the severity of the disease (KL score). Notably, the diminished levels of plasma miR-136 exhibited a significant capability to distinguish OA patients of KL grades 2, 3, and 4 from healthy individuals. Concurrently, serum interleukin (IL)-17 levels, a target gene of miR-136, increased with KOA severity. Together, the above data suggest that decreased plasma miR-136 levels could serve as a potential biomarker for KOA. On the contrary, Beyer et al. [143] demonstrated that the levels of serum let-7e have an inverse correlation with OA severity and the number of knee/hip joint replacements. In their study, Kong et al. [144] reported that the circulating levels of miR-486-5p, miR-320b, miR-122-5p, miR-92a-3p, and miR-19b-3p were elevated in OA patients compared to the control group. A combination of miR-486-5p, miR-122-5p, and miR-19b-3p provided the best diagnostic efficacy, demonstrating a significant correlation with the risk and severity of KOA. Likewise, a study by Ntomou et al. [136] revealed that decreased miR-671-3p, miR-140-3p, and miR-33b-3p serum levels can be used to identify OA patients. These reduced miRNA levels were correlated with the risk and progression of OA.

Interestingly, in a cohort of postmenopausal women, elevated levels of circulating miR-146a-5p and miR-186-5p were associated with existing KOA and the onset of KOA over the following four years, respectively [128], suggesting that circulating miR-136 and the combination of miR-486-5p, miR-122-5p, and miR-19b-3 hold the most significant potential for distinguishing OA patients from healthy individuals [142,144]. Among the validated miRNAs mentioned earlier, miR-146a levels were elevated in the plasma of OA patients compared to controls [118] and in the serum of individuals with prevalent KOA versus healthy controls [128]. The sensitivity and specificity of the most promising circulating miRNAs that can be used as biomarkers for OA were recently thoroughly reviewed by Bottani et al. [68] and are not discussed here.

3.1. Challenges in Using c-miRNAs as Biomarkers for OA: Why Are c-miRNAs Not Being Used in Clinical Applications Yet?

Despite the promising prospect of c-miRNAs serving as prognostic and diagnostic biomarkers for OA and various other bone disorders, these c-miRNAs have not been incorporated into clinical practice to date. Figure 1 presents a SWOT analysis, demonstrating the potential viability of c-miRNAs as biomarkers for monitoring the onset and progression of OA.

Strengths	Weaknesses
• miRNAs are stable in circulation and resistant to degradation. • miRNAs have cell-type specificity, providing precise biological information. • miRNAs can indicate disease stages, making them effective for early detection.	• Challenges in standardizing miRNA extraction and quantification methods. • Inter-individual variability affects the specificity and sensitivity of miRNAs as biomarkers. • Difficulties in identifying target genes of miRNAs.
Opportunities	**Threats**
• Technological advancements such as high-throughput sequencing, artificial intelligence bioinformatics tools can help discover new miRNAs. • Development of novel diagnostic tools or therapeutic strategies based on miRNAs. • Novel methodologies for c-miRNA normalization are being developed.	• Regulatory hurdles for clinical translation of miRNAs. • Competition from other types of biomarkers. • miRNA profiles may overlap in different bone diseases, limiting their specificity.

Figure 1. A SWOT analysis of the potential of c-miRNAs as biomarkers for the monitoring of the development and progress of OA.

3.1.1. Strengths

To achieve the objectives of personalized medicine, new and more accurate biomarkers for the prognosis and diagnosis of OA are sorely needed. An ideal biomarker must meet specific prerequisites. First, it should be easily accessible, which means it can be identified and measured through minimally invasive procedures. The biomarker's specificity to the disease being investigated is another vital factor, coupled with its sensitivity (ideally, it should be detectable even before the clinical symptoms have surfaced, and its quantities should be consistent with the progression of the disease or response to treatment). Lastly, the biomarker should potentially translate from research to clinical practice [145].

As extensively discussed herein, c-miRNAs meet the above criteria, as they are found in circulation, including whole blood, serum, plasma, and other biological fluids, and exhibit stability under extreme conditions [146]. Identifying miRNAs in biological fluids relies on RT-qPCR, which offers highly reproducible outcomes and enhanced specificity and sensitivity [147]. Notably, a number of these c-miRNAs are tissue-specific. Moreover, others can help distinguish patients with OA from healthy individuals or patients suffering from bone-related diseases such as RA [120].

3.1.2. Weaknesses

However, in this study, we have pinpointed several differences and challenges within the research aimed at identifying and validating circulating miRNAs as prognostic and diagnostic biomarkers for OA. Some examples are presented in Table 2 and discussed in the following paragraphs.

Table 2. Examples of discrepancies among studies aiming to evaluate c-miRNAs as biomarkers for OA.

miRNAs	OA [1] Type	Cohort	Sample/ Collection Tube/ Centrifugation (×g/Time/Temp)	Extraction/ Quantification/ Normalization	Ref
146-5p	HOA [2]	• KL [3] grade I and II: n = 28 • Control group: n = 2	Serum Collection tube [4]: N.A. [5] 2000 g/10 min/N.A.	miRCURY RNA Kit/ RT-qRCR/ hsa-miR-103a-3p, -423-5p, & -191-5p	[87]
380 miRNAs	KOA [6]	• KL 2 or 3: n = 27 • Control group: n = 27	Plasma ETDA Tube 1800 g/10 min/RT	miRNeasy kit/ RT-qPCR/ MammU6s	[118]
132	N.A	Cohort of 16 men: • 8 OA patients • 8 healthy controls	Serum/ Collection tube: N.A/ 2000 g/10 min/4 °C	TRIzol/ RT-qRCT/ U6 snRNA	[126]
140	KOA	• KL 1–2: n = 10 • KL 3: n = 10 • KL 4: n = 10 Control group: n = 17	SF [7] & cartilage The collection and centrifugation conditions are not described	microRNA Kit (Norgen)/ RT-qPCR/ U6 snRN	[127]
19 miRNAs (Validation)	KOA	PM women: Screening: KL 2–3 = 10, control group: n = 10 Validation: KL 2–3 = 43, control group: n = 42	Serum The collection and centrifugation conditions are not described	miRCURY Kit/ NGS/ N.A	[128]
136	KOA	• KL 2: n = 22 • KL 3: n = 29 • KL 4: n = 23 Control group n = 29	Plasma Sodium Citrate Centrifugation conditions are not available	RNAVzol LS/ RT-qPCR/ U6 snRNA	[142]
2578 miRNAs (Validation)	KOA	KOA who received celecoxib treatment for six weeks Screening KL 2: n = 4 & KL 3: n = 2 Validation KL 2: n = 159 & KL 3: n = 59	Plasma The collection and centrifugation conditions are not described	TRIzol/ RT-qPCR U6 snRNA	[148]

[1] OA: osteoarthritis; [2] HOA: hip osteoarthritis; [3] KL: Kellgren–Lawrence score; [4] only one type of tube is commonly used for the collection of serum, i.e., without anticoagulant; [5] N.A; not available; [6] KOA: knee osteoarthritis; [7] SF: synovial fluid.

We have identified significant variations in four levels: (i) study design and cohort selection, (ii) preanalytical stage, (iii) analytical stage, and (iv) postanalytical stage, including data analysis and normalization (Figure 2). These variations may lead to inconsistencies among studies aiming to examine the prognostic potential of specific miRNAs in OA. An example is illustrated in Table 1 by comparing the studies conducted by Borgonio Cuadra et al. [118] and Murata et al. [120], both examining the role of miR16 in KOA. Interestingly, while Borgonio Cuadra et al. [118] reported that c-miR-16 is downregulated in KOA patients, Murata et al. [120] reported that the same miR was upregulated. As illustrated in Table 1, the two studies used different methods to extract, quantify, and normalize miRNAs.

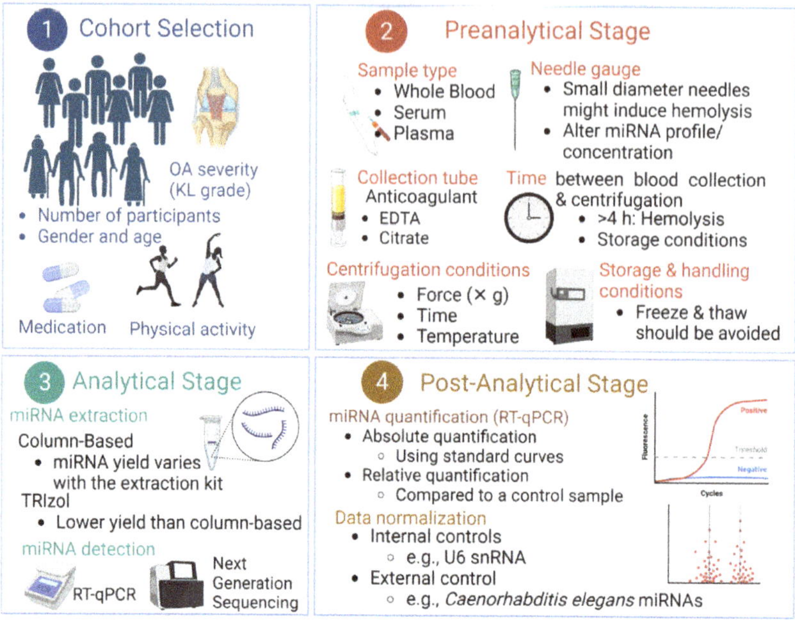

Figure 2. The transition of c-miRNAs as biomarkers for OA from basic research to clinical practice is hampered due to several obstacles and the lack of standardized protocols at four distinct levels: cohort selection, preanalytical, analytical, and post-analytical stages, as discussed in the text.

As discussed in the following paragraphs, the lack of standardized protocols and guidelines has significantly contributed to the delay in transitioning c-miRNAs from basic research to clinical application.

(i) Study design and cohort selection: For clinical application, the most critical evaluation criteria for c-miRNAs as diagnostic and prognostic biomarkers are high sensitivity and specificity to avoid false-positive or false-negative diagnoses. To this end, it is essential to have a larger sample size to be able to discriminate OA patients from healthy controls or assess the severity of the disease [28]. Given the numerous criteria used in clinical applications, such as age, gender, ethnicity, lifestyle, and medical history, it is essential to avoid studies with limited sample sizes. The severity of OA is another factor that should be taken into account. As mentioned in the Introduction, the severity of KOA is assessed using a scale that categorizes the severity of KOA into five grades, ranging from grade 0 (normal) to grade 4 (severe) through visual inspection of X-ray or MRI images [5].

Examples of differences in cohort selection among studies are shown in Table 2. For example, the number of participants in some studies was low, with only male or female participants. In addition, some studies did not report the severity of OA (e.g., KL score). As also shown in Table 2, one study involved participants who received a specific OA drug. Notably, one study did not report the type of OA (i.e., knee, hip, hand, etc.) or the severity of the disease.

Another factor that must be considered is the significant role inflammation plays in OA development, particularly during the disease's early stages (early OA) [149]. Previous studies have shown that inflammatory processes during early stages can contribute to cartilage degradation and changes in the subchondral bone [150–152]. Consequently, the severity of the disease (e.g., as indicated by the KL grade) should be considered when recruiting participants. It is essential to note that the inflammation status in OA can differ based on the severity of the disease. Recently, we proposed that the observed variability in exosomal miRNA levels in patient samples compared to healthy individuals might be partly due to interactions between exosomes carrying c-miRNAs and endothelial receptors. These receptors are often overexpressed under pathological conditions such as inflammation [153].

The participant's physical activity (PA) level is another crucial factor affecting miRNAs concentration and profile in the blood. Whether regularly exercising or sedentary, a person's PA state significantly affects the expression of miRNAs across various tissues and organs. Consequently, it also impacts the profile of c-miRNAs [154]. Numerous miRNA alterations in human skeletal muscle and circulation have been recorded in response to acute and chronic exercise [155]. Studies have shown changes in c-miRNA levels post-exercise, such as the downregulation of miR-16 and miR-93 and a decrease in miR-222. Notably, miR-16 targets the vascular endothelial growth factor/vascular endothelial growth factor receptor pathway, which plays a role in adapting to different types of physical activity. In contrast, miR-93 modulates mitogen- and stress-activated protein kinase 2 (Msk2), a histone kinase activated by muscle contraction [156]. Interestingly, specific miRNAs required for skeletal muscle function, including miR-23a,-133a, -146a, -206, -78b, and -486, showed significant changes within the muscle after a single bout of acute resistance exercise. However, these miRNAs did not show apparent alterations in the plasma post exercise. This observation suggests that c-miRNA levels may not accurately reflect the miRNA responses within skeletal muscle after exercise [157].

Finally, specific drugs such as heparin, aspirin, and antiplatelet therapies may affect the profile and/or concentration of miRNAs [158,159]. In detail, heparin, an anticoagulant frequently used in myocardial infarction treatments, is known to inhibit qPCR-based ncRNA quantification [160]. Administering heparin before blood collection has also been demonstrated to affect the levels of miRNAs, while endogenous heparin also impacts miRNA quantification [161]. Aspirin and antiplatelet treatments also influence miRNA profiles. For example, aspirin alters the expression of platelet-derived miRNAs, such as miR-19b and miR-92a, as well as others, including miR-155, miR-21, miR-98, miR-191, miR-126, miR-223, and miR-150 [159,162]. Other antiplatelet therapies impact the levels of some miRNAs, including miR-191, miR-223, miR-126, and miR-150 [160]. Consequently, special consideration must be taken during patient recruitment for miRNA quantification studies, especially regarding the administration of heparin and antiplatelet therapies before blood collection. It is noteworthy that treating blood samples with the enzyme heparinase can reverse heparin's effects [140]. This more profound insight into the interaction between these drugs and miRNAs emphasizes the complexity of their role in diagnostic and therapeutic applications [163].

(ii) Preanalytical stage: Several preanalytical factors may affect the profiles and levels of c-miRNAs. We recently reviewed the main preanalytical factors affecting the profile and concentration of miRNAs in circulation when they are examined as potential biomarkers for cardiovascular diseases (CVDs) [163], which are also summarized herein. One of the most critical factors is the selection of the blood fraction (whole blood, plasma, or serum), sample collection (e.g., needle gauge), anticoagulant (for plasma collection), centrifugation conditions, and handling/storage conditions of the samples (Figure 2). These factors significantly impact miRNA profiles and are usually overlooked. Furthermore, based on the results of this work and our previous work [163] and that of others [68], the sample source is one of the most critical aspects in determining c-miRNA concentrations [164] for various diseases, including OA. Plasma is often favored over serum as a source of c-mRNAs, since the coagulation process can release RNA molecules, potentially altering the genuine profile of c-miRNAs. However, plasma may contain cellular components such as apoptotic or lysed cells (e.g., red blood cells-RBCs and platelets) that may contribute miRNAs. Furthermore, anticoagulants like citrate and heparin citrate, which are used to isolate plasma, can inhibit downstream methodologies, including RT-qRCR. Therefore, serum is often deemed the optimal fraction for detecting c-miRNAs [165,166]. Furthermore, using whole blood as a source of miRNAs should be avoided because cellular fraction could also contribute miRNAs [167]. Therefore, we suggest that results from studies employing different blood fractions and types of blood tubes not be compared directly. Most crucially, only miRNAs that are not marginally up- or downregulated are likely to be suitable as clinical biomarkers. However, as presented in Tables 1 and 2, various blood fractions have been used in different studies investigating the potential of c-miRNAs as biomarkers for OA. Furthermore, in some studies, the type of collection tube used and the centrifugation conditions were not reported (Tables 1 and 2).

The extraction method/protocol is another factor affecting the yield and profile of c-miRNAs. As illustrated in Tables 1 and 2, the extraction of miRNAs/c-miRNAs in the studies was carried out with either TRIzol-based methods or commercial-column-based extraction kits.

Despite efforts to compare and optimize these methods, interlaboratory differences persist across isolation methods, yielding varied miRNA quantities and qualities (reviewed in [153]). For example, studies have found that TRIzol-based methods negatively impact RNA quality and that column-based methods are preferred. Similarly, two column-based RNA extraction kits (miRNeasy Mini and mirVana PARIS) reported higher RNA yields than the TRIzol-based extraction method [168]. To avoid these discrepancies, the use of equal volumes of starting material (serum or plasma) has been suggested instead of using the same amount of total RNA to ensure accurate results [167].

Thus, standardizing both sample collection and handling protocols would be beneficial to reduce the bias that can affect the preanalytical phase of miRNA validation [169]. Different research groups have endeavored to standardize the steps involved in this stage, encompassing the collection, processing, and storage of blood samples, to mitigate potential inconsistencies and variations in c-miRNA profiles. However, a universally accepted and reliable method for the processing of blood for c-miRNA analysis has not been established. Consequently, scientific publications must thoroughly describe the procedures for blood collection and processing [170]. Recently, guidelines and preventive measures for the collection and handling of blood samples have been proposed by different groups [171,172]. The recommended steps involve (i) utilizing EDTA-treated tubes for blood collection (for plasma samples), (ii) transporting blood samples to the laboratory with minimal agitation and at low temperature, (iii) immediate centrifugation of the samples, (iv) careful removal of plasma followed by prompt aliquoting, (v) storage of plasma/serum samples at −80 °C, and (vi) using pre-aliquoted samples to avoid repetitive freeze–thaw cycles. It is crucial that blood samples are processed within a restricted time frame after collection to minimize miRNA expression contamination by lysed RBCs, platelets, and leukocytes and that plasma and serum samples are cautiously recovered, aliquoted, and frozen within 24 h. Furthermore, they must be stored at −80 °C [171]. Measures should also be taken to reduce contamination by cellular material and hemolyzed samples, such as employing an additional centrifugation step after plasma/serum recovery to eliminate intact blood cells [164]. In vivo hemolysis may be unavoidable, but in vitro hemolysis can be significantly mitigated by adhering to guidelines set forth by the Early Detection Research Network (ERDN) [173]. Lastly, it is essential to refrain from combining different specimen types, such as plasma and serum, within one study, as the uniform expression of miRNAs across these fractions is not guaranteed [163].

(iii) Analytical stage: The platform used for miRNA evaluation introduces a notable source of error in the analytical phase. Common platforms include next-generation sequencing (NGS), microarrays, and RT-qPCR, each with unique pros and cons [174,175]. As illustrated in Tables 1 and 2, most studies aimed at validating miRNAs as biomarkers for OA have employed RT-qPCR, which is cost effective and fast but limited by low throughput. Moreover, the efficiency of each method depends on the quality of the starting material. Notably, c-miRNA levels isolated with different protocols vary significantly, and direct comparisons should be avoided [176]. Therefore, the analytical protocols and platform must be unchanged [146].

(iv) Data analysis and normalization: In the post-analytical stage, reference gene selection and normalization strategy are key challenges in miRNA quantification due to the lack of a standardized methodology. RT-qPCR data for miRNA expression can be normalized using single or multiple endogenous or exogenous reference genes or the averaged expression value of all measured miRNAs (reviewed in [68,163]). As illustrated in Tables 1 and 2, studies aimed at validating miRNA as biomarkers for OA have used a variety of normalization strategies. Given the variables affecting miRNA quantification in the preanalytical, analytical, and post-analytical stages, it is vital to establish detailed, standardized guidelines for consistent and comparable miRNA expression data across studies and labs. Establishing universal guidelines and protocols is critical for c-miRNAs to become clinically valid biomarkers. Overall, normalization is crucial when determining c-miRNAs expression levels. Although various reference genes have been proposed, further studies are needed to identify the most reliable normalization method. This might vary depending on the miRNA release route (e.g., microparticles or protein-bound) [177]. Establishing an optimal endogenous control for each type of cardiovascular disease is essential, as specific c-miRNA expression profiles and/or levels may vary. Several studies suggest the use of multiple reference genes or a suitable combination thereof and a standard concentration of spike-in miRNAs for normalization. All samples should be simultaneously processed using identical starting volumes [178].

3.1.3. Opportunities

Despite the limitations mentioned above, the potential of c-miRNAs as prognostic and diagnostic biomarkers for OA is highly recognized, especially given that the field of c-miRNA research is relatively new. Recent evidence has revealed that miRNAs contribute to the development of OA via various mechanisms [179]. As mentioned above, c-miRNAs have the potential for early detection of OA and to discriminate OA patients from healthy individuals, as well as from patients suffering from other types of bone diseases. Therefore, specific c-miRNAs could be used as indicators for earlier therapeutic intervention, potentially slowing disease progression or preventing severe damage. The early identification of at-risk patients would provide an opportunity to implement lifestyle changes or therapeutic interventions that could modify the disease course. From a personalized medicine perspective, miRNA profiles could facilitate personalized treatment strategies, enabling patient-centered care. Importantly, the unique miRNAs profile of patients can help in tailoring treatments for OA [180]. Detecting c-miRNAs would improve personalized medicine since one of the current clinical challenges is developing minimally invasive approaches. They show remarkable stability in bodily fluid and consistent expression profiles, with the potential to serve as biomarkers for changes in physiological and pathological conditions [181]. The expression signatures of c-miRNAs are emerging as novel biomarkers of numerous diseases, including CVDs, diabetes, cancer, musculoskeletal disorders, etc. [182,183].

Furthermore, the rapid advancement in technologies such as NGS and bioinformatics tools could facilitate the discovery and analysis of new c-miRNAs associated with OA. These tools can help better understand the disease mechanisms and potentially uncover new therapeutic targets and/or strategies [184,185]. Novel methodologies for the normalization of c-miRNAs concentration in biological fluids are also being developed [163]. Moreover, recent advances in artificial intelligence (AI) could facilitate the unraveling of the crucial role of miRNAs in the pathophysiology of OA. For example, in their recent study, Liu et al. [186] shed light on the role of circRNA–miRNA–mRNA regulatory networks in OA using AI tools. Finally, exploring the roles of c-miRNAs and regulating the expression or function of specific c-miRNA expression might offer a potential therapeutic strategy for the treatment of OA. For example, miRNA mimics or inhibitors could be used to modulate the expression of target genes in OA [82,187].

3.1.4. Threats

Finally, we identified specific threats (Figure 1) in utilizing c-miRNAs as biomarkers for OA. Research on circulating miRNAs is still in its early stages, and the human body potentially contains thousands of different miRNAs. These can have intricate interactions with each other and other types of molecules, including other ncRNAs [188]. Maintaining high analytical standards and adopting robust study designs are prerequisites for the generation of reliable data and avoiding the pitfalls of early genetic association studies. As previously discussed, the lack of standard protocols and guidelines for miRNA extraction and quantification significantly hampers the reproducibility and accuracy of results [163]. This lack of standardized methods has led to discrepancies among studies investigating the potential of miRNAs as OA biomarkers [163]. Additionally, before the clinical implementation of miRNAs as biomarkers, they must undergo rigorous clinical trial testing and gain approval from regulatory authorities. This process can be time-consuming and costly, and there is no approval guarantee for a specific biomarker [189].

Moreover, other circulating biomolecules related to bone cartilage and bone turnover (e.g., cartilage oligomeric matrix protein and procollagen type II C-terminal propeptide) are under evaluation as OA biomarkers [190]. The concentrations of these biomarkers can be easily determined using commercially available kits, without laborious extraction steps, and normalization of results is straightforward [191]. Lastly, it is important to note that miRNA profiles may overlap among different bone diseases (e.g., osteoporosis), which can limit their specificity [192].

4. Conclusions

This review focused on examining the potential of c-miRNAs as prognostic and diagnostic biomarkers for OA, a global endemic and debilitating disease. Previously thought to be simply the result of "wear and tear", OA is now understood to involve a complex interplay of local and systemic factors [193]. The development of disease knowledge and treatment innovation in OA has historically been slow, partly due to the alleged lack of valid and responsive biomarkers to evaluate efficacy, which, in turn, depends on the slow evolution of our understanding of complex joint tissue biology.

An approach for optimizing OA treatment lies in identifying effective early diagnostic biomarkers. C-miRNAs fulfill the criteria of ideal biomarkers, as they are stable in biological fluids, including plasma and serum, and can easily be detected using PCR-based techniques (Figure 3). Moreover, the role of miRNAs in OA progression is increasingly recognized, and the development of therapies that modify the disease process and alleviate symptoms is of particular interest [82]. Since OA is the primary cause of joint damage and irreversible disability, early diagnosis and aggressive treatment management of OA are necessary to improve the disease prognosis [69]. Emerging evidence has revealed that the identification of biomarkers, including c-miRNAs, could improve OA diagnosis and treatment [69].

The challenge lies in identifying specific miRNA patterns that can differentiate between various conditions, including distinguishing OA patients from healthy individuals, assessing the severity of the disease, and discriminating OA patients from those with other bone diseases. However, it is worth noting that despite these overlaps, unique patterns of miRNAs could still be identified in specific diseases when a broad enough panel of miRNAs is examined. That could potentially increase the specificity and utility of miRNAs as biomarkers for different diseases. Furthermore, combining miRNA profiles with other biomarkers or clinical information could enhance specificity.

Despite the challenges associated with using c-miRNAs as biomarkers for OA, their potential to aid in early diagnosis and inform treatment strategies is immense. As we unravel the complex interplay of miRNAs and other biological processes, the opportunities for their application in disease management and treatment only grow. Hopefully, with the evolution of our understanding and the development of robust analytical methods, c-miRNAs will serve as a linchpin in personalized medicine, thereby transforming how we diagnose, treat, and ultimately prevent OA.

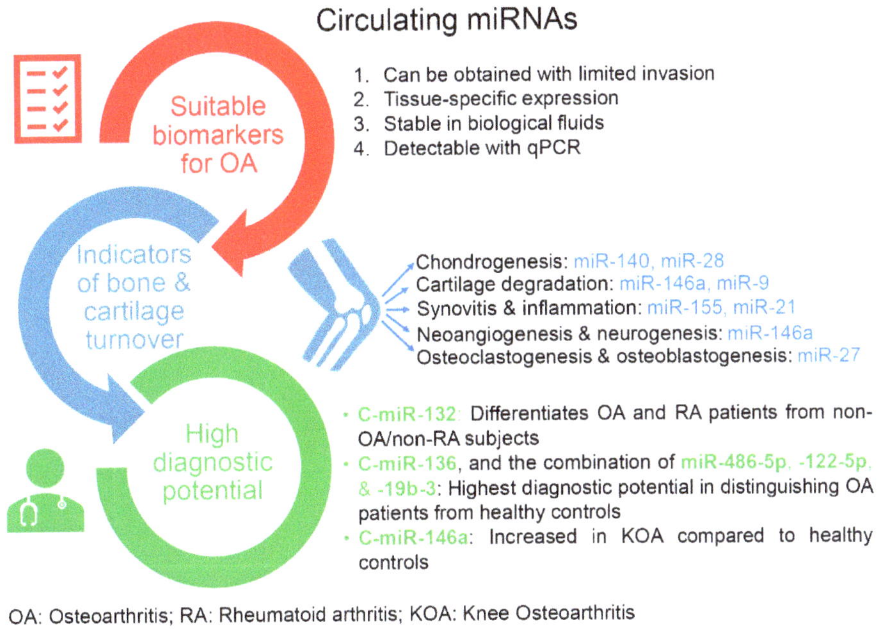

Figure 3. Circulating miRNAs (c-miRs) as promising prognostic and diagnostic biomarkers for osteoarthritis (OA). They are considered promising because they (i) fulfill the criteria of suitable biomarkers, (ii) can serve as indicators of bone and cartilage turnover, and (iii) exhibit high diagnostic potential. The miRNAs shown in this figure are indicative examples; their roles in the development of OA and potential as biomarkers for OA are extensively discussed in the text.

Author Contributions: Conceptualization, K.F. and C.P.; methodology, M.P. and C.P.; software, K.F. and C.P.; validation, K.F. and M.P.; formal analysis, M.P. and C.P.; investigation and resources, C.P.; data curation, M.P. and C.P.; writing—original draft preparation, C.P.; writing—review and editing, K.F., M.P. and C.P.; visualization, C.P.; supervision, C.P.; project administration, C.P.; funding acquisition, C.P. All authors have read and agreed to the published version of the manuscript.

Funding: This research received no external funding.

Data Availability Statement: No new data were created in this work.

Conflicts of Interest: The authors declare no conflict of interest.

References

1. Dahaghin, S.; Bierma-Zeinstra, S.M.; Ginai, A.Z.; Pols, H.A.; Hazes, J.M.; Koes, B.W. Prevalence and pattern of radiographic hand osteoarthritis and association with pain and disability (the Rotterdam study). *Ann. Rheum. Dis.* **2005**, *64*, 682–687. [CrossRef]
2. Lu, K.; Ma, F.; Yi, D.; Yu, H.; Tong, L.; Chen, D. Molecular signaling in temporomandibular joint osteoarthritis. *J. Orthop. Translat.* **2022**, *32*, 21–27. [CrossRef]
3. Yunus, M.H.M.; Nordin, A.; Kamal, H. Pathophysiological perspective of osteoarthritis. *Medicina* **2020**, *56*, 614. [CrossRef]
4. Felson, D.T. Clinical practice. Osteoarthritis of the knee. *N. Engl. J. Med.* **2006**, *354*, 841–848. [CrossRef]
5. Kellgren, J.H.; Lawrence, J.S. Radiological assessment of osteo-arthrosis. *Ann. Rheum. Dis.* **1957**, *16*, 494–502. [CrossRef]
6. Swagerty, D.L., Jr.; Hellinger, D. Radiographic assessment of osteoarthritis. *Am. Fam. Physician* **2001**, *64*, 279–286.
7. Guilak, F. Biomechanical factors in osteoarthritis. *Best Pract. Res. Clin. Rheumatol.* **2011**, *25*, 815–823. [CrossRef]
8. Goldring, M.B. Articular cartilage degradation in osteoarthritis. *HSS J.* **2012**, *8*, 7–9. [CrossRef]

9. Fernandes, G.S.; Parekh, S.M.; Moses, J.; Fuller, C.; Scammell, B.; Batt, M.E.; Zhang, W.; Doherty, M. Prevalence of knee pain, radiographic osteoarthritis and arthroplasty in retired professional footballers compared with men in the general population: A cross-sectional study. *Br. J. Sports Med.* **2018**, *52*, 678–683. [CrossRef]
10. Matthews, G.L.; Hunter, D.J. Emerging drugs for osteoarthritis. *Expert Opin. Emerg. Drugs* **2011**, *16*, 479–491. [CrossRef]
11. Bijlsma, J.W.; Berenbaum, F.; Lafeber, F.P. Osteoarthritis: An update with relevance for clinical practice. *Lancet* **2011**, *377*, 2115–2126. [CrossRef]
12. Roemer, F.W.; Demehri, S.; Omoumi, P.; Link, T.M.; Kijowski, R.; Saarakkala, S.; Crema, M.D.; Guermazi, A. State of the art: Imaging of osteoarthritis—Revisited 2020. *Radiology* **2020**, *296*, 5–21. [CrossRef]
13. Group, B.D.W. Biomarkers and surrogate endpoints: Preferred definitions and conceptual framework. *Clin. Pharmacol. Ther.* **2001**, *69*, 89–95. [CrossRef]
14. Lotz, M.; Martel-Pelletier, J.; Christiansen, C.; Brandi, M.L.; Bruyère, O.; Chapurlat, R.; Collette, J.; Cooper, C.; Giacovelli, G.; Kanis, J.A.; et al. Value of biomarkers in osteoarthritis: Current status and perspectives. *Ann. Rheum. Dis.* **2013**, *72*, 1756–1763. [CrossRef]
15. Karsdal, M.A.; Henriksen, K.; Bay-Jensen, A.C. Biochemical markers in osteoarthritis with lessons learned from osteoporosis. *Clin. Exp. Rheumatol.* **2019**, *37* (Suppl. 120), 73–87.
16. Scanzello, C.R.; Plaas, A.; Crow, M.K. Innate immune system activation in osteoarthritis: Is osteoarthritis a chronic wound? *Curr. Opin. Rheumatol.* **2008**, *20*, 565–572. [CrossRef]
17. Ghafouri-Fard, S.; Poulet, C.; Malaise, M.; Abak, A.; Mahmud Hussen, B.; Taheriazam, A.; Taheri, M.; Hallajnejad, M. The emerging role of non-Coding RNAs in osteoarthritis. *Front. Immunol.* **2021**, *12*, 773171. [CrossRef]
18. Ali, S.A.; Peffers, M.J.; Ormseth, M.J.; Jurisica, I.; Kapoor, M. The non-coding RNA interactome in joint health and disease. *Nat. Rev. Rheumatol.* **2021**, *17*, 692–705. [CrossRef]
19. Zhou, L.; Wan, Y.; Cheng, Q.; Shi, B.; Zhang, L.; Chen, S. The expression and diagnostic value of LncRNA H19 in the blood of patients with osteoarthritis. *Iran. J. Public Health* **2020**, *49*, 1494–1501. [CrossRef]
20. O'Brien, J.; Hayder, H.; Zayed, Y.; Peng, C. Overview of microRNA biogenesis, mechanisms of actions, and circulation. *Front. Endocrinol.* **2018**, *9*, 402. [CrossRef]
21. Rego-Pérez, I.; Durán-Sotuela, A.; Ramos-Louro, P.; Blanco, F.J. Genetic biomarkers in osteoarthritis: A quick overview. *Fac Rev* **2021**, *10*, 78. [CrossRef] [PubMed]
22. Li, J.; Yang, X.; Chu, Q.; Xie, L.; Ding, Y.; Xu, X.; Timko, M.P.; Fan, L. Multi-omics molecular biomarkers and database of osteoarthritis. *Database* **2022**, *2022*, baac052. [CrossRef] [PubMed]
23. Bay-Jensen, A.C.; Mobasheri, A.; Thudium, C.S.; Kraus, V.B.; Karsdal, M.A. Blood and urine biomarkers in osteoarthritis—An update on cartilage associated type II collagen and aggrecan markers. *Curr. Opin. Rheumatol.* **2022**, *34*, 54–60. [CrossRef] [PubMed]
24. Kraus, V.B.; Karsdal, M.A. Osteoarthritis: Current molecular biomarkers and the way rorward. *Calcif. Tissue Int.* **2021**, *109*, 329–338. [CrossRef] [PubMed]
25. Ardekani, A.M.; Naeini, M.M. The role of microRNAs in human diseases. *Avicenna J. Med. Biotechnol.* **2010**, *2*, 161–179. [PubMed]
26. Nimse, S.B.; Sonawane, M.D.; Song, K.S.; Kim, T. Biomarker detection technologies and future directions. *Analyst* **2016**, *141*, 740–755. [CrossRef]
27. Ho, P.T.B.; Clark, I.M.; Le, L.T.T. MicroRNA-based diagnosis and therapy. *Int. J. Mol. Sci.* **2022**, *23*, 7167. [CrossRef]
28. Condrat, C.E.; Thompson, D.C.; Barbu, M.G.; Bugnar, O.L.; Boboc, A.; Cretoiu, D.; Suciu, N.; Cretoiu, S.M.; Voinea, S.C. miRNAs as biomarkers in disease: Latest findings regarding their role in diagnosis and prognosis. *Cells* **2020**, *9*, 276.
29. Ha, M.; Kim, V.N. Regulation of microRNA biogenesis. *Nat. Rev. Mol. Cell Biol.* **2014**, *15*, 509–524. [CrossRef]
30. Saliminejad, K.; Khorram Khorshid, H.R.; Soleymani Fard, S.; Ghaffari, S.H. An overview of microRNAs: Biology, functions, therapeutics, and analysis methods. *J. Cell Physiol.* **2019**, *234*, 5451–5465. [CrossRef]
31. Bartel, D.P. MicroRNAs: Target recognition and regulatory functions. *Cell* **2009**, *136*, 215–233. [CrossRef]
32. Li, S.; Patel, D.J. Drosha and Dicer: Slicers cut from the same cloth. *Cell Res.* **2016**, *26*, 511–512. [CrossRef] [PubMed]
33. Han, J.; Lee, Y.; Yeom, K.H.; Kim, Y.K.; Jin, H.; Kim, V.N. The Drosha-DGCR8 complex in primary microRNA processing. *Genes Dev.* **2004**, *18*, 3016–3027. [CrossRef]
34. Yi, R.; Qin, Y.; Macara, I.G.; Cullen, B.R. Exportin-5 mediates the nuclear export of pre-microRNAs and short hairpin RNAs. *Genes Dev.* **2003**, *17*, 3011–3016. [CrossRef] [PubMed]
35. Köhler, A.; Hurt, E. Exporting RNA from the nucleus to the cytoplasm. *Nat. Rev. Mol. Cell Biol.* **2007**, *8*, 761–773. [CrossRef]
36. Park, J.E.; Heo, I.; Tian, Y.; Simanshu, D.K.; Chang, H.; Jee, D.; Patel, D.J.; Kim, V.N. Dicer recognizes the 5' end of RNA for efficient and accurate processing. *Nature* **2011**, *475*, 201–205. [CrossRef] [PubMed]
37. Wilson, R.C.; Tambe, A.; Kidwell, M.A.; Noland, C.L.; Schneider, C.P.; Doudna, J.A. Dicer-TRBP complex formation ensures accurate mammalian microRNA biogenesis. *Mol. Cell.* **2015**, *57*, 397–407. [CrossRef]
38. Kwak, P.B.; Tomari, Y. The N domain of Argonaute drives duplex unwinding during RISC assembly. *Nat. Struct. Mol. Biol.* **2012**, *19*, 145–151. [CrossRef]
39. Tang, G. siRNA and miRNA: An insight into RISCs. *Trends Biochem. Sci.* **2005**, *30*, 106–114. [CrossRef]
40. Pratt, A.J.; MacRae, I.J. The RNA-induced silencing complex: A versatile gene-silencing machine. *J. Biol. Chem.* **2009**, *284*, 17897–17901. [CrossRef]

41. Pillai, R.S.; Bhattacharyya, S.N.; Filipowicz, W. Repression of protein synthesis by miRNAs: How many mechanisms? *Trends Cell Biol.* **2007**, *17*, 118–126. [CrossRef]
42. Cheng, A.M.; Byrom, M.W.; Shelton, J.; Ford, L.P. Antisense inhibition of human miRNAs and indications for an involvement of miRNA in cell growth and apoptosis. *Nucleic Acids Res.* **2005**, *33*, 1290–1297. [CrossRef] [PubMed]
43. Yang, C.H.; Yue, J.; Fan, M.; Pfeffer, L.M. IFN induces miR-21 through a signal transducer and activator of transcription 3-dependent pathway as a suppressive negative feedback on IFN-induced apoptosis. *Cancer Res.* **2010**, *70*, 8108–8116. [CrossRef]
44. Si, M.L.; Zhu, S.; Wu, H.; Lu, Z.; Wu, F.; Mo, Y.Y. miR-21-mediated tumor growth. *Oncogene* **2007**, *26*, 2799–2803. [CrossRef] [PubMed]
45. Jenike, A.E.; Halushka, M.K. miR-21: A non-specific biomarker of all maladies. *Biomark. Res.* **2021**, *9*, 18. [CrossRef]
46. Jiang, A.; Dong, C.; Li, B.; Zhang, Z.; Chen, Y.; Ning, C.; Wu, W.; Liu, H. MicroRNA-206 regulates cell proliferation by targeting G6PD in skeletal muscle. *FASEB J.* **2019**, *33*, 14083–14094. [CrossRef]
47. Zhang, Y.J.; Xu, F.; Zhang, Y.J.; Li, H.B.; Han, J.C.; Li, L. miR-206 inhibits non small cell lung cancer cell proliferation and invasion by targeting SOX9. *Int. J. Clin. Exp. Med.* **2015**, *8*, 9107–9113. [PubMed]
48. Yuan, S.; Liu, Z.; Yu, S.; Wang, X.; Shi, J. CCND2 and miR-206 as potential biomarkers in the clinical diagnosis of thyroid carcinoma by fine-needle aspiration cytology. *World J. Surg. Oncol.* **2023**, *21*, 22. [CrossRef]
49. Song, G.; Zhang, Y.; Wang, L. MicroRNA-206 targets notch3, activates apoptosis, and inhibits tumor cell migration and focus formation. *J. Biol. Chem.* **2009**, *284*, 31921–31927. [CrossRef]
50. Ren, X.L.; He, G.Y.; Li, X.M.; Men, H.; Yi, L.Z.; Lu, G.F.; Xin, S.N.; Wu, P.X.; Li, Y.L.; Liao, W.T.; et al. MicroRNA-206 functions as a tumor suppressor in colorectal cancer by targeting FMNL2. *J. Cancer Res. Clin. Oncol.* **2016**, *142*, 581–592. [CrossRef] [PubMed]
51. Du, L.; Huang, G.H.; Mou, K.J.; Xiang, Y.; Tang, J.H.; Xu, W.; Xia, S.L.; Zhao, J.N.; Lv, S.Q. MiR-206 is down-regulated and suppresses cell proliferation by targeting FOXP1 in brain gliomas. *Int. J. Clin. Exp. Pathol.* **2018**, *11*, 3405–3415. [PubMed]
52. Weber, J.A.; Baxter, D.H.; Zhang, S.; Huang, D.Y.; Huang, K.H.; Lee, M.J.; Galas, D.J.; Wang, K. The MicroRNA spectrum in 12 body fluids. *Clin. Chem.* **2010**, *56*, 1733–1741. [CrossRef]
53. Turchinovich, A.; Weiz, L.; Langheinz, A.; Burwinkel, B. Characterization of extracellular circulating microRNA. *Nucleic Acids Res.* **2011**, *39*, 7223–7233. [CrossRef]
54. Gallo, A.; Tandon, M.; Alevizos, I.; Illei, G.G. The majority of microRNAs detectable in serum and saliva is concentrated in exosomes. *PLoS ONE* **2012**, *7*, e30679. [CrossRef] [PubMed]
55. Mao, S.; Sun, Q.; Xiao, H.; Zhang, C.; Li, L. Secreted miR-34a in astrocytic shedding vesicles enhanced the vulnerability of dopaminergic neurons to neurotoxins by targeting Bcl-2. *Protein Cell* **2015**, *6*, 529–540. [CrossRef]
56. Jella, K.K.; Nasti, T.H.; Li, Z.; Malla, S.R.; Buchwald, Z.S.; Khan, M.K. Exosomes, their biogenesis and role in inter-cellular communication, tumor microenvironment and cancer immunotherapy. *Vaccines* **2018**, *6*, 69. [CrossRef]
57. Safdar, A.; Saleem, A.; Tarnopolsky, M.A. The potential of endurance exercise-derived exosomes to treat metabolic diseases. *Nat. Rev. Endocrinol.* **2016**, *12*, 504–517. [CrossRef] [PubMed]
58. Cheng, L.; Sharples, R.A.; Scicluna, B.J.; Hill, A.F. Exosomes provide a protective and enriched source of miRNA for biomarker profiling compared to intracellular and cell-free blood. *J. Extracell. Vesicles* **2014**, *3*. [CrossRef] [PubMed]
59. Mathivanan, S.; Lim, J.W.; Tauro, B.J.; Ji, H.; Moritz, R.L.; Simpson, R.J. Proteomics analysis of A33 immunoaffinity-purified exosomes released from the human colon tumor cell line LIM1215 reveals a tissue-specific protein signature. *Mol. Cell Proteom.* **2010**, *9*, 197–208. [CrossRef]
60. Li, C.; Ni, Y.-Q.; Xu, H.; Xiang, Q.-Y.; Zhao, Y.; Zhan, J.-K.; He, J.-Y.; Li, S.; Liu, Y.-S. Roles and mechanisms of exosomal non-coding RNAs in human health and diseases. *Signal Transduct. Target. Ther.* **2021**, *6*, 383. [CrossRef]
61. Castaño, C.; Novials, A.; Párrizas, M. Exosomes and diabetes. *Diabetes Metab. Res. Rev.* **2019**, *35*, e3107. [CrossRef]
62. Zhang, D.; Lee, H.; Zhu, Z.; Minhas, J.K.; Jin, Y. Enrichment of selective miRNAs in exosomes and delivery of exosomal miRNAs in vitro and in vivo. *Am. J. Physiol. Lung Cell Mol. Physiol.* **2017**, *312*, L110–L121. [CrossRef]
63. Cheung, K.C.P.; Jiao, M.; Xingxuan, C.; Wei, J. Extracellular vesicles derived from host and gut microbiota as promising nanocarriers for targeted therapy in osteoporosis and osteoarthritis. *Front. Pharmacol.* **2022**, *13*, 1051134. [CrossRef] [PubMed]
64. Arroyo, J.D.; Chevillet, J.R.; Kroh, E.M.; Ruf, I.K.; Pritchard, C.C.; Gibson, D.F.; Mitchell, P.S.; Bennett, C.F.; Pogosova-Agadjanyan, E.L.; Stirewalt, D.L.; et al. Argonaute2 complexes carry a population of circulating microRNAs independent of vesicles in human plasma. *Proc. Natl. Acad. Sci. USA* **2011**, *108*, 5003–5008. [CrossRef] [PubMed]
65. Tabet, F.; Vickers, K.C.; Cuesta Torres, L.F.; Wiese, C.B.; Shoucri, B.M.; Lambert, G.; Catherinet, C.; Prado-Lourenco, L.; Levin, M.G.; Thacker, S.; et al. HDL-transferred microRNA-223 regulates ICAM-1 expression in endothelial cells. *Nat. Commun.* **2014**, *5*, 3292. [CrossRef] [PubMed]
66. Wagner, J.; Riwanto, M.; Besler, C.; Knau, A.; Fichtlscherer, S.; Röxe, T.; Zeiher, A.M.; Landmesser, U.; Dimmeler, S. Characterization of levels and cellular transfer of circulating lipoprotein-bound microRNAs. *Arterioscler. Thromb. Vasc. Biol.* **2013**, *33*, 1392–1400. [CrossRef]
67. Huang, W. MicroRNAs: Biomarkers, Diagnostics, and Therapeutics. In *Bioinformatics in MicroRNA Research*; Huang, J., Borchert, G.M., Dou, D., Huan, J., Lan, W., Tan, M., Wu, B., Eds.; Springer: New York, NY, USA, 2017; pp. 57–67.
68. Bottani, M.; Banfi, G.; Lombardi, G. The clinical potential of circulating miRNAs as biomarkers: Present and future applications for diagnosis and prognosis of age-associated bone diseases. *Biomolecules* **2020**, *10*, 589. [CrossRef]

69. Li, Z.; Chen, Z.; Wang, X.; Li, Z.; Sun, H.; Wei, J.; Zeng, X.; Cao, X.; Wan, C. Integrated analysis of miRNAs and gene expression profiles reveals potential biomarkers for osteoarthritis. *Front. Genet.* **2022**, *13*, 814645. [CrossRef]
70. McAlinden, A.; Im, G.I. MicroRNAs in orthopaedic research: Disease associations, potential therapeutic applications, and perspectives. *J. Orthop. Res.* **2018**, *36*, 33–51. [CrossRef]
71. Liu, J.; Dang, L.; Wu, X.; Li, D.; Ren, Q.; Lu, A.; Zhang, G. microRNA-mediated regulation of bone remodeling: A brief review. *JBMR Plus* **2019**, *3*, e10213. [CrossRef]
72. Wei, J.; Shi, Y.; Zheng, L.; Zhou, B.; Inose, H.; Wang, J.; Guo, X.E.; Grosschedl, R.; Karsenty, G. miR-34s inhibit osteoblast proliferation and differentiation in the mouse by targeting SATB2. *J. Cell Biol.* **2012**, *197*, 509–521. [CrossRef]
73. Inose, H.; Ochi, H.; Kimura, A.; Fujita, K.; Xu, R.; Sato, S.; Iwasaki, M.; Sunamura, S.; Takeuchi, Y.; Fukumoto, S.; et al. A microRNA regulatory mechanism of osteoblast differentiation. *Proc. Natl. Acad. Sci. USA* **2009**, *106*, 20794–20799. [CrossRef] [PubMed]
74. Williams, A.; Henao-Mejia, J.; Harman, C.C.; Flavell, R.A. miR-181 and metabolic regulation in the immune system. *Cold Spring Harb. Symp. Quant. Biol.* **2013**, *78*, 223–230. [CrossRef] [PubMed]
75. Inoue, K.; Deng, Z.; Chen, Y.; Giannopoulou, E.; Xu, R.; Gong, S.; Greenblatt, M.B.; Mangala, L.S.; Lopez-Berestein, G.; Kirsch, D.G.; et al. Bone protection by inhibition of microRNA-182. *Nat. Commun.* **2018**, *9*, 4108. [CrossRef]
76. Yang, N.; Wang, G.; Hu, C.; Shi, Y.; Liao, L.; Shi, S.; Cai, Y.; Cheng, S.; Wang, X.; Liu, Y.; et al. Tumor necrosis factor α suppresses the mesenchymal stem cell osteogenesis promoter miR-21 in estrogen deficiency-induced osteoporosis. *J. Bone Miner. Res.* **2013**, *28*, 559–573. [CrossRef] [PubMed]
77. Sugatani, T.; Vacher, J.; Hruska, K.A. A microRNA expression signature of osteoclastogenesis. *Blood* **2011**, *117*, 3648–3657. [CrossRef] [PubMed]
78. Hu, C.H.; Sui, B.D.; Du, F.Y.; Shuai, Y.; Zheng, C.X.; Zhao, P.; Yu, X.R.; Jin, Y. miR-21 deficiency inhibits osteoclast function and prevents bone loss in mice. *Sci. Rep.* **2017**, *7*, 43191. [CrossRef] [PubMed]
79. Yang, J.X.; Xie, P.; Li, Y.S.; Wen, T.; Yang, X.C. Osteoclast-derived miR-23a-5p-containing exosomes inhibit osteogenic differentiation by regulating Runx2. *Cell. Signal.* **2020**, *70*, 109504. [CrossRef]
80. Eisenberg, I.; Alexander, M.S.; Kunkel, L.M. miRNAS in normal and diseased skeletal muscle. *J. Cell Mol. Med.* **2009**, *13*, 2–11. [CrossRef]
81. Kobayashi, T.; Lu, J.; Cobb, B.S.; Rodda, S.J.; McMahon, A.P.; Schipani, E.; Merkenschlager, M.; Kronenberg, H.M. Dicer-dependent pathways regulate chondrocyte proliferation and differentiation. *Proc. Natl. Acad. Sci. USA* **2008**, *105*, 1949–1954. [CrossRef]
82. Sondag, G.R.; Haqqi, T.M. The role of microRNAs and their targets in osteoarthritis. *Curr. Rheumatol. Rep.* **2016**, *18*, 56. [CrossRef] [PubMed]
83. Li, X.; Gibson, G.; Kim, J.-S.; Kroin, J.; Xu, S.; van Wijnen, A.J.; Im, H.-J. MicroRNA-146a is linked to pain-related pathophysiology of osteoarthritis. *Gene* **2011**, *480*, 34–41. [CrossRef] [PubMed]
84. Wang, X.; Guo, Y.; Wang, C.; Yu, H.; Yu, X.; Yu, H. MicroRNA-142-3p inhibits chondrocyte apoptosis and inflammation in osteoarthritis by targeting HMGB1. *Inflammation* **2016**, *39*, 1718–1728. [CrossRef] [PubMed]
85. Li, Z.C.; Han, N.; Li, X.; Li, G.; Liu, Y.Z.; Sun, G.X.; Wang, Y.; Chen, G.T.; Li, G.F. Decreased expression of microRNA-130a correlates with TNF-α in the development of osteoarthritis. *Int. J. Clin. Exp. Pathol.* **2015**, *8*, 2555–2564.
86. Miyaki, S.; Sato, T.; Inoue, A.; Otsuki, S.; Ito, Y.; Yokoyama, S.; Kato, Y.; Takemoto, F.; Nakasa, T.; Yamashita, S.; et al. MicroRNA-140 plays dual roles in both cartilage development and homeostasis. *Genes Dev.* **2010**, *24*, 1173–1185. [CrossRef]
87. Skrzypa, M.; Szala, D.; Gablo, N.; Czech, J.; Pajak, J.; Kopanska, M.; Trzeciak, M.; Gargasz, K.; Snela, S.; Zawlik, I. miRNA-146a-5p is upregulated in serum and cartilage samples of patients with osteoarthritis. *Pol. Przegl. Chir.* **2019**, *91*, 1–5. [CrossRef] [PubMed]
88. Nakasa, T.; Miyaki, S.; Okubo, A.; Hashimoto, M.; Nishida, K.; Ochi, M.; Asahara, H. Expression of microRNA-146 in rheumatoid arthritis synovial tissue. *Arthritis Rheum.* **2008**, *58*, 1284–1292. [CrossRef]
89. Saba, R.; Sorensen, D.L.; Booth, S.A. MicroRNA-146a: A dominant, negative regulator of the innate immune response. *Front. Immunol.* **2014**, *5*, 578. [CrossRef] [PubMed]
90. Song, J.; Ahn, C.; Chun, C.H.; Jin, E.J. A long non-coding RNA, GAS5, plays a critical role in the regulation of miR-21 during osteoarthritis. *J. Orthop. Res.* **2014**, *32*, 1628–1635. [CrossRef]
91. Kumarswamy, R.; Volkmann, I.; Thum, T. Regulation and function of miRNA-21 in health and disease. *RNA Biol.* **2011**, *8*, 706–713. [CrossRef]
92. Francis-West, P.H.; Abdelfattah, A.; Chen, P.; Allen, C.; Parish, J.; Ladher, R.; Allen, S.; MacPherson, S.; Luyten, F.P.; Archer, C.W. Mechanisms of GDF-5 action during skeletal development. *Development* **1999**, *126*, 1305–1315. [CrossRef] [PubMed]
93. Yu, X.M.; Meng, H.Y.; Yuan, X.L.; Wang, Y.; Guo, Q.Y.; Peng, J.; Wang, A.Y.; Lu, S.B. MicroRNAs' involvement in osteoarthritis and the prospects for treatments. *Evid. Based Complement. Alternat. Med.* **2015**, *2015*, 236179. [CrossRef]
94. Endisha, H.; Rockel, J.; Jurisica, I.; Kapoor, M. The complex landscape of microRNAs in articular cartilage: Biology, pathology, and therapeutic targets. *JCI Insight* **2018**, *3*, e121630. [CrossRef] [PubMed]
95. Dai, L.; Zhang, X.; Hu, X.; Zhou, C.; Ao, Y. Silencing of microRNA-101 prevents IL-1β-induced extracellular matrix degradation in chondrocytes. *Arthritis Res. Ther.* **2012**, *14*, R268. [CrossRef] [PubMed]
96. Wang, Q.; Wang, W.; Zhang, F.; Deng, Y.; Long, Z. NEAT1/miR-181c regulates osteopontin (OPN)-mediated synoviocyte proliferation in osteoarthritis. *J. Cell Biochem.* **2017**, *118*, 3775–3784. [CrossRef]

97. Steck, E.; Boeuf, S.; Gabler, J.; Werth, N.; Schnatzer, P.; Diederichs, S.; Richter, W. Regulation of H19 and its encoded microRNA-675 in osteoarthritis and under anabolic and catabolic in vitro conditions. *J. Mol. Med.* **2012**, *90*, 1185–1195. [CrossRef]
98. Kang, Y.; Song, J.; Kim, D.; Ahn, C.; Park, S.; Chun, C.H.; Jin, E.J. PCGEM1 stimulates proliferation of osteoarthritic synoviocytes by acting as a sponge for miR-770. *J. Orthop. Res.* **2016**, *34*, 412–418. [CrossRef]
99. Yang, C.-R.; Shih, K.-S.; Liou, J.-P.; Wu, Y.-W.; Hsieh, I.N.; Lee, H.-Y.; Lin, T.-C.; Wang, J.-H. Denbinobin upregulates miR-146a expression and attenuates IL-1β-induced upregulation of ICAM-1 and VCAM-1 expressions in osteoarthritis fibroblast-like synoviocytes. *J. Mol. Med.* **2014**, *92*, 1147–1158. [CrossRef]
100. Tardif, G.; Hum, D.; Pelletier, J.P.; Duval, N.; Martel-Pelletier, J. Regulation of the IGFBP-5 and MMP-13 genes by the microRNAs miR-140 and miR-27a in human osteoarthritic chondrocytes. *BMC Musculoskelet. Disord.* **2009**, *10*, 148. [CrossRef]
101. Gao, Y.; He, Y.; Ding, J.; Wu, K.; Hu, B.; Liu, Y.; Wu, Y.; Guo, B.; Shen, Y.; Landi, D.; et al. An insertion/deletion polymorphism at miRNA-122-binding site in the interleukin-1alpha 3′ untranslated region confers risk for hepatocellular carcinoma. *Carcinogenesis* **2009**, *30*, 2064–2069. [CrossRef]
102. Lu, X.; Lin, J.; Jin, J.; Qian, W.; Weng, X. Hsa-miR-15a exerts protective effects against osteoarthritis by targeting aggrecanase-2 (ADAMTS5) in human chondrocytes. *Int. J. Mol. Med.* **2016**, *37*, 509–516. [CrossRef] [PubMed]
103. Park, S.J.; Cheon, E.J.; Lee, M.H.; Kim, H.A. MicroRNA-127-5p regulates matrix metalloproteinase 13 expression and interleukin-1β-induced catabolic effects in human chondrocytes. *Arthritis Rheum.* **2013**, *65*, 3141–3152. [CrossRef]
104. Liu, H.; Yan, L.; Li, X.; Li, D.; Wang, G.; Shen, N.N.; Li, J.J.; Wang, B. MicroRNA expression in osteoarthritis: A meta-analysis. *Clin. Exp. Med.* **2023**, 1–13. [CrossRef]
105. Duan, L.; Liang, Y.; Xu, X.; Xiao, Y.; Wang, D. Recent progress on the role of miR-140 in cartilage matrix remodelling and its implications for osteoarthritis treatment. *Arthritis Res. Ther.* **2020**, *22*, 194. [CrossRef]
106. Boehme, K.A.; Rolauffs, B. Onset and progression of human osteoarthritis-can growth factors, inflammatory cytokines, or differential miRNA expression concomitantly induce proliferation, ECM degradation, and inflammation in articular cartilage? *Int. J. Mol. Sci.* **2018**, *19*, 2282. [CrossRef] [PubMed]
107. Xie, Y.; Zinkle, A.; Chen, L.; Mohammadi, M. Fibroblast growth factor signalling in osteoarthritis and cartilage repair. *Nat. Rev. Rheumatol.* **2020**, *16*, 547–564. [CrossRef]
108. Felka, T.; Rothdiener, M.; Bast, S.; Uynuk-Ool, T.; Zouhair, S.; Ochs, B.G.; De Zwart, P.; Stoeckle, U.; Aicher, W.K.; Hart, M.L.; et al. Loss of spatial organization and destruction of the pericellular matrix in early osteoarthritis in vivo and in a novel in vitro methodology. *Osteoarthr. Cartil.* **2016**, *24*, 1200–1209. [CrossRef]
109. Jann, J.; Gascon, S.; Roux, S.; Faucheux, N. Influence of the TGF-β superfamily on osteoclasts/osteoblasts balance in physiological and pathological bone conditions. *Int. J. Mol. Sci.* **2020**, *21*, 7597. [CrossRef]
110. Huang, J.; Zhao, L.; Chen, D. Growth factor signalling in osteoarthritis. *Growth Factors* **2018**, *36*, 187–195. [CrossRef]
111. Yoo, K.H.; Thapa, N.; Chwae, Y.J.; Yoon, S.H.; Kim, B.J.; Lee, J.O.; Jang, Y.N.; Kim, J. Transforming growth factor-β family and stem cell-derived exosome therapeutic treatment in osteoarthritis. *Int. J. Mol. Med.* **2022**, *49*, 62. [CrossRef]
112. Li, J.; Huang, J.; Dai, L.; Yu, D.; Chen, Q.; Zhang, X.; Dai, K. miR-146a, an IL-1β responsive miRNA, induces vascular endothelial growth factor and chondrocyte apoptosis by targeting Smad4. *Arthritis Res. Ther.* **2012**, *14*, R75. [CrossRef] [PubMed]
113. Meng, Y.B.; Li, X.; Li, Z.Y.; Zhao, J.; Yuan, X.B.; Ren, Y.; Cui, Z.D.; Liu, Y.D.; Yang, X.J. microRNA-21 promotes osteogenic differentiation of mesenchymal stem cells by the PI3K/β-catenin pathway. *J. Orthop. Res.* **2015**, *33*, 957–964. [CrossRef] [PubMed]
114. Qin, Y.; Guan, J.; Zhang, C. Mesenchymal stem cells: Mechanisms and role in bone regeneration. *Postgrad. Med. J.* **2014**, *90*, 643–647. [CrossRef]
115. Shirazi, S.; Huang, C.-C.; Kang, M.; Lu, Y.; Ravindran, S.; Cooper, L.F. The importance of cellular and exosomal miRNAs in mesenchymal stem cell osteoblastic differentiation. *Sci. Rep.* **2021**, *11*, 5953. [CrossRef]
116. Jones, S.W.; Watkins, G.; Le Good, N.; Roberts, S.; Murphy, C.L.; Brockbank, S.M.V.; Needham, M.R.C.; Read, S.J.; Newham, P. The identification of differentially expressed microRNA in osteoarthritic tissue that modulate the production of TNF-alpha and MMP13. *Osteoarthr. Cartil.* **2009**, *17*, 464–472. [CrossRef] [PubMed]
117. Gu, R.; Liu, N.; Luo, S.; Huang, W.; Zha, Z.; Yang, J. MicroRNA-9 regulates the development of knee osteoarthritis through the NF-kappaB1 pathway in chondrocytes. *Medicine* **2016**, *95*, e4315. [CrossRef]
118. Borgonio Cuadra, V.M.; González-Huerta, N.C.; Romero-Córdoba, S.; Hidalgo-Miranda, A.; Miranda-Duarte, A. Altered expression of circulating microRNA in plasma of patients with primary osteoarthritis and in silico analysis of their pathways. *PLoS ONE* **2014**, *9*, e97690. [CrossRef]
119. Iliopoulos, D.; Malizos, K.N.; Oikonomou, P.; Tsezou, A. Integrative microRNA and proteomic approaches identify novel osteoarthritis genes and their collaborative metabolic and inflammatory networks. *PLoS ONE* **2008**, *3*, e3740. [CrossRef]
120. Murata, K.; Yoshitomi, H.; Tanida, S.; Ishikawa, M.; Nishitani, K.; Ito, H.; Nakamura, T. Plasma and synovial fluid microRNAs as potential biomarkers of rheumatoid arthritis and osteoarthritis. *Arthritis Res. Ther.* **2010**, *12*, R86. [CrossRef]
121. Li, L.; Jia, J.; Liu, X.; Yang, S.; Ye, S.; Yang, W.; Zhang, Y. MicroRNA-16-5p controls development of osteoarthritis by targeting SMAD3 in chondrocytes. *Curr. Pharm. Des.* **2015**, *21*, 5160–5167. [CrossRef]
122. Zhang, Y.; Jia, J.; Yang, S.; Liu, X.; Ye, S.; Tian, H. MicroRNA-21 controls the development of osteoarthritis by targeting GDF-5 in chondrocytes. *Exp. Mol. Med.* **2014**, *46*, e79. [CrossRef]
123. Yassin, A.M.; AbuBakr, H.O.; Abdelgalil, A.I.; Farid, O.A.; El-Behairy, A.M.; Gouda, E.M. Circulating miR-146b and miR-27b are efficient biomarkers for early diagnosis of Equidae osteoarthritis. *Sci. Rep.* **2023**, *13*, 7966. [CrossRef]

124. Li, Y.H.; Tavallaee, G.; Tokar, T.; Nakamura, A.; Sundararajan, K.; Weston, A.; Sharma, A.; Mahomed, N.N.; Gandhi, R.; Jurisica, I.; et al. Identification of synovial fluid microRNA signature in knee osteoarthritis: Differentiating early- and late-stage knee osteoarthritis. *Osteoarthr. Cartil.* **2016**, *24*, 1577–1586. [CrossRef] [PubMed]
125. Chang, T.; Xie, J.; Li, H.; Li, D.; Liu, P.; Hu, Y. MicroRNA-30a promotes extracellular matrix degradation in articular cartilage via downregulation of Sox9. *Cell Prolif.* **2016**, *49*, 207–218. [CrossRef] [PubMed]
126. Zhang, W.; Hu, C.; Zhang, C.; Luo, C.; Zhong, B.; Yu, X. MiRNA-132 regulates the development of osteoarthritis in correlation with the modulation of PTEN/PI3K/AKT signaling. *BMC Geriatr.* **2021**, *21*, 175. [CrossRef]
127. Si, H.-b.; Zeng, Y.; Zhou, Z.-k.; Pei, F.-x.; Lu, Y.-r.; Cheng, J.-q.; Shen, B. Expression of miRNA-140 in chondrocytes and synovial fluid of knee joints in patients with osteoarthritis. *Chin. Med. Sci.* **2016**, *31*, 207–212. [CrossRef] [PubMed]
128. Rousseau, J.C.; Millet, M.; Croset, M.; Sornay-Rendu, E.; Borel, O.; Chapurlat, R. Association of circulating microRNAs with prevalent and incident knee osteoarthritis in women: The OFELY study. *Arthritis Res. Ther.* **2020**, *22*, 2. [CrossRef]
129. Lin, Z.; Tian, X.Y.; Huang, X.X.; He, L.L.; Xu, F. microRNA-186 inhibition of PI3K-AKT pathway via SPP1 inhibits chondrocyte apoptosis in mice with osteoarthritis. *J. Cell Physiol.* **2019**, *234*, 6042–6053. [CrossRef]
130. Akhtar, N.; Rasheed, Z.; Ramamurthy, S.; Anbazhagan, A.N.; Voss, F.R.; Haqqi, T.M. MicroRNA-27b regulates the expression of matrix metalloproteinase 13 in human osteoarthritis chondrocytes. *Arthritis Rheum.* **2010**, *62*, 1361–1371. [CrossRef]
131. Yamasaki, K.; Nakasa, T.; Miyaki, S.; Ishikawa, M.; Deie, M.; Adachi, N.; Yasunaga, Y.; Asahara, H.; Ochi, M. Expression of MicroRNA-146a in osteoarthritis cartilage. *Arthritis Rheum.* **2009**, *60*, 1035–1041. [CrossRef]
132. Pacholewska, A.; Mach, N.; Mata, X.; Vaiman, A.; Schibler, L.; Barrey, E.; Gerber, V. Novel equine tissue miRNAs and breed-related miRNA expressed in serum. *BMC Genom.* **2016**, *17*, 831. [CrossRef]
133. Castanheira, C.; Balaskas, P.; Falls, C.; Ashraf-Kharaz, Y.; Clegg, P.; Burke, K.; Fang, Y.; Dyer, P.; Welting, T.J.M.; Peffers, M.J. Equine synovial fluid small non-coding RNA signatures in early osteoarthritis. *BMC Vet. Res.* **2021**, *17*, 26. [CrossRef]
134. Sui, C.; Zhang, L.; Hu, Y. MicroRNA-let-7a inhibition inhibits LPS-induced inflammatory injury of chondrocytes by targeting IL6R. *Mol. Med. Rep.* **2019**, *20*, 2633–2640. [CrossRef]
135. Feng, L.; Feng, C.; Wang, C.X.; Xu, D.Y.; Chen, J.J.; Huang, J.F.; Tan, P.L.; Shen, J.M. Circulating microRNA let-7e is decreased in knee osteoarthritis, accompanied by elevated apoptosis and reduced autophagy. *Int. J. Mol. Med.* **2020**, *45*, 1464–1476. [CrossRef]
136. Ntoumou, E.; Tzetis, M.; Braoudaki, M.; Lambrou, G.; Poulou, M.; Malizos, K.; Stefanou, N.; Anastasopoulou, L.; Tsezou, A. Serum microRNA array analysis identifies miR-140-3p, miR-33b-3p and miR-671-3p as potential osteoarthritis biomarkers involved in metabolic processes. *Clin. Epigenetics* **2017**, *9*, 127. [CrossRef]
137. Lao, T.D.; Le, T.A.H. Data Integration Reveals the Potential Biomarkers of Circulating MicroRNAs in Osteoarthritis. *Diagnostics* **2021**, *11*, 412. [CrossRef]
138. He, Q.; Ren, X.; Chen, J.; Li, Y.; Tang, X.; Wen, X.; Yang, X.; Zhang, J.; Wang, Y.; Ma, J.; et al. miR-16 targets fibroblast growth factor 2 to inhibit NPC cell proliferation and invasion via PI3K/AKT and MAPK signaling pathways. *Oncotarget* **2016**, *7*, 3047–3058. [CrossRef]
139. Chou, C.K.; Chi, S.Y.; Huang, C.H.; Chou, F.F.; Huang, C.C.; Liu, R.T.; Kang, H.Y. IRAK1, a target of miR-146b, reduces cell aggressiveness of human papillary thyroid carcinoma. *J. Clin. Endocrinol. Metab.* **2016**, *101*, 4357–4366. [CrossRef]
140. Li, Y.; Zhang, H.; Dong, Y.; Fan, Y.; Li, Y.; Zhao, C.; Wang, C.; Liu, J.; Li, X.; Dong, M.; et al. MiR-146b-5p functions as a suppressor miRNA and prognosis predictor in non-small cell lung cancer. *J. Cancer* **2017**, *8*, 1704–1716. [CrossRef]
141. Liu, X.; Liu, L.; Zhang, H.; Shao, Y.; Chen, Z.; Feng, X.; Fang, H.; Zhao, C.; Pan, J.; Zhang, H.; et al. MiR-146b accelerates osteoarthritis progression by targeting alpha-2-macroglobulin. *Aging* **2019**, *11*, 6014–6028. [CrossRef]
142. Wan, L.; Zhao, Q.; Niu, G.; Xiang, T.; Ding, C.; Wang, S. Plasma miR-136 can be used to screen patients with knee osteoarthritis from healthy controls by targeting IL-17. *Exp. Ther. Med.* **2018**, *16*, 3419–3424. [CrossRef] [PubMed]
143. Beyer, C.; Zampetaki, A.; Lin, N.Y.; Kleyer, A.; Perricone, C.; Iagnocco, A.; Distler, A.; Langley, S.R.; Gelse, K.; Sesselmann, S.; et al. Signature of circulating microRNAs in osteoarthritis. *Ann. Rheum. Dis.* **2015**, *74*, e18. [CrossRef] [PubMed]
144. Kong, R.; Gao, J.; Si, Y.; Zhao, D. Combination of circulating miR-19b-3p, miR-122-5p and miR-486-5p expressions correlates with risk and disease severity of knee osteoarthritis. *Am. J. Transl. Res.* **2017**, *9*, 2852–2864.
145. Taylor, C.R. Introduction to Predictive Biomarkers: Definitions and Characteristics. In *Predictive Biomarkers in Oncology: Applications in Precision Medicine*; Badve, S., Kumar, G.L., Eds.; Springer International Publishing: Cham, Switzerland, 2019; pp. 3–18.
146. Hackl, M.; Heilmeier, U.; Weilner, S.; Grillari, J. Circulating microRNAs as novel biomarkers for bone diseases—Complex signatures for multifactorial diseases? *Mol. Cell Endocrinol.* **2016**, *432*, 83–95. [CrossRef]
147. Bustin, S.A.; Nolan, T. Pitfalls of quantitative real-time reverse-transcription polymerase chain reaction. *J. Biomol. Tech.* **2004**, *15*, 155–166. [PubMed]
148. Dong, Z.; Jiang, H.; Jian, X.; Zhang, W. Change of miRNA expression profiles in patients with knee osteoarthritis before and after celecoxib treatment. *J. Clin. Lab. Anal.* **2019**, *33*, e22648. [CrossRef]
149. Goldring, M.B.; Otero, M. Inflammation in osteoarthritis. *Curr. Opin. Rheumatol.* **2011**, *23*, 471–478. [CrossRef]
150. Sellam, J.; Berenbaum, F. The role of synovitis in pathophysiology and clinical symptoms of osteoarthritis. *Nat. Rev. Rheumatol.* **2010**, *6*, 625–635. [CrossRef] [PubMed]
151. Berenbaum, F. Osteoarthritis as an inflammatory disease (osteoarthritis is not osteoarthrosis!). *Osteoarthr. Cartil.* **2013**, *21*, 16–21. [CrossRef]

152. Scanzello, C.R.; Goldring, S.R. The role of synovitis in osteoarthritis pathogenesis. *Bone* **2012**, *51*, 249–257. [CrossRef]
153. Felekkis, K.; Pieri, M.; Papaneophytou, C. Variability in the levels of exosomal miRNAs among human subjects could be explained by differential interactions of exosomes with the endothelium. *IUBMB Life* **2021**, *73*, 1400–1405. [CrossRef] [PubMed]
154. Lombardi, G.; Perego, S.; Sansoni, V.; Banfi, G. Circulating miRNA as fine regulators of the physiological responses to physical activity: Pre-analytical warnings for a novel class of biomarkers. *Clin. Biochem.* **2016**, *49*, 1331–1339. [CrossRef] [PubMed]
155. Ultimo, S.; Zauli, G.; Martelli, A.M.; Vitale, M.; McCubrey, J.A.; Capitani, S.; Neri, L.M. Influence of physical exercise on microRNAs in skeletal muscle regeneration, aging and diseases. *Oncotarget* **2018**, *9*, 17220–17237. [CrossRef] [PubMed]
156. Horak, M.; Novak, J.; Bienertova-Vasku, J. Muscle-specific microRNAs in skeletal muscle development. *Dev. Biol.* **2016**, *410*, 1–13. [CrossRef]
157. D'Souza, R.F.; Markworth, J.F.; Aasen, K.M.M.; Zeng, N.; Cameron-Smith, D.; Mitchell, C.J. Acute resistance exercise modulates microRNA expression profiles: Combined tissue and circulatory targeted analyses. *PLoS ONE* **2017**, *12*, e0181594. [CrossRef]
158. Boeckel, J.N.; Thomé, C.E.; Leistner, D.; Zeiher, A.M.; Fichtlscherer, S.; Dimmeler, S. Heparin selectively affects the quantification of microRNAs in human blood samples. *Clin. Chem.* **2013**, *59*, 1125–1127. [CrossRef]
159. Russo, I.; Penna, C.; Musso, T.; Popara, J.; Alloatti, G.; Cavalot, F.; Pagliaro, P. Platelets, diabetes and myocardial ischemia/reperfusion injury. *Cardiovasc. Diabetol.* **2017**, *16*, 71. [CrossRef]
160. Willeit, P.; Zampetaki, A.; Dudek, K.; Kaudewitz, D.; King, A.; Kirkby, N.S.; Crosby-Nwaobi, R.; Prokopi, M.; Drozdov, I.; Langley, S.R.; et al. Circulating microRNAs as novel biomarkers for platelet activation. *Circ. Res.* **2013**, *112*, 595–600. [CrossRef]
161. Boileau, A.; Lino Cardenas, C.L.; Lindsay, M.E.; Devaux, Y. Endogenous Heparin Interferes with Quantification of MicroRNAs by RT-qPCR. *Clin. Chem.* **2018**, *64*, 863–865. [CrossRef]
162. Khazeei Tabari, M.A.; Mishan, M.A.; Moradi, M.; Khandan, M.; Khoshhal, H.; Mahrooz, A.; Bagheri, A. Noncoding RNA roles in pharmacogenomic responses to aspirin: New molecular mechanisms for an old drug. *Biomed. Res. Int.* **2021**, *2021*, 6830560. [CrossRef]
163. Felekkis, K.; Papaneophytou, C. Challenges in using circulating micro-RNAs as biomarkers for cardiovascular diseases. *Int. J. Mol. Sci.* **2020**, *21*, 561. [CrossRef] [PubMed]
164. Wang, K.; Yuan, Y.; Cho, J.-H.; McClarty, S.; Baxter, D.; Galas, D.J. Comparing the microRNA spectrum between serum and plasma. *PLoS ONE* **2012**, *7*, e41561. [CrossRef] [PubMed]
165. Hastings, M.L.; Palma, J.; Duelli, D.M. Sensitive PCR-based quantitation of cell-free circulating microRNAs. *Methods* **2012**, *58*, 144–150. [CrossRef]
166. Kim, D.J.; Linnstaedt, S.; Palma, J.; Park, J.C.; Ntrivalas, E.; Kwak-Kim, J.Y.; Gilman-Sachs, A.; Beaman, K.; Hastings, M.L.; Martin, J.N.; et al. Plasma components affect accuracy of circulating cancer-related microRNA quantitation. *J. Mol. Diagn.* **2012**, *14*, 71–80. [CrossRef]
167. Sohel, M.H. Extracellular/circulating microRNAs: Release mechanisms, functions and challenges. *Adv. Life Sci.* **2016**, *10*, 175–186. [CrossRef]
168. Sourvinou, I.S.; Markou, A.; Lianidou, E.S. Quantification of circulating miRNAs in plasma: Effect of preanalytical and analytical parameters on their isolation and stability. *J. Mol. Diagn.* **2013**, *15*, 827–834. [CrossRef]
169. Cheng, H.H.; Yi, H.S.; Kim, Y.; Kroh, E.M.; Chien, J.W.; Eaton, K.D.; Goodman, M.T.; Tait, J.F.; Tewari, M.; Pritchard, C.C. Plasma processing conditions substantially influence circulating microRNA biomarker levels. *PLoS ONE* **2013**, *8*, e64795. [CrossRef]
170. Ono, S.; Lam, S.; Nagahara, M.; Hoon, D.S.B. Circulating microRNA biomarkers as liquid biopsy for cancer patients: Pros and cons of current assays. *J. Clin. Med.* **2015**, *4*, 1890–1907. [CrossRef]
171. Glinge, C.; Clauss, S.; Boddum, K.; Jabbari, R.; Jabbari, J.; Risgaard, B.; Tomsits, P.; Hildebrand, B.; Kääb, S.; Wakili, R.; et al. Stability of circulating blood-based microRNAs—Pre-analytic methodological considerations. *PLoS ONE* **2017**, *12*, e0167969. [CrossRef]
172. Ruiz-Godoy, L.; Enríquez-Cárcamo, V.; Suárez-Roa, L.; Lopez-Castro, M.; Santamaría, A.; Orozco-Morales, M.; Colín-González, A. Identification of specific pre-analytical quality control markers in plasma and serum samples. *Anal. Methods* **2019**, *11*, 2259–2271. [CrossRef]
173. Tuck, M.K.; Chan, D.W.; Chia, D.; Godwin, A.K.; Grizzle, W.E.; Krueger, K.E.; Rom, W.; Sanda, M.; Sorbara, L.; Stass, S.; et al. Standard operating procedures for serum and plasma collection: Early detection research network consensus statement standard operating procedure integration working group. *J. Proteome Res.* **2009**, *8*, 113–117. [CrossRef] [PubMed]
174. Mestdagh, P.; Hartmann, N.; Baeriswyl, L.; Andreasen, D.; Bernard, N.; Chen, C.; Cheo, D.; D'Andrade, P.; DeMayo, M.; Dennis, L.; et al. Evaluation of quantitative miRNA expression platforms in the microRNA quality control (miRQC) study. *Nat. Methods* **2014**, *11*, 809–815. [CrossRef]
175. Koshiol, J.; Wang, E.; Zhao, Y.; Marincola, F.; Landi, M.T. Strengths and limitations of laboratory procedures for microRNA detection. *Cancer Epidemiol. Biomarkers Prev.* **2010**, *19*, 907–911. [CrossRef] [PubMed]
176. Li, Y.; Kowdley, K.V. Method for microRNA isolation from clinical serum samples. *Anal. Biochem.* **2012**, *431*, 69–75. [CrossRef] [PubMed]
177. Lee, I.; Baxter, D.; Lee, M.Y.; Scherler, K.; Wang, K. The Importance of Standardization on Analyzing Circulating RNA. *Mol. Diagn. Ther.* **2017**, *21*, 259–268. [CrossRef]
178. Cheng, G. Circulating miRNAs: Roles in cancer diagnosis, prognosis and therapy. *Adv. Drug Deliv. Rev.* **2015**, *81*, 75–93. [CrossRef]

179. Li, Y.P.; Wei, X.C.; Li, P.C.; Chen, C.W.; Wang, X.H.; Jiao, Q.; Wang, D.M.; Wei, F.Y.; Zhang, J.Z.; Wei, L. The role of miRNAs in cartilage homeostasis. *Curr. Genom.* **2015**, *16*, 393–404. [CrossRef]
180. Mc Cormack, B.A.; González-Cantó, E.; Agababyan, C.; Espinoza-Sánchez, N.A.; Tomás-Pérez, S.; Llueca, A.; Marí-Alexandre, J.; Götte, M.; Gilabert-Estellés, J. miRNAs in the rra of personalized medicine: From biomarkers to therapeutics. *Int. J. Mol. Sci.* **2021**, *22*, 8154. [CrossRef]
181. Weiland, M.; Gao, X.H.; Zhou, L.; Mi, Q.S. Small RNAs have a large impact: Circulating microRNAs as biomarkers for human diseases. *RNA Biol.* **2012**, *9*, 850–859. [CrossRef]
182. Zampetaki, A.; Willeit, P.; Drozdov, I.; Kiechl, S.; Mayr, M. Profiling of circulating microRNAs: From single biomarkers to re-wired networks. *Cardiovasc. Res.* **2012**, *93*, 555–562. [CrossRef]
183. Avci, C.B.; Baran, Y. Use of microRNAs in personalized medicine. *Methods Mol. Biol.* **2014**, *1107*, 311–325. [CrossRef] [PubMed]
184. Liu, J.; Jennings, S.F.; Tong, W.; Hong, H. Next generation sequencing for profiling expression of miRNAs: Technical progress and applications in drug development. *J. Biomed. Sci. Eng.* **2011**, *4*, 666–676. [CrossRef] [PubMed]
185. Satam, H.; Joshi, K.; Mangrolia, U.; Waghoo, S.; Zaidi, G.; Rawool, S.; Thakare, R.P.; Banday, S.; Mishra, A.K.; Das, G.; et al. Next-generation sequencing technology: Current trends and advancements. *Biology* **2023**, *12*, 997. [CrossRef] [PubMed]
186. Liu, X.; Xiao, H.; Peng, X.; Chai, Y.; Wang, S.; Wen, G. Identification and comprehensive analysis of circRNA–miRNA–mRNA regulatory networks in osteoarthritis. *Front. Immunol.* **2023**, *13*, 1050743. [CrossRef] [PubMed]
187. Zhang, M.; Lygrisse, K.; Wang, J. Role of microRNA in osteoarthritis. *J. Arthritis* **2017**, *6*, 239. [CrossRef]
188. Bartel, D.P. Metazoan microRNAs. *Cell* **2018**, *173*, 20–51. [CrossRef]
189. Scott, S.A. Personalizing medicine with clinical pharmacogenetics. *Genet. Med.* **2011**, *13*, 987–995. [CrossRef]
190. Watt, F.E. Osteoarthritis biomarkers: Year in review. *Osteoarthr. Cartil.* **2018**, *26*, 312–318. [CrossRef]
191. Mazières, B.; Garnero, P.; Guéguen, A.; Abbal, M.; Berdah, L.; Lequesne, M.; Nguyen, M.; Salles, J.P.; Vignon, E.; Dougados, M. Molecular markers of cartilage breakdown and synovitis at baseline as predictors of structural progression of hip osteoarthritis. The ECHODIAH Cohort. *Ann. Rheum. Dis.* **2006**, *65*, 354–359. [CrossRef]
192. Jones, T.L.; Esa, M.S.; Li, K.H.C.; Krishnan, S.R.G.; Elgallab, G.M.; Pearce, M.S.; Young, D.A.; Birrell, F.N. Osteoporosis, fracture, osteoarthritis & sarcopenia: A systematic review of circulating microRNA association. *Bone* **2021**, *152*, 116068. [CrossRef]
193. Abramoff, B.; Caldera, F.E. Osteoarthritis: Pathology, diagnosis, and treatment options. *Med. Clin. N. Am.* **2020**, *104*, 293–311. [CrossRef] [PubMed]

International Journal of
Molecular Sciences

MDPI

Review

Expression and Regulatory Mechanisms of MicroRNA in Cholesteatoma: A Systematic Review

Karolina Dżaman [1], Katarzyna Czerwaty [1], Torsten E. Reichert [2], Mirosław J. Szczepański [1,3,*] and Nils Ludwig [2]

[1] Department of Otolaryngology, The Medical Centre of Postgraduate Education, 01-813 Warsaw, Poland; otolaryngologia@brodnowski.pl (K.D.); katarzynaczerwaty@gmail.com (K.C.)
[2] Department of Oral and Maxillofacial Surgery, University Hospital Regensburg, 93053 Regensburg, Germany; torsten.reichert@ukr.de (T.E.R.); nils.ludwig@ukr.de (N.L.)
[3] Department of Biochemistry, Medical University of Warsaw, 02-097 Warsaw, Poland
* Correspondence: mszczepanski@wum.edu.pl

Abstract: Cholesteatoma is a temporal bone disease characterized by dysfunctions of keratinocytes. MicroRNAs (miRNAs) are evolutionary conserved noncoding RNAs that regulate mRNA expression. They can be packaged into exosomes and transported to target cells that can be used in the future therapy of cholesteatoma. This study aimed to collect knowledge on the role of miRNAs and exosomal miRNAs in cholesteatoma and was conducted according to the PRISMA guidelines for systematic reviews. Four databases were screened: Pubmed/MEDLINE, Web of Science, Scopus, and the Cochrane Library. The last search was run on the 6th of June 2023. We included full-text original studies written in English, which examined miRNAs in cholesteatoma. The risk of bias was assessed using the Office of Health Assessment and Translation (OHAT) Risk of Bias Rating Tool, modified for the needs of this review. We identified 118 records and included 18 articles. Analyses revealed the downregulation of exosomal miR-17 as well as miR-10a-5p, miR-125b, miR-142-5p, miR34a, miR-203a, and miR-152-5p and the overexpression of exosomal miR-106b-5p as well as miR-1297, miR-26a-5p, miR-199a, miR-508-3p, miR-21-3p, miR-584-5p, and miR-16-1-3p in cholesteatoma. The role of differentially expressed miRNAs in cholesteatoma, including cell proliferation, apoptosis, the cell cycle, differentiation, bone resorption, and the remodeling process, was confirmed, making them a potential therapeutic target in this disease.

Keywords: cholesteatoma; microRNA; exosomes; small extracellular vesicles

1. Introduction

Cholesteatoma is a well-demarcated non-neoplastic cystic formation composed of stratified keratinizing squamous epithelium, developing in the temporal bone. In cholesteatoma, the hyperproliferation, migration, and altered differentiation of keratinocytes are observed, which are associated with invasive and destructive growth [1].

The incidence rate of cholesteatoma is estimated at 5–9.2 per 100,000 per year [2–6]. The familial occurrence of cholesteatoma is the basis for investigating the genetic background of the disease [7,8], although a family history in this disease is nevertheless quite rare and can only explain a limited number of cases [9]. To date, the higher incidence of cholesteatoma has been associated with various genetic syndromes characterized by craniofacial anomalies, such as Down syndrome [10,11] or Turner syndrome [12,13]. There is a classic clinical classification of cholesteatoma into two phenotypes: congenital and acquired [14]. Middle ear cholesteatomas can be also classified according to the growth pattern into anterior epitympanic, posterior epitympanic, posterior mesotympanic, two routes (both the pars flaccida and the pars tensa are involved), and undetermined [15].

Four theories of the origin of acquired cholesteatoma were presented, including the theories of invagination, immigration, squamous metaplasia, and basal cell hyperplasia [16]. The growth and proliferation of cholesteatoma are closely linked to the upregulation of growth factors, such as epidermal growth factor (EGF) and keratinocyte growth factor (KGF) and its receptors [17–19], and cytokines, including IL-1, IL-6, and tumor necrosis factor α (TNF-α) [20]. Cholesteatoma growth can cause both sensorineural and conductive hearing loss, as well as other symptoms, such as dizziness or facial nerve paralysis, and can lead to other complications that include meningitis, mastoiditis, bacterial labyrinthitis, sigmoid sinus thrombophlebitis, or brain abscess.

Currently, the only effective method of treating cholesteatoma is the complete removal and repair of the damaged middle ear structures, and various surgical accesses are used for this purpose [21–23]. Unfortunately, recurrence after surgery is still common, and risk factors for recurrence include young age, cholesteatoma localized to the mastoid, stapes, or incus erosion [3]. It was also observed that surgical outcomes are different when comparing rural and urban cohorts [24]. Therefore, there is an urgent need to develop nonsurgical, alternative treatments based on molecular mechanisms. For this reason, it is very important to explore the cellular and molecular mechanisms underlying the pathogenesis of middle ear cholesteatoma.

MicroRNAs (miRNAs) are among small, evolutionary conserved, noncoding RNAs whose function is to regulate the activity of mRNA expression, thus affecting various biological processes. The importance of miRNA-directed gene regulation is becoming increasingly apparent as more miRNAs and their regulatory targets and functions are discovered. A single miRNA can regulate many targets [25]. MiRNAs can be packaged into exosomes and transported to different cells. This process allows for the transfer of genetic information between cells and can have significant functional implications; therefore, they may be used in the future therapy of cholesteatoma.

The role of certain miRNAs in differentiation, development, apoptosis, and oncogenesis in various cancers has been recognized. However, miRNAs have also been shown to play an important role in the development of various otological disorders, such as progressive sensorineural hearing loss, age-related hearing loss, noise-induced hearing loss, cholesteatoma, schwannomas, and inner ear inflammation [26].

In this systematic review, we collect and analyze the results of all studies on the role of miRNAs and exosomal miRNAs in cholesteatoma pathogenesis and their potential usefulness in the nonsurgical treatment of cholesteatoma.

2. Materials and Methods

This systematic review was conducted and written in accordance with the updated 2020 Preferred Reporting Items for Systematic Reviews and MetaAnalyses Statement (PRISMA) [27,28].

The PubMed/MEDLINE, Web of Science, Scopus, and Cochrane Library databases were searched systematically. The last search was initiated on the 6th of June 2023 with the language limited to English. No publication time limits were imposed. The detailed search strategy is presented in Table 1. Subsequently, the automatic duplicate identifier function built in EndNote X9 (Clarivate Analytics, London, UK) was used.

Table 1. Databases used and corresponding search lines.

Databases	Number of Hits	Search Lines
Scopus	83	TITLE-ABS-KEY ((microRNA) OR (miRNA)) AND (cholesteatoma) Limited to 1. Document type: Article 2. Language: English
Web of Science	18	(AB = ((microRNA) OR (miRNA))) AND (AB = (cholesteatoma))
Pubmed/Medline	17	("Cholesteatoma" [Mesh]) AND "MicroRNAs" [Mesh]
Cochrane	0	#1 "cholesteatoma" [Mesh] #2 "microRNAs" [Mesh] #1 AND #2

Both types of studies, observational and experimental, were included and assessed. Two reviewers (KD and KC) independently screened the studies by title and abstract. Following this selection, full texts were checked. Studies that met the selection criteria were included (Table 2).

Table 2. Inclusion and exclusion criteria for the studies included in the review.

	Inclusion Criteria	Exclusion Criteria
Topic	Studies concerning miRNA/cholesteatoma	Studies not related to miRNA/cholesteatoma
Study type	Original articles	Reviews, case reports, book chapters, expert opinions, letters to the editor, conference reports, posters
Study status	Completed, published	Unfinished, unpublished
Language	Full text available in English	Language other than English or only abstract available in English
Quality	Good-quality research studies	Poor-quality research studies

Only original papers according to the study design were eligible for inclusion. Any apparent discrepancies resulting from the selection process were discussed and resolved by the co-authors (MS and NL or by discussion). Figure 1 provides an overview of the selection process summarized in the PRISMA flowchart.

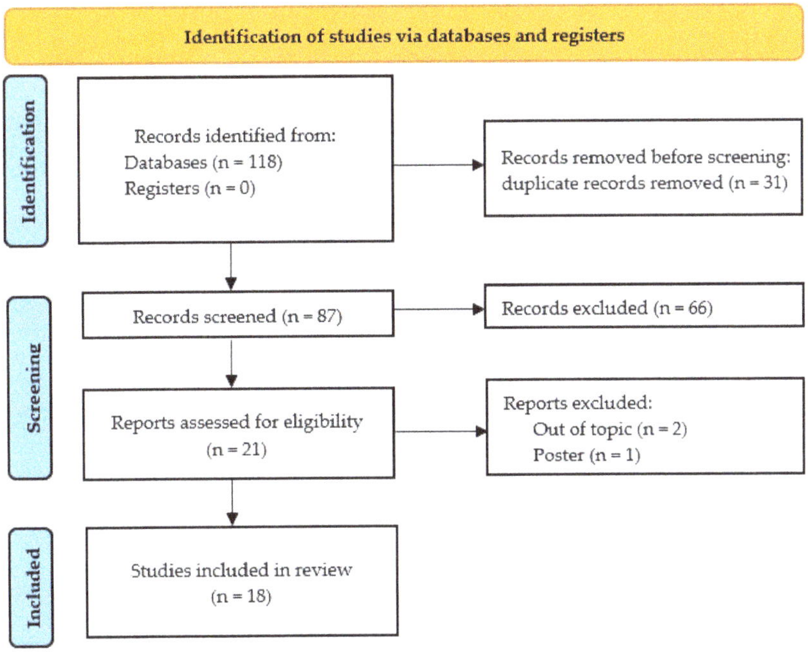

Figure 1. The PRISMA 2020 flow diagram for new systematic reviews, which included searches of databases and registers only.

After assessing the eligibility of all included studies, data for each study were extracted individually by two reviewers. The following results are presented in this review: first author, year of publication, country of origin, the purpose of the study, the miRNAs investigated, and finally the main results on the role of miRNAs in cholesteatoma.

The risk of bias was assessed independently by two authors (KD and KC) using the Office of Health Assessment and Translation (OHAT) Risk of Bias Rating Tool for Human and Animal Studies, modified for the needs of this review [29]. Any disagreement was discussed and resolved by the fourth and fifth co-authors (MS and NL). In general, the studies included were of high to moderate quality. The data are presented in Supplementary Table S2.

3. Results and Discussion

3.1. Search Results, Study Characteristics, and Study Quality

In total, 118 articles were retrieved through the database search. After applying the search strategy described above, a total of 18 articles were identified that met all inclusion criteria [30–47]. A flow diagram of the detailed study selection process is presented in Figure 1.

The articles were analyzed for their basic data and the main results are described in Table S1. The included studies were published between 2009 and 2023, more than half of them in the last five years, reflecting the rapid development in miRNA research in recent years. Almost all of the included studies were conducted in China, except for one study by Friedland et al. [33], which originated in the USA. All included studies used cholesteatoma tissues for analysis. Additionally, the majority of studies included in vitro experiments on cell lines [30,32,34–39,42–44,46,47]. In the Zheng study [46], a mouse model was also used for in vivo testing.

The quality of the included studies was estimated using the OHAT Risk of Bias Rating Tool for Human and Animal Studies, modified for the needs of this review [29], and it showed that the quality of most studies was intermediate or high. The results are presented in Table S2.

3.2. Differentially Expressed MicroRNAs in Acquired Middle Ear Cholesteatoma

In the included studies, miRNA expression profiling analysis was performed between acquired middle ear cholesteatoma and normal skin to identify miRNAs that may be involved in the aetiopathogenesis of middle ear cholesteatoma. Selected miRNAs with significantly increased or decreased expression in cholesteatoma tissue compared to normal skin are shown in Table 3. In Xie's study [40], the miRNA microarray technology revealed 44 miRNAs with increased expressions (miR-21-3p, miR-584-5p, miR-16-1-3p, etc.) and 175 miRNAs with decreased expressions (miR-10a-5p, miR-152-5p, miR-203b-5p, etc.) in cholesteatoma tissues with a 2-fold change compared to normal skin. Subsequently, qRT-PCR validation showed that miR-21-3p and miR-16-1-3p had significantly higher expressions, while miR-10a-5p showed a markedly reduced expression in middle ear cholesteatoma tissues compared to normal skin. Finally, the Gene Ontology (GO) and Kyoto Encyclopedia Genes and Genomes (KEGG) pathway analyses provided clues that these differentially expressed miRNAs may play an important role in the aetiopathogenesis of middle ear cholesteatoma, including cell proliferation, apoptosis, the cell cycle, differentiation, bone resorption, and the remodeling process.

Differences in miRNA expression may have a potential role in the pathogenesis of cholesteatoma and be a future therapeutic target. The role of some miRNAs has been identified in cholesteatoma and the results of the available studies are presented below. More cholesteatoma research is needed to extend these early discovered relationships.

Table 3. Differentially expressed miRNAs in cholesteatoma tissue.

Study	miRNA	Expression in Cholesteatoma Tissue
Yang, J. [41]	miR-10a-5p	downregulated
Zhu, X. [47]	miR-1297, miR-26a-5p	downregulated
Zang, J. [44]	miR-125b	downregulated

Table 3. *Cont.*

Study	miRNA	Expression in Cholesteatoma Tissue
Sui, R. [39]	miR-142-5p	downregulated
Yao, L. [42]	miR-199a	upregulated
Liu, D. [38]	miR-508-3p	upregulated
Hu, Y. [35]	miR-22-3p, miR-125a-5p	upregulated
Gong, N. [34]	exosomal miR-17	downregulated
Zang, J. [43]	miR-203a	downregulated
Xie, S. [40]	miR-21-3p, miR-584-5p, miR-16-1-3p, miR-338-5p, miR-320b, miR-181a-3p, miR-181a-5p, miR-181b-5p, miR-335-3p, miR-155-5p, miR-224-3p, etc.	upregulated
	miR-10a-5p, miR-152-5p, miR-203b-5p, miR-30a-5p, miR-1297, miR-539-3p, miR-9-3p, miR-769-3p, etc.	downregulated
Li, Y. [37]	exosomal miR-106b-5p	downregulated
Chen, X. [31]	miR-21, let-7a miRNA	upregulated
Friedland, D.R. [33]	miR-21	upregulated

Differences in miRNA expression may have a potential role in the pathogenesis of cholesteatoma and be a future therapeutic target. The role of some miRNAs has been identified in cholesteatoma and the results of the available studies are presented below. More cholesteatoma research is needed to extend these early discovered relationships.

3.3. *Role of Exosomal MicroRNAs in Cholesteatoma*

3.3.1. MicroRNA-17 Carried by Small Extracellular Vesicles

It was identified that miRNAs may regulate osteoclasts in bone diseases, such as bone loss, metastasis, and osteosarcoma [48]. Furthermore, the role of osteoclasts in bone resorption in cholesteatoma has been confirmed [49,50]. The balance between bone resorption by osteoclasts and bone formation by osteoblasts and the communication between these cells are crucial for bone homeostasis [51,52].

Gong et al. [34] demonstrated that miRNA-17 carried by small extracellular vesicles (sEVs) isolated from keratinocytes of patients with middle ear cholesteatoma can increase receptor activator of nuclear factor-κB ligand (RANKL) expression in fibroblasts and induce osteoclast differentiation. Hereby, RANKL is known to induce osteoclast differentiation [52,53]. Using an in vitro co-culture system consisting of keratinocytes, fibroblasts, and osteoclast precursors, keratinocytes have been shown to stimulate osteoclast differentiation through the induction of RANKL in fibroblasts [52]. Gong et al. [34] observed that fibroblasts treated with keratinocyte-derived sEVs (Ker-Exo) from cholesteatoma patients increased RANKL expression and promoted osteoclast differentiation, and this was dependent on the sEV component miRNA-17, which was downregulated in Ker-Exo in cholesteatoma. These new findings provide a foundation for further research into the effect of miRNAs carried by sEVs on osteoclast differentiation in cholesteatoma, which may be used in the future to treat cholesteatoma and reduce bone destruction in this disease.

3.3.2. MicroRNA-106b-5p Carried by Small Extracellular Vesicles

Proliferating tissue in cholesteatoma requires an increased blood supply; therefore, angiogenesis appears to be a necessary condition for cholesteatoma expansion. The presence and distribution of blood vessels in cholesteatoma were investigated and revealed the presence of increased vascularization in the subepithelial connective tissue of cholesteatoma (perimatrix), which probably plays a role in maintaining continued abnormal growth [54–56]. Analysis of the angiogenic growth factors in middle ear cholesteatoma revealed their altered expression compared to middle ear mucosa or auditory meatal skin [55,57–63]. Therefore, the possibility of using antiangiogenic molecules in the adjuvant treatment of cholesteatoma surgery has been proposed [62].

Moreover, it was confirmed that miRNAs carried by sEVs can participate in stimulating angiogenesis [64–66]. Therefore, miRNAs carried by sEVs were one of the factors studied in cholesteatoma angiogenesis. sEVs (size range of 30–150 nm) released into the extracellular space carry cell-specific cargos of proteins, lipids, and nucleic acids (including miRNAs) that can be transferred and participate in intercellular communication [67]. One source of miRNAs in human cholesteatoma is perimatrix fibroblasts (hCPFs), which stimulate microvascular endothelial cells to proliferate and migrate and alter their gene expression patterns by releasing various angiogenic factors [37]. Li et al. [37] were the first to study the function of hCPF-derived sEVs (hCPF-Exo). Their results indicate that hCPF-Exo transports low-expressed miR-106b-5p into endothelial cells and promotes angiogenesis by overexpressing angiopoietin 2.

In conclusion, to develop effective antiangiogenic therapies, it is necessary to understand the mechanisms underlying abnormal angiogenesis.

3.4. Other MicroRNAs in Cholesteatoma

3.4.1. MicroRNA-21

The main miRNA investigated in cholesteatoma is miR-21. Its expression is altered in many cancers and other pathologies, including cardiomyopathies [68]. The upregulation of miR-21 in colorectal cancer was shown to be associated with the downregulation of phosphatase tension homolog (PTEN) protein expression, which affects proliferation, antiapoptosis, cell cycle progression, and the invasion of colorectal cancer cells [69]. Furthermore, the tumor suppressor PTEN modulates cell cycle progression and cell survival and its expression was found to be significantly lower in cholesteatoma [70,71], which may account for the impaired inhibition in cholesteatoma. Moreover, programmed cell death 4 (PDCD4) plays a role in the pathogenesis of many cancers, including skin cancer [72–74]. However, IL-6 and signal transducers and activators of transcription 3 (STAT3) are upstream activators of miR-21, whose role in cholesteatoma epithelial hyperproliferation has also been confirmed [71,75].

Some investigators [33,76] observed a more than 4-fold higher expression of miR-21 and a reduction in the downstream targets of miR-21, PTEN, and PDCD4 in cholesteatoma compared to normal skin. Chen et al. [31] confirmed these results and additionally reported that miR-21 was particularly elevated in pediatric patients, potentially leading to greater tumor cell proliferation and cholesteatoma invasion, which is reflected in clinical practice, where the disease process is more aggressive and invasive in children than adults [77]. Moreover, PTEN, PDCD4, and high-mobility group AT-hook 2 (HMGA2) protein levels were significantly decreased in pediatric versus adult cholesteatoma patients [31].

Others [30] provided pieces of evidence that miR-21 promotes the proliferation and invasion of cholesteatoma keratinocytes. The number of proliferative cholesteatoma keratinocytes increased after transfection with miR-21 mimics, compared to miRNA controls and miR-21 inhibitors, indicating that miR-21 can promote the growth of cholesteatoma keratinocytes. The inhibitory effect of miR-21 on the apoptosis of cholesteatoma keratinocytes was shown by changing the percentage of apoptotic cells—decreasing it after transfection with miR-21 mimetics and increasing it after transfection with miR-21 inhibitors. The percentage of migrated cholesteatoma keratinocytes transfected with miR-21 mimics was 4.82 times higher when compared to cells transfected with miR-21 inhibitors, indicating that miR-21 promotes the migration and invasion of cholesteatoma keratinocytes.

In a recent study, Chen et al. [32] demonstrated that miR-21 inhibition to some extent slows keratinocyte proliferation and induces the apoptosis of cholesteatoma keratinocytes by inducing cell cycle arrest in the G0/G1 phase through a mechanism associated with the negative regulation of PTEN and PDCD4 expression. Specifically, it was sequentially demonstrated that the proliferation of CK cells (cultured keratinocytes) in the miR-21 inhibition group was significantly lower than in the negative and blind control group, the percentage of CK cells in the G0/G1 phase in the miR-21 inhibition group was significantly higher than in the negative and blind control groups, and protein and mRNA expression levels of PTEN and PDCD4 in CK in the miR-21 group were significantly higher than in the negative and blind control groups.

To sum up, the results of the above studies on cholesteatoma showed a higher level of miR-21 expression, which was associated with the downregulation of two cholesteatoma suppressors, PTEN and PDCD4, which was particularly noticeable for pediatric cholesteatoma. Therefore, the researchers point to the potential use of miR-21 as a therapeutic target in the disease.

3.4.2. MicroRNA-508-3p and Hsa_circ_0000007

Circular RNAs (circRNAs) are covalently closed RNA molecules, which are stable, conserved, and expressed in a tissue-specific manner [78]. CircRNA has been discovered to play a role in the pathogenesis of many cancers [78,79], along with other diseases, such as cardiovascular [80], renal [81], and skin diseases [82], diabetes mellitus [83], and neurological and neuropsychiatric disorders [84,85]. The circRNA expression profile was examined in cholesteatoma and revealed 101 upregulated and 254 downregulated circRNAs in cholesteatoma [86]. CircRNAs can regulate gene expression (transcription and translation) and act as miRNA sponges [79].

It was demonstrated that the increased expression of miR-508-3p in cholesteatoma tissues and cells is inversely correlated with hsa_circ_0000007 expression [38]. Hsa_circ_0000007 expression was significantly lower in cholesteatoma than in normal skin. miR-508-3p appeared to be the targeted miRNA downstream of hsa_circ_0000007; it was upregulated in cholesteatoma, and there was a statistically negative correlation between miR-508-30 and hsa_circ_0000007 in cholesteatoma tissue. The miR-508-3p inhibitor elevated the level of pro-apoptotic factor B-cell lymphoma-2 (Bcl-2), increased PTEN expression, and impeded class I phosphoinositide 3-kinase (PIK3) and protein kinase B (Akt) expressions. Therefore, the miR-508-3p mimic decreased PTEN expression, increased the expression of PIK3 and phosphorylated Akt (p-Akt), and promoted the proliferation of middle ear cholesteatoma cells. The PIK3/Akt signaling pathway plays a key role in cell growth, survival, proliferation, metabolism, and motility in various cancer phenotypes and PIK3/Akt signaling pathway inhibitors are a highly effective treatment strategy [87,88].

In conclusion, the findings point to the targeted regulation of the PTEN/PI3K/Akt signaling pathway by miR-508-3p in the cholesteatoma pathogenesis, while miR-508-3p overexpression in cholesteatoma is probably mediated by the regulation of upstream hsa_circ_0000007.

3.4.3. Let-7a MicroRNA

Let-7a miRNA is categorized as a tumor suppressor and targets oncogenes, including the HMGA2 [89,90]. However, in some rare cases, let-7a miRNA can also act as an oncogene, increasing cancer invasion and chemoresistance [91]. Furthermore, its expression is also altered in cardiovascular diseases and inflammatory processes [92].

The level of let-7a miRNA was significantly elevated in cholesteatoma tissue compared to normal skin, particularly in pediatric patients [31]. HMGA2 protein levels were significantly decreased in pediatric versus adult cholesteatoma patients, which may correspond to an increased apoptosis of keratinocytes and decreased proliferation of cholesteatoma cells [31].

It was observed that let-7a miRNA inhibited the growth of cholesteatoma keratinocytes by reducing keratinocyte proliferation by promoting cell cycle arrest in the G0/G1 phase and inducing cell apoptosis. The authors demonstrated also that let-7a miRNA induces the cell apoptosis, migration, and invasion of cholesteatoma keratinocytes. Finally, let-7a miRNA downregulated miR-21 expression in cholesteatoma keratinocytes, which was also confirmed by the upregulation of miR-21 in the cholesteatoma keratinocytes transfected with the let-7a inhibitor.

To sum up, the results of those studies indicate that let-7a plays a positive role in inhibiting cholesteatoma and can control the proliferation and apoptosis of cholesteatoma keratinocytes through both miR-21 and HMGA2 downregulation.

3.4.4. MicroRNA-125 and Circ_0074491

Another miRNA whose role has been explored in cholesteatoma pathogenesis is miE-125. It has been linked to the regulation of tumorigenesis and tumor development, and its downstream targets include transcription factors such as STAT3, cytokines such as IL-6 and TGFβ, and suppressor protein p53 [93,94]. MiR-125b was found to be significantly downregulated in psoriatic skin, which may contribute to the extensive proliferation and inappropriate differentiation of keratinocytes observed in this disease [95,96]. STAT3 target genes are involved in proliferation, survival, self-renewal, invasion, and angiogenesis [97,98]. The blocking of the JAK2/STAT3 signaling pathway inhibits IL-17-induced vascular endothelial growth factor (VEGF) expression, leading to the inhibition of angiogenesis in inflammatory skin diseases [99]. The expression of STAT3 was found to be significantly higher in the cholesteatoma epithelium than in the normal epithelium of the external auditory canal or retroauricular skin, indicating its potential role in the mechanisms of hyperproliferation and growth in cholesteatoma [100,101].

Zang et al. [44] observed downregulation of miR-125b and upregulation of STAT3, cyclin D1, survivin, and VEGF in cholesteatoma tissues, which is consistent with the results of previous studies [17,57,102,103]. These reports imply that the inhibition of miR-125b expression in cholesteatoma may contribute to high proliferation and low apoptosis through an increased expression of STAT3, which could be used as a potential therapeutic target for intratympanic pharmacotherapy of cholesteatoma [17].

Hu et al. [35] presented the results of a study in which circ_0074491 was defined as a decoy for miR-22-3p and miR-125a-5p in cholesteatoma keratinocytes and was downregulated in cholesteatoma specimens. Furthermore, circ_0074491 knockdown reduced cell cycle arrest and apoptosis; facilitated cell proliferation, migration, colony formation, and the invasion of cholesteatoma keratinocytes; and activated the PIK3/Akt pathway via miR-22-3p and miR-125a-5p in cholesteatoma keratinocytes. Moreover, both miR-22-3p and miR-125a-5p inhibitors reversed the impacts of circ_0074491 silencing on the proliferation, apoptosis, migration, and invasion of cholesteatoma keratinocytes. In conclusion, circ_0074491 was shown to regulate cholesteatoma progression through the PIK3/Akt pathway and binding to miR-22-3p and miR-125a-5p.

Finally, most researchers confirmed the protective role of miR-125 in cholesteatoma, especially through the impact on the apoptosis pathway, and found it to be significantly downregulated in patients with this disease.

3.4.5. MicroRNA-10a-5p

MiR-10a-5p is another miRNA that plays a role in cancer and was studied in cholesteatoma. A lower expression of miR-10a-5p has been linked to the progression of hepatocellular carcinoma [104], melanoma [105], and ovarian cancer [106]. However, miR-10a-5p was also found to be upregulated in the affected skin of atopic dermatitis patients and in proliferating keratinocytes [107]. Xie et al. [40] identified the phosphatidylinositol-4,5-bisphosphonate 3-kinase catalytic subunit α (PIK3CA) as an important miR-10a-5p target gene. PIK3CA is a catalytic subunit of PIK3 that plays a role in the EGFR/PIK3/Akt/cyclinD1 signaling pathway, which is active in cholesteatoma and may have a key function in epithelial hyperproliferation [17].

In another study [41], middle ear cholesteatoma tissues showed significantly decreased miR-10a-5p expression and increased PIK3CA expression compared to normal posterior ear skin tissues ($p < 0.05$). The results of target gene prediction revealed that PIK3CA may be an miR-10a-5p target gene. Additionally, miR-10a-5p and PIK3CA expression levels were significantly negatively correlated in middle ear cholesteatoma tissues ($r = -0.926$, $p < 0.001$). These results indicate that miR-10a-5p can inhibit cholesteatoma proliferation and differentiation through the negative regulation of its target gene, PIK3CA.

3.4.6. MicroRNA-802

MiR-802 is another miRNA that has been studied in cholesteatoma. Previously, its role was confirmed in mediating the pathogenesis of a variety of tumors [108], obesity [109,110], osteoporosis [111], and inflammatory bowel disease [112].

Li et al. [36], for the first time, demonstrated the role of the NF-κB/miR-802/PTEN signaling pathway in cholesteatoma. In this study, cholesteatoma tissues showed a high activation of NF-κB and upregulation of proinflammatory cytokines such as TNFα, IL-1b, and IL-6. The researchers observed the upregulation of miR-802 in keratinocytes treated with TNFα, IL-1b, or IL-6 and revealed that the miR-802 promoter contained a functional NF-κB/P65 binding site. The authors demonstrated that miR-802 can promote keratinocyte proliferation and cell cycle progression. They also noticed that keratinocytes transfected with an has-miR-802 mimetic showed a significantly increased proliferation rate, a significantly reduced percentage of cells in the G1/G0 phase, and an increased percentage of cells in the S phase. Additionally, the inhibition of miR-802 reduced the above effects. Finally, the authors performed a computational analysis using miRWalk software and found that the gene encoding PTEN harbored an miR-802 binding site and suggested that miR-802 could directly inhibit PTEN expression by targeting its 30-untranslated region (UTR).

3.4.7. MicroRNA-1297, MicroRNA-26a-5p, and MicroRNA-203a

MiR-1297 and miR-26a-5p are other miRNAs that have been studied in cholesteatoma and are widely investigated in oncology. The inhibitory effect of miR-1297 on cancer progression has been observed in various malignant entities [113–124]. In addition to its role in tumorigenesis, miR-26a-5p similarly plays an important role in pulmonary fibrosis [125] and Alzheimer's disease [126]. However, the included studies provided new information on the role of miR-1297 and miR-26a-5p concerning the pathogenesis of cholesteatoma. miR-1297 and miR-26a-5p were shown to inhibit the progression of cholesteatoma keratinocytes by targeting the B-cell-specific Moloney murine leukemia virus insertion site 1 (BMI1) [47]. BMI1 is a component of the polycomb 70 repressive complex 1, which is a transcriptional repressor that increases cell survival [127]. The role of BMI1 has been identified in various cancers [128–133] and was shown to be linked to poor prognosis. Additionally, BMI1 levels were found to be elevated in transformed keratinocytes, skin tumors, and psoriasis and were shown to play a role in keratinocyte survival [127,134]. It was noted that BMI1 is expressed mainly in the basal and suprabasal layers of the epithelium in normal skin [43,134,135], but in cholesteatoma it was found in almost all layers of the epithelium, and the intensity of staining was comparably stronger [43].

A significantly elevated mRNA expression of BMI1 and protein level of BMI1 were demonstrated in cholesteatoma tumor tissues relative to normal retroauricular skin specimens [43,47]. The downregulation of BMI1 was shown to significantly inhibit the proliferation of human immortalized keratinocytes, indicating a pro-oncogenic role for BMI1 in cholesteatomas, which is consistent with observations in other malignant entities [47]. The prediction result obtained with DianaTools software showed that BMI1 directly interacts with miR-1297 and miR-26a-5p in human immortalized keratinocytes.

Another miRNA targeting BMI1 is miR-203a. The role of miR-203 as a tumor suppressor has been demonstrated in a number of cancers [136–138]. It is known that miR-203 promotes epidermal differentiation by limiting the proliferative potential [139] and is upregulated in psoriatic lesions [95], while the upregulation of miR-203 expression by oleic acid treatment accelerates human keratinocyte differentiation [140]. Therefore, its role in the growth and proliferation of cholesteatoma was investigated [43], leading to the observation that miR-203a is downregulated and BMI1 upregulated in cholesteatoma. Subsequently, the authors demonstrated in the dual-luciferase reporter assay (DLRA) that BMI1 is a direct target of miR-203a. Furthermore, reduced miR-203a expression increased BMI1 expression; promoted the proliferation, colony formation, and migration of HaCaT cells (a spontaneously transformed aneuploid immortal keratinocyte cell line from adult human skin); and inhibited apoptosis. Moreover, p-Akt was significantly increased in cholesteatoma tissues and was positively correlated with BMI1, which is consistent with other studies [141,142]. Finally, BMI1 suppression decreased p-Akt expression in HaCaT cells and subsequent miR-203a inhibition reversed this effect [43].

Taken together, all the studies confirmed that all three miRs, miR-1297, miR-26a-5p, and miR-203a, inhibit the progression of cholesteatoma keratinocytes by targeting BMI1. The new reports reveal the need for further research to explore specific therapeutic methods for cholesteatoma patients through a combination of nanotechnology and molecular drugs targeting those miRNAs or BMI1.

3.4.8. MicroRNA-199a

MiR-199a was shown to be involved in the proliferation, migration, invasion, and apoptosis of cancer cells from different primary tumors [143]. Its role has also been confirmed in other noncancerous diseases, such as generalized epilepsy [144] or osteoporosis [145].

Yao et al. [42] reported that miR-199a was significantly upregulated in cholesteatoma tissues compared to the normal retroauricular skin tissues, which facilitated the proliferation, migration, and invasion of cholesteatoma keratinocytes in vitro, while miR-199a downregulation caused the opposite effects. The authors demonstrated a negative regulatory relationship between miR-199a and PNRC1 and found miR-199a binding to the 3′-UTR of PNRC1.

To sum up, the upregulation of miR-199a in cholesteatoma potentially plays a role in the development of cholesteatoma and could be used as a biomarker or therapeutic target.

3.4.9. MicroRNA-142-5p

Another study focused on the role of miR-142 in cholesteatoma pathogenesis. miR-142 is expressed in many tissues and was found to play an important role in inflammatory and immune responses [146–149]. The differential expression patterns of miR-142-5p were observed in cancer tissues [150–154]. A downstream target gene of miR-142-5p is cyclin-dependent kinase 5 (CDK5) [39], whose role has been found in the development and progression of many cancers [155–157] and neurodegenerative [158,159] and inflammatory disorders [160–162]. Sui et al. [39] demonstrated that miR-142-5p directly inhibits the CDK5-mediated upregulation of inflammatory cytokines in acquired middle ear cholesteatoma, making it a potential therapeutic target in this disease.

3.4.10. Micro-RNA-34

MiR-34 is characterized by antitumor properties, is induced by the tumor suppressor p53, and can be used to control tumor progression [163,164]. Thus, one of the included studies revealed the function of miR-34 in cholesteatoma [42]. The researchers conducted an experimental study using nanoparticles in which therapeutic targets were achieved by regulating miR-34a expression. They showed that free rubone and rubon-containing drug nanoparticles (RC NPs) can significantly inhibit the proliferation and migration ability of cholesteatoma cells in children's middle ear cholesteatoma. The authors demonstrated that drug-loaded nanoparticles could deliver rubone into cells and upregulate miR-34a levels and downregulate mRNA expression levels of Bcl-2, CDK6, and cyclin D1 in cells. The expression level of miR-34a in the free rubone group was slightly higher than that in the

RC NPs group, which was due to the slow and controlled release of nanoparticles. The study indicated the potential of these novel forms of treatment in both cholesteatoma and also other oncologic diseases.

3.5. Strengths and Limitations of the Included Studies

Research on the association between cholesteatoma and microRNAs is still in the exploratory phase. The main advantage of the included studies is that they provide discoveries in the field of understanding the etiopathogenesis of cholesteatoma, which may open up the possibility of developing new microparticle-based treatment strategies for this disease. It should be noted that some of the studies included cholesteatoma samples collected in the pediatric population, which also provide better insight into the pathogenesis of this more aggressive form of the disease [31,32,46]. Moreover, new studies involving miRNAs carried by sEVs were also among the included studies, which provided new insight into the role of these nanoparticles in cholesteatoma [34,37].

However, many of the included studies analyzed a small number of clinical samples (most studies include cholesteatoma tissues from less than 30 patients), and the population from which tissues were collected was not homogeneous in terms of age or other parameters. In some studies, it was not specified precisely whether the samples were only from acquired cholesteatoma patients. Another limitation is the lack of information on the number of people from whom the tissues analyzed in some studies were collected [32–34,37]. Future studies should use a larger number of samples and ensure the homogeneity of the groups to avoid factors that could affect the results of the study. The pathogenesis of acquired and congenital cholesteatoma can differ, so it is important to emphasize which form of this disease the study is concerned with. Larger, more homogenized cohorts would provide more robust and generalizable data on differentially expressed miRNAs. Another problem is the limited validation in the included studies. Many findings are based on initial microarray screens that have not been extensively replicated in independent datasets. Further validation is needed to confirm key microRNAs and their roles. Unfortunately, microRNA mechanisms of action in promoting or inhibiting cholesteatoma progression are still unclear, and further work is needed to clarify how they affect keratinocyte behaviors.

Retroauricular skin has been used as a control material in many of the included studies [31,33,35–38,40–45,47]. There are reports that the selection of control group material in cholesteatoma studies can be important for the outcome of the studies, with deep meatal skin specimens being more similar in protein profile to cholesteatoma than retroauricular skin [165,166]. Moreover, canal skin is exposed to infection, which can stimulate cytokine expression and changes in protein expression, which can also affect research results [33].

Most studies show only correlational analyses established between miRNA levels and target mRNAs/proteins. Experimental manipulations are often lacking to demonstrate causality.

The study by Zhu et al. [47] provides new insights into the involvement of miR-1297/BMI1 and miR-26a-5p/BMI1 regulatory mechanisms in the pathogenesis of cholesteatoma, but only in vitro experiments were performed, which have some limitations. Animal studies are needed to confirm the relationships discovered.

The study by Li et al. [37] has its limitations because it is an in vitro study in which miR-106b-5p carried by sEVs derived from cholesteatoma hCPFs is just one of the sEV miRNA cargo components that promote angiogenesis in middle ear cholesteatoma, and the possible influence of other pro-angiogenic factors of the sEV cargo cannot be ignored. Therefore, further studies should be performed in an animal model of middle ear cholesteatoma to investigate the suppressive effect of sEV-associated miR-106b-5p on cholesteatoma growth in vivo. Animal models could help determine the true significance of microRNAs in cholesteatoma pathogenesis.

Recent studies suggest that abnormal miRNA expression may contribute to cholesteatoma pathogenesis, though causal links remain unclear based on preliminary evidence. Targeting relevant microRNAs could potentially offer adjunctive treatments to supplement current surgical treatments like type 1 tympanoplasty [167,168]. However, microRNA-based therapies for cholesteatoma remain in the early stages of research, with limited validation in large cohorts and animal models. Further rigorous investigation is needed before these approaches could mature sufficiently to realistically improve outcomes of procedures like type 1 tympanoplasty in clinical practice.

4. Conclusions

Cholesteatoma is a disease that significantly impairs function, and its complications can be life-threatening. The pathogenesis of this disease is still under investigation, and knowledge in this field has recently been supplemented by studying the role of certain miRNAs and circRNAs. These are very important discoveries as they might contribute to the identification of pharmacological targets for the treatment of cholesteatoma, but there is a need for further validation and causative evidence. The presence of miRNAs in exosomes suggests a mechanism by which cells can communicate and modulate gene expression in a paracrine or endocrine manner. However, further large-scale validation studies, the experimental manipulation of microRNA levels, the elucidation of mechanisms, and in vivo testing are needed to substantiate these initial correlations and guide potential therapeutic applications. Comprehensive microRNA profiling in larger cohorts to identify robust disease-associated miRNAs are needed. Furthermore, future research should include integrated analyses of miRNAs, target mRNAs, and proteins to map regulatory networks. Moreover, the specific mechanisms by which miRNAs regulate cholesteatoma development remain largely unknown. Future work should contain functional studies manipulating microRNA levels and assessing impacts on keratinocyte behaviors and clarify how miRNAs influence keratinocyte proliferation, apoptosis, differentiation, etc. Additionally, there is limited functional testing in vitro and a lack of in vivo animal models to determine the true significance of miRNAs in disease pathogenesis. Investigations of miRNA roles in growth factor and cytokine signaling dysregulation are required. Finally, future work should also consider the development and testing of miRNA-based therapies in animal models. Hopefully, in the future, RNA- and protein-based therapies using nanoparticles will be possible for the nonsurgical or adjunctive treatment of cholesteatoma.

Supplementary Materials: The following supporting information can be downloaded at: https://www.mdpi.com/article/10.3390/ijms241512277/s1.

Author Contributions: Conceptualization, K.D. and M.J.S.; methodology, K.D. and K.C.; investigation, K.D. and K.C.; data curation, K.D. and K.C.; writing—original draft preparation, K.D., K.C., T.E.R., M.J.S. and N.L.; writing—review and editing, M.J.S. and N.L.; visualization, K.D. and K.C.; supervision, M.J.S.; project administration, M.J.S. All authors have read and agreed to the published version of the manuscript.

Funding: This work was supported by the National Science Centre, Poland, grant 2017/25/B/NZ5/02949 to MJS.

Institutional Review Board Statement: Not applicable.

Informed Consent Statement: Not applicable.

Data Availability Statement: Not applicable.

Conflicts of Interest: The authors declare no conflict of interest.

References

1. Kim, H.J.; Tinling, S.P.; Chole, R.A. Expression patterns of cytokeratins in cholesteatomas: Evidence of increased migration and proliferation. *J. Korean Med. Sci.* **2002**, *17*, 381–388. [CrossRef]
2. Li, J.; Jufas, N.; Forer, M.; Patel, N. Incidence and trends of middle ear cholesteatoma surgery and mastoidectomy in Australia-A national hospital morbidity database analysis. *Laryngoscope Investig. Otolaryngol.* **2022**, *7*, 210–218. [CrossRef]
3. Britze, A.; Møller, M.L.; Ovesen, T. Incidence, 10-year recidivism rate and prognostic factors for cholesteatoma. *J. Laryngol. Otol.* **2017**, *131*, 319–328. [CrossRef]

4. Shibata, S.; Murakami, K.; Umeno, Y.; Komune, S. Epidemiological study of cholesteatoma in Fukuoka City. *J. Laryngol. Otol.* **2015**, *129* (Suppl. 2), S6–S11. [CrossRef]
5. Kemppainen, H.O.; Puhakka, H.J.; Laippala, P.J.; Sipilä, M.M.; Manninen, M.P.; Karma, P.H. Epidemiology and aetiology of middle ear cholesteatoma. *Acta Otolaryngol.* **1999**, *119*, 568–572. [CrossRef] [PubMed]
6. Homøe, P.; Rosborg, J. Family cluster of cholesteatoma. *J. Laryngol. Otol.* **2007**, *121*, 65–67. [CrossRef] [PubMed]
7. Jennings, B.A.; Prinsley, P.; Philpott, C.; Willis, G.; Bhutta, M.F. The genetics of cholesteatoma. A systematic review using narrative synthesis. *Clin. Otolaryngol.* **2018**, *43*, 55–67. [CrossRef] [PubMed]
8. Collins, R.; Ta, N.H.; Jennings, B.A.; Prinsley, P.; Philpott, C.M.; Steel, N.; Clark, A. Cholesteatoma and family history: An international survey. *Clin. Otolaryngol.* **2020**, *45*, 500–505. [CrossRef] [PubMed]
9. Bonnard, Å.; Engmér Berglin, C.; Wincent, J.; Eriksson, P.O.; Westman, E.; Feychting, M.; Mogensen, H. The Risk of Cholesteatoma in Individuals With First-degree Relatives Surgically Treated for the Disease. *JAMA Otolaryngol. Head Neck Surg.* **2023**, *149*, 390–396. [CrossRef]
10. Poliner, A.; Mahomva, C.; Williams, C.; Alfonso, K.; Anne, S.; Musso, M.; Liu, Y.C. Prevalence and surgical management of cholesteatoma in Down Syndrome children. *Int. J. Pediatr. Otorhinolaryngol.* **2022**, *157*, 111126. [CrossRef]
11. Spinner, A.; Munjuluru, A.; Wootten, C.T. Prevalence of Cholesteatoma in Children With Down Syndrome Receiving Treatment at Pediatric Health Care Facilities. *JAMA Otolaryngol. Head Neck Surg.* **2020**, *146*, 864–865. [CrossRef] [PubMed]
12. Zanetti, D.; Di Lella, F.; Negri, M.; Vincenti, V. Surgical management of middle ear cholesteatoma in children with Turner syndrome: A multicenter experience. *Acta Biomed.* **2018**, *89*, 382–388. [CrossRef] [PubMed]
13. Dorney, I.; Otteson, T.; Kaelber, D.C. Middle ear cholesteatoma prevalence in over 3,600 children with Turner Syndrome. *Int. J. Pediatr. Otorhinolaryngol.* **2022**, *161*, 111289. [CrossRef] [PubMed]
14. Meyerhoff, W.L.; Truelson, J. Cholesteatoma staging. *Laryngoscope* **1986**, *96*, 935–939. [CrossRef]
15. Rosito, L.S.; Netto, L.F.; Teixeira, A.R.; da Costa, S.S. Classification of Cholesteatoma According to Growth Patterns. *JAMA Otolaryngol. Head Neck Surg.* **2016**, *142*, 168–172. [CrossRef]
16. Kuo, C.L. Etiopathogenesis of Acquired Cholesteatoma: Prominent Theories and Recent Advances in Biomolecular Research. *Laryngoscope* **2015**, *125*, 234–240. [CrossRef]
17. Liu, W.; Ren, H.; Ren, J.; Yin, T.; Hu, B.; Xie, S.; Dai, Y.; Wu, W.; Xiao, Z.; Yang, X.; et al. The role of EGFR/PI3K/Akt/cyclinD1 signaling pathway in acquired middle ear cholesteatoma. *Mediat. Inflamm.* **2013**, *2013*, 651207. [CrossRef]
18. Yamamoto-Fukuda, T.; Takahashi, H.; Koji, T. Expression of keratinocyte growth factor (KGF) and its receptor in a middle-ear cavity problem. *Int. J. Pediatr. Otorhinolaryngol.* **2012**, *76*, 76–81. [CrossRef]
19. Yamamoto-Fukuda, T.; Akiyama, N. Keratinocyte growth factor signaling promotes stem/progenitor cell proliferation under p63 expression during middle ear cholesteatoma formation. *Curr. Opin. Otolaryngol. Head Neck Surg.* **2020**, *28*, 291–295. [CrossRef]
20. Kuczkowski, J.; Sakowicz-Burkiewicz, M.; Iżycka-Świeszewska, E.; Mikaszewski, B.; Pawełczyk, T. Expression of tumor necrosis factor-α, interleukin-1α, interleukin-6 and interleukin-10 in chronic otitis media with bone osteolysis. *ORL J. Otorhinolaryngol. Relat. Spec.* **2011**, *73*, 93–99. [CrossRef]
21. Mulazimoglu, S.; Meco, C. Endoscopic diving technique for hearing preservation in managing labyrinth-invading cholesteatomas. *Eur. Arch. Otorhinolaryngol.* **2023**, *280*, 1639–1646. [CrossRef]
22. Salem, J.; Bakundukize, J.; Milinis, K.; Sharma, S.D. Mastoid obliteration versus canal wall down or canal wall up mastoidectomy for cholesteatoma: Systematic review and meta-analysis. *Am. J. Otolaryngol.* **2023**, *44*, 103751. [CrossRef] [PubMed]
23. Shakya, D.; Nepal, A. Transcanal Endoscopic Retrograde Mastoidectomy for Cholesteatoma: A Prospective Study. *Ear Nose Throat J.* **2023**, *102*, Np269–Np276. [CrossRef] [PubMed]
24. Kennedy, K.L.; Connolly, K.M.; Albert, C.L.; Goldman, J.L.; Cash, E.D.; Severtson, M.A. Postoperative Recurrent Cholesteatoma in Rural Versus Urban Populations. *Otol. Neurotol.* **2021**, *42*, e459–e463. [CrossRef] [PubMed]
25. Bartel, D.P. MicroRNAs: Genomics, biogenesis, mechanism, and function. *Cell* **2004**, *116*, 281–297. [CrossRef]
26. Mahmoudian-Sani, M.R.; Mehri-Ghahfarrokhi, A.; Ahmadinejad, F.; Hashemzadeh-Chaleshtori, M.; Saidijam, M.; Jami, M.S. MicroRNAs: Effective elements in ear-related diseases and hearing loss. *Eur. Arch. Otorhinolaryngol.* **2017**, *274*, 2373–2380. [CrossRef]
27. Page, M.J.; McKenzie, J.E.; Bossuyt, P.M.; Boutron, I.; Hoffmann, T.C.; Mulrow, C.D.; Shamseer, L.; Tetzlaff, J.M.; Akl, E.A.; Brennan, S.E.; et al. The PRISMA 2020 statement: An updated guideline for reporting systematic reviews. *J. Clin. Epidemiol.* **2021**, *134*, 178–189. [CrossRef] [PubMed]
28. Page, M.J.; Moher, D.; Bossuyt, P.M.; Boutron, I.; Hoffmann, T.C.; Mulrow, C.D.; Shamseer, L.; Tetzlaff, J.M.; Akl, E.A.; Brennan, S.E.; et al. PRISMA 2020 explanation and elaboration: Updated guidance and exemplars for reporting systematic reviews. *BMJ* **2021**, *372*, n160. [CrossRef] [PubMed]
29. Eick, S.M.; Goin, D.E.; Chartres, N.; Lam, J.; Woodruff, T.J. Assessing risk of bias in human environmental epidemiology studies using three tools: Different conclusions from different tools. *Syst. Rev.* **2020**, *9*, 249. [CrossRef]
30. Chen, X.; Li, X.; Qin, Z. MicroRNA-21 promotes the proliferation and invasion of cholesteatoma keratinocytes. *Acta Oto-Laryngol.* **2016**, *136*, 1261–1266. [CrossRef]
31. Chen, X.; Qin, Z. Post-transcriptional regulation by microRNA-21 and let-7a microRNA in paediatric cholesteatoma. *J. Int. Med. Res.* **2011**, *39*, 2110–2118. [CrossRef]

32. Chen, X.; Xiao, F.; Dong, S.; Chen, Y.; Wang, J. MiR-21 inhibits the proliferation of childhood cholesteatoma glioma cells by negatively regulating the expressions of PTEN and PDCD4. *Trop. J. Pharm. Res.* **2021**, *20*, 1119–1124. [CrossRef]
33. Friedland, D.R.; Eernisse, R.; Erbe, C.; Gupta, N.; Cioffi, J.A. Cholesteatoma growth and proliferation: Posttranscriptional regulation by microRNA-21. *Otol. Neurotol.* **2009**, *30*, 998–1005. [CrossRef]
34. Gong, N.; Zhu, W.; Xu, R.; Teng, Z.; Deng, C.; Zhou, H.; Xia, M.; Zhao, M. Keratinocytes-derived exosomal miRNA regulates osteoclast differentiation in middle ear cholesteatoma. *Biochem. Biophys. Res. Commun.* **2020**, *525*, 341–347. [CrossRef]
35. Hu, Y.; Qian, X. Hsa_circ_0074491 regulates the malignance of cholesteatoma keratinocytes by modulating the PI3K/Akt pathway by binding to miR-22-3p and miR-125a-5p: An observational study. *Medicine* **2021**, *100*, e27122. [CrossRef]
36. Li, N.; Qin, Z.B. Inflammation-induced miR-802 promotes cell proliferation in cholesteatoma. *Biotechnol. Lett.* **2014**, *36*, 1753–1759. [CrossRef] [PubMed]
37. Li, Y.; Liang, J.; Hu, J.; Ren, X.; Sheng, Y. Down-regulation of exosomal miR-106b-5p derived from cholesteatoma perimatrix fibroblasts promotes angiogenesis in endothelial cells by overexpression of Angiopoietin 2. *Cell Biol. Int.* **2018**, *42*, 1300–1310. [CrossRef] [PubMed]
38. Liu, D.; Ma, X. Mir-508-3p promotes proliferation and inhibits apoptosis of middle ear cholesteatoma cells by targeting pten/pi3k/akt pathway. *Int. J. Med. Sci.* **2021**, *18*, 3224–3235. [CrossRef] [PubMed]
39. Sui, R.; Shi, W.; Han, S.; Fan, X.; Zhang, X.; Wang, N.; Zhang, H.; Xu, A.; Liu, C. MiR-142-5p directly targets cyclin-dependent kinase 5-mediated upregulation of the inflammatory process in acquired middle ear cholesteatoma. *Mol. Immunol.* **2022**, *141*, 236–245. [CrossRef]
40. Xie, S.; Liu, X.; Pan, Z.; Chen, X.; Peng, A.; Yin, T.; Ren, J.; Liu, W. Microarray analysis of differentially-expressed microRNAs in acquired middle ear cholesteatoma. *Int. J. Med. Sci.* **2018**, *15*, 1547–1554. [CrossRef]
41. Yang, J.; Yan, W.; Tang, S.; Huang, Z.; Ye, M.; Lu, Z.; Liu, Q. Expression and Correlation Research of MicroRNA10a-5p and PIK3CA in Middle Ear Cholesteatoma. *J. Int. Adv. Otol.* **2023**, *19*, 212–216. [CrossRef]
42. Yao, L.; Zhang, W.; Zheng, J.; Lu, X.; Zhang, F. MiR-199a Targeting PNRC1 to Promote Keratinocyte Proliferation and Invasion in Cholesteatoma. *BioMed Res. Int.* **2021**, *2021*, 1442093. [CrossRef]
43. Zang, J.; Hui, L.; Yang, N.; Yang, B.; Jiang, X. Downregulation of MiR-203a disinhibits bmi1 and promotes growth and proliferation of keratinocytes in cholesteatoma. *Int. J. Med. Sci.* **2018**, *15*, 447–455. [CrossRef]
44. Zang, J.; Yang, B.; Feng, S.; Jiang, X. Low expression of microRNA-125b enhances the expression of STAT3 and contributes to cholesteatoma growth. *Arch. Med. Sci.* **2022**, *18*, 1596–1606. [CrossRef] [PubMed]
45. Zhang, W.; Chen, X.; Qin, Z. MicroRNA let-7a suppresses the growth and invasion of cholesteatoma keratinocytes. *Mol. Med. Rep.* **2015**, *11*, 2097–2103. [CrossRef] [PubMed]
46. Zheng, H.; Wang, W.; Li, S.; Han, L. The Effect of Zbxz23ir-21 NANO(nanomaterials) Delivery Vector on Apoptosis and PTEN(phosphatase and tensin homolog deleted on chromosome ten)/PI3K(Intracellular phosphatidylinositol kinase)/AKT(related to the A and C kinase) in Children with CHOLESTEATOMA in Middle Ear. *Bioengineered* **2021**, *12*, 8809–8821. [CrossRef]
47. Zhu, X.; Ye, F.; Hao, S.; Yu, Q.; Wang, Y.; Lou, W.; Zhao, K.; Li, H. MiR-1297 and MiR-26a-5p Inhibit Cell Progression of Keratinocytes in Cholesteatoma Depending on the Regulation of BMI1. *Biotechnol. Bioprocess Eng.* **2022**, *27*, 79–88. [CrossRef]
48. Doghish, A.S.; Elballal, M.S.; Elazazy, O.; Elesawy, A.E.; Shahin, R.K.; Midan, H.M.; Sallam, A.M.; Elbadry, A.M.M.; Mohamed, A.K.I.; Ishak, N.W.; et al. miRNAs as potential game-changers in bone diseases: Future medicinal and clinical uses. *Pathol. Res. Pract.* **2023**, *245*, 154440. [CrossRef]
49. Kuczkowski, J.; Brzoznowski, W.; Nowicki, T. Bone Damage in Chronic Otitis Media. *Ear Nose Throat J.* **2022**, *101*, 428–429. [CrossRef]
50. Imai, R.; Sato, T.; Iwamoto, Y.; Hanada, Y.; Terao, M.; Ohta, Y.; Osaki, Y.; Imai, T.; Morihana, T.; Okazaki, S.; et al. Osteoclasts Modulate Bone Erosion in Cholesteatoma via RANKL Signaling. *J. Assoc. Res. Otolaryngol.* **2019**, *20*, 449–459. [CrossRef]
51. Chen, X.; Wang, Z.; Duan, N.; Zhu, G.; Schwarz, E.M.; Xie, C. Osteoblast-osteoclast interactions. *Connect Tissue Res.* **2018**, *59*, 99–107. [CrossRef]
52. Iwamoto, Y.; Nishikawa, K.; Imai, R.; Furuya, M.; Uenaka, M.; Ohta, Y.; Morihana, T.; Itoi-Ochi, S.; Penninger, J.M.; Katayama, I.; et al. Intercellular Communication between Keratinocytes and Fibroblasts Induces Local Osteoclast Differentiation: A Mechanism Underlying Cholesteatoma-Induced Bone Destruction. *Mol. Cell Biol.* **2016**, *36*, 1610–1620. [CrossRef]
53. Yoon, W.J.; Kim, K.N.; Heo, S.J.; Han, S.C.; Kim, J.; Ko, Y.J.; Kang, H.K.; Yoo, E.S. Sargachromanol G inhibits osteoclastogenesis by suppressing the activation NF-κB and MAPKs in RANKL-induced RAW 264.7 cells. *Biochem. Biophys. Res. Commun.* **2013**, *434*, 892–897. [CrossRef] [PubMed]
54. Bujía, J.; Holly, A.; Stammberger, M.; Sudhoff, H. Angiogenesis in cholesteatoma of the middle ear. *Acta Otorrinolaringol. Esp.* **1996**, *47*, 187–192. [PubMed]
55. Sudhoff, H.; Dazert, S.; Gonzales, A.M.; Borkowski, G.; Park, S.Y.; Baird, A.; Hildmann, H.; Ryan, A.F. Angiogenesis and angiogenic growth factors in middle ear cholesteatoma. *Am. J. Otol.* **2000**, *21*, 793–798.
56. Olszewska, E.; Chodynicki, S.; Chyczewski, L. Apoptosis in the pathogenesis of cholesteatoma in adults. *Eur. Arch. Oto-Rhino-Laryngol.* **2006**, *263*, 409–413. [CrossRef]

57. Fukudome, S.; Wang, C.; Hamajima, Y.; Ye, S.; Zheng, Y.; Narita, N.; Sunaga, H.; Fujieda, S.; Hu, X.; Feng, L.; et al. Regulation of the angiogenesis of acquired middle ear cholesteatomas by inhibitor of DNA binding transcription factor. *JAMA Otolaryngol. Head Neck Surg.* **2013**, *139*, 273–278. [CrossRef]
58. Starke, R.D.; Ferraro, F.; Paschalaki, K.E.; Dryden, N.H.; McKinnon, T.A.; Sutton, R.E.; Payne, E.M.; Haskard, D.O.; Hughes, A.D.; Cutler, D.F.; et al. Endothelial von Willebrand factor regulates angiogenesis. *Blood* **2011**, *117*, 1071–1080. [CrossRef]
59. Samuelson Bannow, B.; Recht, M.; Négrier, C.; Hermans, C.; Berntorp, E.; Eichler, H.; Mancuso, M.E.; Klamroth, R.; O'Hara, J.; Santagostino, E.; et al. Factor VIII: Long-established role in haemophilia A and emerging evidence beyond haemostasis. *Blood Rev.* **2019**, *35*, 43–50. [CrossRef]
60. Acharya, S.S.; Kaplan, R.N.; Macdonald, D.; Fabiyi, O.T.; DiMichele, D.; Lyden, D. Neoangiogenesis contributes to the development of hemophilic synovitis. *Blood* **2011**, *117*, 2484–2493. [CrossRef] [PubMed]
61. Costa, J.R.; Rego, A.R.; Soares, T.; Sousa, C.A.E.; Coutinho, M.B. Changes in Coagulation Study and Risk of Developing Cholesteatoma: Is There a Link? *J. Audiol. Otol.* **2023**, *27*, 30–36. [CrossRef] [PubMed]
62. Reis Rego, Â.; Santos, M.; Coutinho, M.; Feliciano, T.; Almeida e Sousa, C. Is von Willebrand disease linked to cholesteatoma aetiology? *Med. Hypotheses* **2017**, *100*, 43–45. [CrossRef]
63. Yoshikawa, M.; Kojima, H.; Yaguchi, Y.; Okada, N.; Saito, H.; Moriyama, H. Cholesteatoma Fibroblasts Promote Epithelial Cell Proliferation through Overexpression of Epiregulin. *PLoS ONE* **2013**, *8*, e66725. [CrossRef] [PubMed]
64. van Balkom, B.W.; de Jong, O.G.; Smits, M.; Brummelman, J.; den Ouden, K.; de Bree, P.M.; van Eijndhoven, M.A.; Pegtel, D.M.; Stoorvogel, W.; Würdinger, T.; et al. Endothelial cells require miR-214 to secrete exosomes that suppress senescence and induce angiogenesis in human and mouse endothelial cells. *Blood* **2013**, *121*, 3997–4006. [CrossRef]
65. Qu, Q.; Liu, L.; Cui, Y.; Liu, H.; Yi, J.; Bing, W.; Liu, C.; Jiang, D.; Bi, Y. miR-126-3p containing exosomes derived from human umbilical cord mesenchymal stem cells promote angiogenesis and attenuate ovarian granulosa cell apoptosis in a preclinical rat model of premature ovarian failure. *Stem Cell Res. Ther.* **2022**, *13*, 352. [CrossRef]
66. Cabello, P.; Torres-Ruiz, S.; Adam-Artigues, A.; Forés-Martos, J.; Martínez, M.T.; Hernando, C.; Zazo, S.; Madoz-Gúrpide, J.; Rovira, A.; Burgués, O.; et al. miR-146a-5p Promotes Angiogenesis and Confers Trastuzumab Resistance in HER2+ Breast Cancer. *Cancers* **2023**, *15*, 2138. [CrossRef]
67. Dżaman, K.; Czerwaty, K. Roles of Exosomes in Chronic Rhinosinusitis: A Systematic Review. *Int. J. Mol. Sci.* **2022**, *23*, 1284. [CrossRef]
68. Surina, S.; Fontanella, R.A.; Scisciola, L.; Marfella, R.; Paolisso, G.; Barbieri, M. miR-21 in Human Cardiomyopathies. *Front. Cardiovasc. Med.* **2021**, *8*, 767064. [CrossRef]
69. Wu, Y.; Song, Y.; Xiong, Y.; Wang, X.; Xu, K.; Han, B.; Bai, Y.; Li, L.; Zhang, Y.; Zhou, L. MicroRNA-21 (Mir-21) Promotes Cell Growth and Invasion by Repressing Tumor Suppressor PTEN in Colorectal Cancer. *Cell Physiol. Biochem.* **2017**, *43*, 945–958. [CrossRef]
70. Yune, T.Y.; Byun, J.Y. Expression of PTEN and phosphorylated Akt in human cholesteatoma epithelium. *Acta Otolaryngol.* **2009**, *129*, 501–506. [CrossRef] [PubMed]
71. Liu, D.; Ma, X. Expression and significance of PTEN, P-ERK and P-AKT in the middle ear cholesteatoma. *Lin Chuang Er Bi Yan Hou Tou Jing Wai Ke Za Zhi* **2014**, *28*, 238–242, 245.
72. Cai, Q.; Yang, H.S.; Li, Y.C.; Zhu, J. Dissecting the Roles of PDCD4 in Breast Cancer. *Front. Oncol.* **2022**, *12*, 855807. [CrossRef] [PubMed]
73. Mao, X.H.; Chen, M.; Wang, Y.; Cui, P.G.; Liu, S.B.; Xu, Z.Y. MicroRNA-21 regulates the ERK/NF-κB signaling pathway to affect the proliferation, migration, and apoptosis of human melanoma A375 cells by targeting SPRY1, PDCD4, and PTEN. *Mol. Carcinog.* **2017**, *56*, 886–894. [CrossRef]
74. Kim, J.Y.; Lee, H.; Kim, E.K.; Lee, W.M.; Hong, Y.O.; Hong, S.A. Low PDCD4 Expression Is Associated With Poor Prognosis of Colorectal Carcinoma. *Appl. Immunohistochem. Mol. Morphol.* **2021**, *29*, 685–692. [CrossRef] [PubMed]
75. Xie, S.; Xiang, Y.; Wang, X.; Ren, H.; Yin, T.; Ren, J.; Liu, W. Acquired cholesteatoma epithelial hyperproliferation: Roles of cell proliferation signal pathways. *Laryngoscope* **2016**, *126*, 1923–1930. [CrossRef]
76. Friedland, D.R.; Eernisse, R.; Erbe, C.; Gupta, N.; Cioffi, J.A. MicroRNA regulation of cholesteatoma growth microrna regulation of cholesteatoma growth. *Laryngoscope* **2009**, *119*, S114. [CrossRef]
77. Orobello, N.; Harrington, C.; Reilly, B.K. Updates in paediatric cholesteatoma. *Curr. Opin. Otolaryngol. Head Neck Surg.* **2022**, *30*, 422–425. [CrossRef] [PubMed]
78. Li, J.; Sun, D.; Pu, W.; Wang, J.; Peng, Y. Circular RNAs in Cancer: Biogenesis, Function, and Clinical Significance. *Trends Cancer* **2020**, *6*, 319–336. [CrossRef]
79. Li, W.; Liu, J.Q.; Chen, M.; Xu, J.; Zhu, D. Circular RNA in cancer development and immune regulation. *J. Cell Mol. Med.* **2022**, *26*, 1785–1798. [CrossRef]
80. Altesha, M.A.; Ni, T.; Khan, A.; Liu, K.; Zheng, X. Circular RNA in cardiovascular disease. *J. Cell Physiol.* **2019**, *234*, 5588–5600. [CrossRef]
81. Jin, J.; Sun, H.; Shi, C.; Yang, H.; Wu, Y.; Li, W.; Dong, Y.H.; Cai, L.; Meng, X.M. Circular RNA in renal diseases. *J. Cell Mol. Med.* **2020**, *24*, 6523–6533. [CrossRef] [PubMed]
82. Wu, X.; Xiao, Y.; Ma, J.; Wang, A. Circular RNA: A novel potential biomarker for skin diseases. *Pharmacol. Res.* **2020**, *158*, 104841. [CrossRef] [PubMed]

83. Zhang, Y.L.; Zhang, S.Y.; Hu, J.; Zhang, H.H. Circular RNA in Diabetes and its Complications. *Zhongguo Yi Xue Ke Xue Yuan Xue Bao* **2022**, *44*, 521–528. [CrossRef]
84. Li, T.R.; Jia, Y.J.; Wang, Q.; Shao, X.Q.; Lv, R.J. Circular RNA: A new star in neurological diseases. *Int. J. Neurosci.* **2017**, *127*, 726–734. [CrossRef]
85. Singh, M.; Dwibedy, S.L.L.; Biswal, S.R.; Muthuswamy, S.; Kumar, A.; Kumar, S. Circular RNA: A novel and potential regulator in pathophysiology of schizophrenia. *Metab. Brain Dis.* **2022**, *37*, 1309–1316. [CrossRef]
86. Gao, J.; Tang, Q.; Xue, R.; Zhu, X.; Wang, S.; Zhang, Y.; Liu, W.; Gao, Z.; Yang, H. Comprehensive circular RNA expression profiling with associated ceRNA network reveals their therapeutic potential in cholesteatoma. *Oncol. Rep.* **2020**, *43*, 1234–1244. [CrossRef]
87. He, Y.; Sun, M.M.; Zhang, G.G.; Yang, J.; Chen, K.S.; Xu, W.W.; Li, B. Targeting PI3K/Akt signal transduction for cancer therapy. *Signal Transduct Target Ther.* **2021**, *6*, 425. [CrossRef] [PubMed]
88. Stanciu, S.; Ionita-Radu, F.; Stefani, C.; Miricescu, D.; Stanescu, S., II; Greabu, M.; Ripszky Totan, A.; Jinga, M. Targeting PI3K/AKT/mTOR Signaling Pathway in Pancreatic Cancer: From Molecular to Clinical Aspects. *Int. J. Mol. Sci.* **2022**, *23*, 132. [CrossRef]
89. Mao, X.B.; Sheng, T.; Zhuang, L.P.; Wu, C.K.; Zhao, G.G. Expression of miRNA let-7a and HMGA2 and Diagnostic Value of Serum miRNA let-7a Level in Pancreatic Cancer. *Sichuan Da Xue Xue Bao Yi Xue Ban* **2020**, *51*, 540–545. [CrossRef]
90. Motoyama, K.; Inoue, H.; Nakamura, Y.; Uetake, H.; Sugihara, K.; Mori, M. Clinical significance of high mobility group A2 in human gastric cancer and its relationship to let-7 microRNA family. *Clin. Cancer Res.* **2008**, *14*, 2334–2340. [CrossRef]
91. Chirshev, E.; Oberg, K.C.; Ioffe, Y.J.; Unternaehrer, J.J. Let-7 as biomarker, prognostic indicator, and therapy for precision medicine in cancer. *Clin. Transl. Med.* **2019**, *8*, 24. [CrossRef]
92. Bernstein, D.L.; Jiang, X.; Rom, S. let-7 microRNAs: Their Role in Cerebral and Cardiovascular Diseases, Inflammation, Cancer, and Their Regulation. *Biomedicines* **2021**, *9*, 606. [CrossRef]
93. Yin, H.; Sun, Y.; Wang, X.; Park, J.; Zhang, Y.; Li, M.; Yin, J.; Liu, Q.; Wei, M. Progress on the relationship between miR-125 family and tumorigenesis. *Exp. Cell Res.* **2015**, *339*, 252–260. [CrossRef] [PubMed]
94. Wang, Q.W.; Sun, Y.N.; Tan, L.J.; Zhao, J.N.; Zhou, X.J.; Yu, T.J.; Liu, J.T. MiR-125 family improves the radiosensitivity of head and neck squamous cell carcinoma. *Mol. Biol. Rep.* **2023**, *50*, 5307–5317. [CrossRef] [PubMed]
95. Wang, M.J.; Xu, Y.Y.; Huang, R.Y.; Chen, X.M.; Chen, H.M.; Han, L.; Yan, Y.H.; Lu, C.J. Role of an imbalanced miRNAs axis in pathogenesis of psoriasis: Novel perspectives based on review of the literature. *Oncotarget* **2017**, *8*, 5498–5507. [CrossRef] [PubMed]
96. Xu, N.; Brodin, P.; Wei, T.; Meisgen, F.; Eidsmo, L.; Nagy, N.; Kemeny, L.; Ståhle, M.; Sonkoly, E.; Pivarcsi, A. MiR-125b, a microRNA downregulated in psoriasis, modulates keratinocyte proliferation by targeting FGFR2. *J. Investig. Dermatol.* **2011**, *131*, 1521–1529. [CrossRef] [PubMed]
97. Chan, K.S.; Sano, S.; Kiguchi, K.; Anders, J.; Komazawa, N.; Takeda, J.; DiGiovanni, J. Disruption of Stat3 reveals a critical role in both the initiation and the promotion stages of epithelial carcinogenesis. *J. Clin. Investig.* **2004**, *114*, 720–728. [CrossRef]
98. Frank, D.A. STAT3 as a central mediator of neoplastic cellular transformation. *Cancer Lett.* **2007**, *251*, 199–210. [CrossRef]
99. Xu, Y.; Xu, X.; Gao, X.; Chen, H.; Geng, L. Shikonin suppresses IL-17-induced VEGF expression via blockage of JAK2/STAT3 pathway. *Int. Immunopharmacol.* **2014**, *19*, 327–333. [CrossRef]
100. Ho, K.Y.; Huang, H.H.; Hung, K.F.; Chen, J.C.; Chai, C.Y.; Chen, W.T.; Tsai, S.M.; Chien, C.Y.; Wang, H.M.; Wu, Y.J. Cholesteatoma growth and proliferation: Relevance with serpin B3. *Laryngoscope* **2012**, *122*, 2818–2823. [CrossRef]
101. Liu, W.; Xie, S.; Chen, X.; Rao, X.; Ren, H.; Hu, B.; Yin, T.; Xiang, Y.; Ren, J. Activation of the IL-6/JAK/STAT3 signaling pathway in human middle ear cholesteatoma epithelium. *Int. J. Clin. Exp. Pathol.* **2014**, *7*, 709–715. [PubMed]
102. Park, H.R.; Min, S.K.; Min, K.; Jun, S.Y.; Seo, J.; Kim, H.J. Increased expression of p63 and survivin in cholesteatomas. *Acta Otolaryngol.* **2009**, *129*, 268–272. [CrossRef] [PubMed]
103. Hamajima, Y.; Komori, M.; Preciado, D.A.; Choo, D.I.; Moribe, K.; Murakami, S.; Ondrey, F.G.; Lin, J. The role of inhibitor of DNA-binding (Id1) in hyperproliferation of keratinocytes: The pathological basis for middle ear cholesteatoma from chronic otitis media. *Cell Prolif.* **2010**, *43*, 457–463. [CrossRef] [PubMed]
104. Shen, D.; Zhao, H.Y.; Gu, A.D.; Wu, Y.W.; Weng, Y.H.; Li, S.J.; Song, J.Y.; Gu, X.F.; Qiu, J.; Zhao, W. miRNA-10a-5p inhibits cell metastasis in hepatocellular carcinoma via targeting SKA1. *Kaohsiung J. Med. Sci.* **2021**, *37*, 784–794. [CrossRef]
105. Zhu, H.; Kang, M.; Bai, X. TCF21 regulates miR-10a-5p/LIN28B signaling to block the proliferation and invasion of melanoma cells. *PLoS ONE* **2021**, *16*, e0255971. [CrossRef] [PubMed]
106. Liu, L.J.; Sun, X.Y.; Yang, C.X.; Zou, X.Y. MiR-10a-5p restrains the aggressive phenotypes of ovarian cancer cells by inhibiting HOXA1. *Kaohsiung J. Med. Sci.* **2021**, *37*, 276–285. [CrossRef]
107. Vaher, H.; Runnel, T.; Urgard, E.; Aab, A.; Carreras Badosa, G.; Maslovskaja, J.; Abram, K.; Raam, L.; Kaldvee, B.; Annilo, T.; et al. miR-10a-5p is increased in atopic dermatitis and has capacity to inhibit keratinocyte proliferation. *Allergy* **2019**, *74*, 2146–2156. [CrossRef]
108. Gao, T.; Zou, M.; Shen, T.; Duan, S. Dysfunction of miR-802 in tumors. *J. Clin. Lab. Anal.* **2021**, *35*, e23989. [CrossRef]
109. Sun, D.; Chen, J.; Wu, W.; Tang, J.; Luo, L.; Zhang, K.; Jin, L.; Lin, S.; Gao, Y.; Yan, X.; et al. MiR-802 causes nephropathy by suppressing NF-κB-repressing factor in obese mice and human. *J. Cell Mol. Med.* **2019**, *23*, 2863–2871. [CrossRef]

110. Zhang, F.; Ma, D.; Zhao, W.; Wang, D.; Liu, T.; Liu, Y.; Yang, Y.; Liu, Y.; Mu, J.; Li, B.; et al. Obesity-induced overexpression of miR-802 impairs insulin transcription and secretion. *Nat. Commun.* **2020**, *11*, 1822. [CrossRef]
111. Liu, W.; Li, G.; Li, J.; Chen, W. Long noncoding RNA TRG-AS1 protects against glucocorticoid-induced osteoporosis in a rat model by regulating miR-802-mediated CAB39/AMPK/SIRT-1/NF-κB axis. *Hum. Cell* **2022**, *35*, 1424–1439. [CrossRef]
112. Yao, J.; Gao, R.; Luo, M.; Li, D.; Guo, L.; Yu, Z.; Xiong, F.; Wei, C.; Wu, B.; Xu, Z.; et al. miR-802 participates in the inflammatory process of inflammatory bowel disease by suppressing SOCS5. *Biosci. Rep.* **2020**, *40*, BSR20192257. [CrossRef]
113. Wang, Y.; Liu, X.; Wang, L.; Zhang, Z.; Li, Z.; Li, M. Circ_PGPEP1 Serves as a Sponge of miR-1297 to Promote Gastric Cancer Progression via Regulating E2F3. *Dig. Dis. Sci.* **2021**, *66*, 4302–4313. [CrossRef]
114. Zhang, X.; Zhang, M.; Guo, Q.; Hu, X.; Zhao, Z.; Ni, L.; Liu, L.; Wang, X.; Wang, Z.; Tong, D.; et al. MicroRNA-1297 inhibits proliferation and promotes apoptosis in gastric cancer cells by downregulating CDC6 expression. *Anticancer Drugs* **2019**, *30*, 803–811. [CrossRef]
115. Liang, L.; Feng, L.; Wei, B. microRNA-1297 involves in the progression of oral squamous cell carcinoma through PTEN. *Saudi J. Biol. Sci.* **2018**, *25*, 923–927. [CrossRef] [PubMed]
116. Liu, F.; He, Y.; Shu, R.; Wang, S. MicroRNA-1297 regulates hepatocellular carcinoma cell proliferation and apoptosis by targeting EZH2. *Int. J. Clin. Exp. Pathol.* **2015**, *8*, 4972–4980.
117. Wang, Y.; Xue, J.; Kuang, H.; Zhou, X.; Liao, L.; Yin, F. microRNA-1297 Inhibits the Growth and Metastasis of Colorectal Cancer by Suppressing Cyclin D2 Expression. *DNA Cell Biol.* **2017**, *36*, 991–999. [CrossRef]
118. Park, C.R.; Lee, M.; Lee, S.Y.; Kang, D.; Park, S.J.; Lee, D.C.; Koo, H.; Park, Y.G.; Yu, S.L.; Jeong, I.B.; et al. Regulating POLR3G by MicroRNA-26a-5p as a promising therapeutic target of lung cancer stemness and chemosensitivity. *Noncoding RNA Res.* **2023**, *8*, 273–281. [CrossRef]
119. Li, M.; Xiao, Y.; Liu, M.; Ning, Q.; Xiang, Z.; Zheng, X.; Tang, S.; Mo, Z. MiR-26a-5p regulates proliferation, apoptosis, migration and invasion via inhibiting hydroxysteroid dehydrogenase like-2 in cervical cancer cell. *BMC Cancer* **2022**, *22*, 876. [CrossRef]
120. Chen, X.; Wu, G.; Qing, J.; Li, C.; Chen, X.; Shen, J. LINC00240 knockdown inhibits nasopharyngeal carcinoma progress by targeting miR-26a-5p. *J. Clin. Lab. Anal.* **2022**, *36*, e24424. [CrossRef] [PubMed]
121. Cai, B.; Qu, X.; Kan, D.; Luo, Y. miR-26a-5p suppresses nasopharyngeal carcinoma progression by inhibiting PTGS2 expression. *Cell Cycle* **2022**, *21*, 618–629. [CrossRef] [PubMed]
122. Chung, Y.H.; Cheng, Y.T.; Kao, Y.H.; Tsai, W.C.; Huang, G.K.; Chen, Y.T.; Shen, Y.C.; Tai, M.H.; Chiang, P.H. MiR-26a-5p as a useful therapeutic target for upper tract urothelial carcinoma by regulating WNT5A/β-catenin signaling. *Sci. Rep.* **2022**, *12*, 6955. [CrossRef] [PubMed]
123. Wang, Z.; He, S.; Guo, P.; Guo, X.; Zheng, J. MicroRNA-1297 inhibits metastasis and epithelial-mesenchymal transition by targeting AEG-1 in cervical cancer. *Oncol. Rep.* **2017**, *38*, 3121–3129. [CrossRef] [PubMed]
124. Pan, X.; Li, H.; Tan, J.; Weng, X.; Zhou, L.; Weng, Y.; Cao, X. miR-1297 Suppresses Osteosarcoma Proliferation and Aerobic Glycolysis by Regulating PFKFB2. *Onco. Targets Ther.* **2020**, *13*, 11265–11275. [CrossRef]
125. Zhao, J.; Jiang, Q.; Xu, C.; Jia, Q.; Wang, H.; Xue, W.; Wang, Y.; Zhu, Z.; Tian, L. MiR-26a-5p from HucMSC-derived extracellular vesicles inhibits epithelial mesenchymal transition by targeting Adam17 in silica-induced lung fibrosis. *Ecotoxicol. Environ. Saf.* **2023**, *257*, 114950. [CrossRef]
126. Xie, T.; Pei, Y.; Shan, P.; Xiao, Q.; Zhou, F.; Huang, L.; Wang, S. Identification of miRNA-mRNA Pairs in the Alzheimer's Disease Expression Profile and Explore the Effect of miR-26a-5p/PTGS2 on Amyloid-β Induced Neurotoxicity in Alzheimer's Disease Cell Model. *Front. Aging Neurosci.* **2022**, *14*, 909222. [CrossRef]
127. Balasubramanian, S.; Adhikary, G.; Eckert, R.L. The Bmi-1 polycomb protein antagonizes the (-)-epigallocatechin-3-gallate-dependent suppression of skin cancer cell survival. *Carcinogenesis* **2010**, *31*, 496–503. [CrossRef]
128. Ma, N.; Zhao, S.; Yang, W.; Wang, Y. B-cell-specific Moloney murine leukemia virus integration site 1 knockdown impairs adriamycin resistance of gastric cancer cells. *Arab J. Gastroenterol.* **2023**. [CrossRef]
129. Liu, J.Y.; Jiang, Y.N.; Huang, H.; Xu, J.F.; Wu, Y.H.; Wang, Q.; Zhu, Y.; Zheng, B.; Shen, C.; Qian, W.F.; et al. BMI-1 promotes breast cancer proliferation and metastasis through different mechanisms in different subtypes. *Cancer Sci.* **2023**, *114*, 449–462. [CrossRef]
130. Liu, S.; Yang, Y.; Chen, L.; Liu, D.; Dong, H. MicroRNA-154 functions as a tumor suppressor in non-small cell lung cancer through directly targeting B-cell-specific Moloney murine leukemia virus insertion site 1. *Oncol. Lett.* **2018**, *15*, 10098–10104. [CrossRef]
131. Li, J.; Wang, Y.; Ge, J.; Li, W.; Yin, L.; Zhao, Z.; Liu, S.; Qin, H.; Yang, J.; Wang, L.; et al. Doublecortin-Like Kinase 1 (DCLK1) Regulates B Cell-Specific Moloney Murine Leukemia Virus Insertion Site 1 (Bmi-1) and is Associated with Metastasis and Prognosis in Pancreatic Cancer. *Cell Physiol. Biochem.* **2018**, *51*, 262–277. [CrossRef] [PubMed]
132. Espersen, M.L.; Linnemann, D.; Christensen, I.J.; Alamili, M.; Troelsen, J.T.; Høgdall, E. The prognostic value of polycomb group protein B-cell-specific moloney murine leukemia virus insertion site 1 in stage II colon cancer patients. *Apmis* **2016**, *124*, 541–546. [CrossRef] [PubMed]
133. Abobaker, S.; Kulbe, H.; Taube, E.T.; Darb-Esfahani, S.; Richter, R.; Denkert, C.; Jank, P.; Sehouli, J.; Braicu, E.I. Polycomb Protein BMI-1 as a Potential Therapeutic Target in Mucinous Ovarian Cancer. *Anticancer Res.* **2022**, *42*, 1739–1747. [CrossRef] [PubMed]
134. Lee, K.; Adhikary, G.; Balasubramanian, S.; Gopalakrishnan, R.; McCormick, T.; Dimri, G.P.; Eckert, R.L.; Rorke, E.A. Expression of Bmi-1 in epidermis enhances cell survival by altering cell cycle regulatory protein expression and inhibiting apoptosis. *J. Invest. Dermatol.* **2008**, *128*, 9–17. [CrossRef]

135. Reinisch, C.M.; Uthman, A.; Erovic, B.M.; Pammer, J. Expression of BMI-1 in normal skin and inflammatory and neoplastic skin lesions. *J. Cutan Pathol.* **2007**, *34*, 174–180. [CrossRef]
136. Wang, S.L.; Dong, X.W.; Zhao, F.; Li, C.X. MiR-203 inhibits cell proliferation, invasion, and migration of ovarian cancer through regulating RGS17. *J. Biol. Regul. Homeost. Agents* **2021**, *35*, 1109–1115. [CrossRef]
137. Song, S.; Johnson, K.S.; Lujan, H.; Pradhan, S.H.; Sayes, C.M.; Taube, J.H. Nanoliposomal Delivery of MicroRNA-203 Suppresses Migration of Triple-Negative Breast Cancer through Distinct Target Suppression. *Noncoding. RNA* **2021**, *7*, 45. [CrossRef]
138. Altan, Z.; Sahin, Y. miR-203 suppresses pancreatic cancer cell proliferation and migration by modulating DUSP5 expression. *Mol. Cell Probes.* **2022**, *66*, 101866. [CrossRef]
139. Yi, R.; Poy, M.N.; Stoffel, M.; Fuchs, E. A skin microRNA promotes differentiation by repressing 'stemness'. *Nature* **2008**, *452*, 225–229. [CrossRef]
140. Kim, Y.J.; Lee, S.B.; Lee, H.B. Oleic acid enhances keratinocytes differentiation via the upregulation of miR-203 in human epidermal keratinocytes. *J. Cosmet. Dermatol.* **2019**, *18*, 383–389. [CrossRef]
141. Liu, W.; Yin, T.; Ren, J.; Li, L.; Xiao, Z.; Chen, X.; Xie, D. Activation of the EGFR/Akt/NF-κB/cyclinD1 survival signaling pathway in human cholesteatoma epithelium. *Eur. Arch. Otorhinolaryngol.* **2014**, *271*, 265–273. [CrossRef]
142. Huisman, M.A.; De Heer, E.; Grote, J.J. Survival signaling and terminal differentiation in cholesteatoma epithelium. *Acta Otolaryngol.* **2007**, *127*, 424–429. [CrossRef]
143. Wang, Q.; Ye, B.; Wang, P.; Yao, F.; Zhang, C.; Yu, G. Overview of microRNA-199a Regulation in Cancer. *Cancer Manag. Res.* **2019**, *11*, 10327–10335. [CrossRef] [PubMed]
144. Zahra, M.A.; Kamha, E.S.; Abdelaziz, H.K.; Nounou, H.A.; Deeb, H.M.E. Aberrant Expression of Serum MicroRNA-153 and -199a in Generalized Epilepsy and its Correlation with Drug Resistance. *Ann. Neurosci.* **2022**, *29*, 203–208. [CrossRef] [PubMed]
145. Tang, J.; Yu, H.; Wang, Y.; Duan, G.; Wang, B.; Li, W.; Zhu, Z. microRNA-199a counteracts glucocorticoid inhibition of bone marrow mesenchymal stem cell osteogenic differentiation through regulation of Klotho expression in vitro. *Cell Biol. Int.* **2020**, *44*, 2532–2540. [CrossRef]
146. Sharma, S. Immunomodulation: A definitive role of microRNA-142. *Dev. Comp. Immunol.* **2017**, *77*, 150–156. [CrossRef]
147. Berrien-Elliott, M.M.; Sun, Y.; Neal, C.; Ireland, A.; Trissal, M.C.; Sullivan, R.P.; Wagner, J.A.; Leong, J.W.; Wong, P.; Mah-Som, A.Y.; et al. MicroRNA-142 Is Critical for the Homeostasis and Function of Type 1 Innate Lymphoid Cells. *Immunity* **2019**, *51*, 479–490.e476. [CrossRef] [PubMed]
148. Zhou, C.; Wang, P.; Lei, L.; Huang, Y.; Wu, Y. Overexpression of miR-142-5p inhibits the progression of nonalcoholic steatohepatitis by targeting TSLP and inhibiting JAK-STAT signaling pathway. *Aging* **2020**, *12*, 9066–9084. [CrossRef]
149. Shrestha, A.; Mukhametshina, R.T.; Taghizadeh, S.; Vásquez-Pacheco, E.; Cabrera-Fuentes, H.; Rizvanov, A.; Mari, B.; Carraro, G.; Bellusci, S. MicroRNA-142 is a multifaceted regulator in organogenesis, homeostasis, and disease. *Dev. Dyn.* **2017**, *246*, 285–290. [CrossRef]
150. Yao, R.; Xu, L.; Wei, B.; Qian, Z.; Wang, J.; Hui, H.; Sun, Y. miR-142-5p regulates pancreatic cancer cell proliferation and apoptosis by regulation of RAP1A. *Pathol. Res. Pract.* **2019**, *215*, 152416. [CrossRef]
151. Yu, W.; Li, D.; Zhang, Y.; Li, C.; Zhang, C.; Wang, L. MiR-142-5p Acts as a Significant Regulator Through Promoting Proliferation, Invasion, and Migration in Breast Cancer Modulated by Targeting SORBS1. *Technol. Cancer Res. Treat* **2019**, *18*, 1533033819892264. [CrossRef] [PubMed]
152. Yan, J.; Yang, B.; Lin, S.; Xing, R.; Lu, Y. Downregulation of miR-142-5p promotes tumor metastasis through directly regulating CYR61 expression in gastric cancer. *Gastric. Cancer* **2019**, *22*, 302–313. [CrossRef]
153. Cheng, D.; Li, J.; Zhang, L.; Hu, L. miR-142-5p suppresses proliferation and promotes apoptosis of human osteosarcoma cell line, HOS, by targeting PLA2G16 through the ERK1/2 signaling pathway. *Oncol. Lett.* **2019**, *17*, 1363–1371. [CrossRef] [PubMed]
154. Wei, G.; Yuan, Y.; He, X.; Jin, L.; Jin, D. Enhanced plasma miR-142-5p promotes the progression of intrahepatic cholangiocarcinoma via targeting PTEN. *Exp. Ther. Med.* **2019**, *17*, 4190–4196. [CrossRef] [PubMed]
155. NavaneethaKrishnan, S.; Rosales, J.L.; Lee, K.Y. Loss of Cdk5 in breast cancer cells promotes ROS-mediated cell death through dysregulation of the mitochondrial permeability transition pore. *Oncogene* **2018**, *37*, 1788–1804. [CrossRef]
156. Wang, Q.; Yin, R.; Chen, X.; Hu, B.; Jiang, B.; Tang, W.; Zhang, X.; Jin, X.; Ying, M.; Fu, J. Higher Levels of Tumour-Infiltrating Lymphocytes (TILs) are Associated with a Better Prognosis, While CDK5 Plays a Different Role Between Nonmetastatic and Metastatic Colonic Carcinoma. *Cancer Control* **2023**, *30*, 10732748231169396. [CrossRef]
157. Ling, R.; Sheng, Y.; Hu, Y.; Wang, D.; Zhou, Y.; Shu, Y. Comprehensive analysis of CDK5 as a novel biomarker for progression in esophageal cancer. *Esophagus* **2023**, *20*, 502–514. [CrossRef]
158. Shao, X.; Yang, Y.; Chen, J.; Zhao, R.; Xu, L.; Guo, X.; Feng, Y.; Qin, L. Identification of Two CDK5R1-Related Subtypes and Characterization of Immune Infiltrates in Alzheimer's Disease Based on an Integrated Bioinformatics Analysis. *Comput. Math Methods Med.* **2022**, *2022*, 6766460. [CrossRef]
159. Batra, S.; Jahan, S.; Ashraf, A.; Alharby, B.; Jawaid, T.; Islam, A.; Hassan, I. A review on cyclin-dependent kinase 5: An emerging drug target for neurodegenerative diseases. *Int. J. Biol. Macromol.* **2023**, *230*, 123259. [CrossRef]
160. Liu, C.C.; Zhang, H.L.; Zhi, L.L.; Jin, P.; Zhao, L.; Li, T.; Zhou, X.M.; Sun, D.S.; Cheng, G.H.; Xin, Q.; et al. CDK5 Regulates PD-L1 Expression and Cell Maturation in Dendritic Cells of CRSwNP. *Inflammation* **2019**, *42*, 135–144. [CrossRef]
161. Zhang, D.; Lee, H.; Wang, X.; Groot, M.; Sharma, L.; Dela Cruz, C.S.; Jin, Y. A potential role of microvesicle-containing miR-223/142 in lung inflammation. *Thorax* **2019**, *74*, 865–874. [CrossRef] [PubMed]

162. Ji, Y.; Fang, Q.Y.; Wang, S.N.; Zhang, Z.W.; Hou, Z.J.; Li, J.N.; Fu, S.Q. Lnc-RNA BLACAT1 regulates differentiation of bone marrow stromal stem cells by targeting miR-142-5p in osteoarthritis. *Eur. Rev. Med. Pharmacol. Sci.* **2020**, *24*, 2893–2901. [CrossRef] [PubMed]
163. Rokavec, M.; Huang, Z.; Hermeking, H. Meta-analysis of miR-34 target mRNAs using an integrative online application. *Comput. Struct. Biotechnol. J.* **2023**, *21*, 267–274. [CrossRef] [PubMed]
164. Li, S.; Wei, X.; He, J.; Cao, Q.; Du, D.; Zhan, X.; Zeng, Y.; Yuan, S.; Sun, L. The comprehensive landscape of miR-34a in cancer research. *Cancer Metastasis Rev.* **2021**, *40*, 925–948. [CrossRef] [PubMed]
165. Mehta, D.; Daudia, A.; Birchall, J.P.; Banerjee, A.R. The localization of matrix metalloproteinases-8 and -13 in cholesteatoma, deep-meatal and post-auricular skin: A comparative analysis. *Acta Otolaryngol.* **2007**, *127*, 138–142. [CrossRef]
166. Lee, R.J.; Mackenzie, I.C.; Hall, B.K.; Gantz, B.J. The nature of the epithelium in acquired cholesteatoma. *Clin. Otolaryngol. Allied Sci.* **1991**, *16*, 168–173. [CrossRef]
167. Ferlito, S.; Fadda, G.; Lechien, J.R.; Cammaroto, G.; Bartel, R.; Borello, A.; Cavallo, G.; Piccinini, F.; La Mantia, I.; Cocuzza, S.; et al. Type 1 Tympanoplasty Outcomes between Cartilage and Temporal Fascia Grafts: A Long-Term Retrospective Study. *J. Clin. Med.* **2022**, *11*, 7000. [CrossRef]
168. Ferlito, S.; La Mantia, I.; Merlino, F.; Cocuzza, S.; Di Stadio, A.; Cammaroto, G.; Bartel, R.; Fadda, G.; Iannella, G.; Mat, Q.; et al. Long-Term Anatomical and Hearing Outcomes of Canal Wall down Tympanoplasty for Tympano-Mastoid Cholesteatoma: A 20-Year Retrospective Study. *Life* **2022**, *12*, 1745. [CrossRef]

International Journal of *Molecular Sciences*

MDPI

Review

miRNAs as Modern Biomarkers in Asthma Therapy

Natalia Kierbiedź-Guzik [1,*] and Barbara Sozańska [2]

1 14th Paediatric Ward—Pulmonology and Allergology, J. Gromkowski Provincial Specialist Hospital, ul. Koszarowa 5, 51-149 Wrocław, Poland

2 1st Department and Clinic of Paediatrics, Allergology and Cardiology Wrocław Medical University, ul. Chałubińskiego 2a, 50-368 Wrocław, Poland; barbara.sozanska@umw.edu.pl

* Correspondence: natalia.kierbiedz@gmail.com

Abstract: Asthma is a chronic inflammatory disease of the airways characterized by shortness of breath, chest tightness, coughing, and wheezing. For several decades (approximately 30 years), miRNAs and their role in asthma have been of constant interest among scientists. These small, non-coding RNA fragments, 18–25 nucleotides long, regulate gene expression at the post-transcriptional level by binding to the target mRNA. In this way, they affect several biological processes, e.g., shaping airway structures, producing cytokines and immune mediators, and controlling defense mechanisms. Publications confirm their potential role in the diagnosis and monitoring of the disease, but only some articles address the use of miRNAs in the treatment of asthma. The following paper reviews the latest available studies and presents miRNAs as a useful tool for predicting the effectiveness of the included treatment, early diagnosis of exacerbations, and in assessing patient compliance for different groups of drugs used in asthma. The latest known pathways underlying the pathogenesis of the disease, which are associated with a change in miRNA expression, may be precise targets of therapeutic activity in the future.

Keywords: asthma; miRNA; antagomir; biomarkers; response; treatment; glucocorticosteroids; beta-mimetic; biological therapy; pathways; infection

1. Introduction

Asthma is a chronic inflammatory disease of the airways characterized by shortness of breath, chest tightness, wheezing, and coughing, affecting nearly 300 million people worldwide, and an additional 100 million people are expected to be affected by the end of 2025 [1]. Environmental factors and genetic conditions predispose the development of the disease, and their exact pathomechanism is not known [2]. A particular clinical challenge is severe asthma, which, despite the use of high doses of inhaled corticosteroids (ICS) and a second controller and/or oral corticosteroids, still cannot be adequately controlled [3]. In adults, 5–10% are affected by this disease [4], and the exact statistics among children are unknown [5]. This problem is estimated to affect about 5% of the pediatric population [6]. In addition, its various phenotypes can be distinguished, considering the cells involved in the systemic inflammatory process (eosinophilic, neutrophilic, and mixed form) [7,8]. In most cases, severe asthma is characterized by atopy, higher levels of IgE and eosinophils in the blood, increased levels of FeNO (fractional exhaled nitric oxide) in exhaled air, and, usually, positive skin tests after contact with an allergen. Neutrophilic asthma is characterized by the predominance of neutrophils in the blood and sputum and the involvement of Th1 (T helper type 1) and Th17 (T helper type 17) cells in the inflammatory process [9]. It has been proven that Th17 lymphocytes increase neutrophil infiltration, which may cause increased inflammation in the airways and contribute to severe cases of steroid-resistant disease [10,11]. The phenotype of severe asthma is very complex and understanding the molecular basis of its course will allow for greater control of disease symptoms and an improved response to treatment. The hope for understanding these biological pathways is provided by miRNA molecules, which have been an intense subject of scientific interest in recent years. Their broad impact on the processes involved in the pathogenesis of asthma has been confirmed (including the proliferation of smooth muscle cells, increased mucus production, collagen, and overactivity of the respiratory tract) [12]. These are small, non-coding RNA biomolecules, 18–25 nucleotides long, and regulate gene expression at the post-transcriptional level by binding to the target mRNA (messenger RNA) [13]. Due to their role in multiple cellular processes, such as inflammatory and immunomodulatory processes, they are an important future target of therapeutic activities leading to increased and improved sensitivity to steroid therapy [14]. The following paper summarizes the results of recent studies showing miRNAs and their role as molecules in monitoring the effectiveness of therapy, assessing the response to the introduced treatment, and examples of their experimental use in asthma therapy. The paper pays special attention to studies conducted among the pediatric population. However, due to their small number, the review also included studies conducted on groups of adult patients. The potential use of miRNAs in the treatment of asthma is summarized and presented in Figure 1 below.

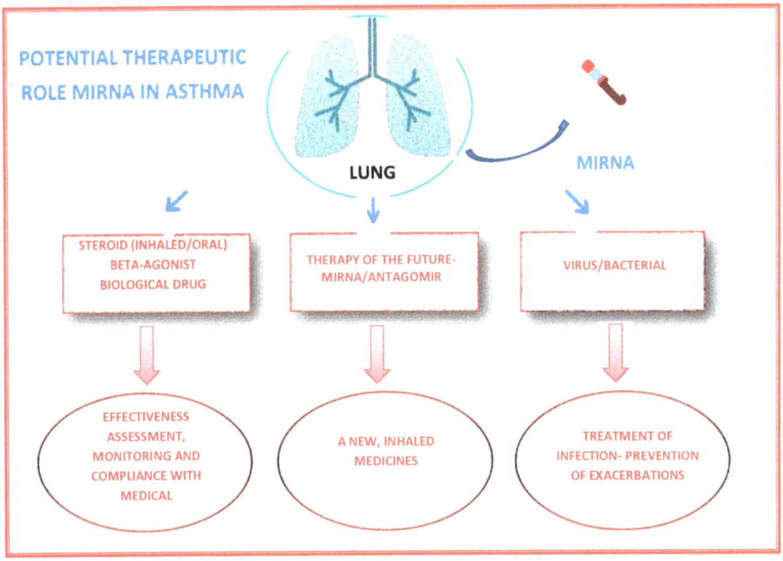

Figure 1. miRNAs and their potential use in asthma.

2. miRNAs and Individual Groups of Drugs Used in Asthma (Glucocorticosteroids, Short-Acting Beta-Agonists, and Biological Therapy)

Currently, many publications show miRNAs as potential asthma diagnostic biomarkers [13,15–17]. Few studies have also drawn attention to the relationship between miRNAs and a particular therapeutic regimen. The following chapter divides the studies according to the main groups of medicines used in asthma and shows their relationship with individual miRNAs, thus confirming their use as predictive biomarkers, and in monitoring and evaluating the effectiveness of the treatment.

The first object of consideration will be inhaled glucocorticosteroids (ICS), the primary asthma drug [18]. They soothe the inflammatory process in the respiratory tract and prevent future disease exacerbations. Their mechanism of action is complex and can be divided into genomic and non-genomic. The first directly relates to the effect on gene transcription, causing their transactivation (increased transcription) or transrepression (reduced transcription). The other activates signaling pathways of kinases, G proteins, and ion channels. Their anti-inflammatory effect associated with the inhibition of cytokine production is used in clinical practice [19]. A diverse response to ICS among patients is constantly observed. A relationship between the response and the change in the concentration of individual miRNA molecules in the blood and their direct impact on given biological processes was noted. Among the pediatric population aged 9.0 ± 2.1 years who suffer from asthma and were using inhaled glucocorticosteroids for 4 years, it was shown that two biomolecules—miRNA-155-5p and miRNA-532-5p—can be used to predict the response to the applied treatment. These miRNAs affect the transcription of NF-κB in airway epithelial cells (nuclear factor kappa-light-chain-enhancer of activated B cells), the expression of which is reduced (transrepression) by dexamethasone. The transcription factor NF-κB regulates many aspects of innate and adaptive immune function and is a crucial mediator of the inflammatory response. Consequently, its excessive activation leads to the initiation of pathological processes of many inflammatory diseases, including asthma [20]. miRNA-155-5p inhibits NF-κB transrepression, while miRNA-532-5p increases it. Interestingly, with the increase in the concentration of miRNA155-5p in the blood, the FEV1% (forced expiratory volume during the first second of expiration) improves, and patients have a better response to ICS, while the opposite effect was noted for miRNA-532-5p [21]. Why is it that miRNA-155-5p causes a better therapeutic response even though it intensifies the synthesis of the pro-inflammatory factor NF-κB? The results of other publications may provide a solution to this problem. It has been shown that the expression of miRNA-155-5p is predominantly in the serum in healthy individuals [22], and in asthmatics, it is in the bronchial epithelial cells [23]. Therefore, it can be assumed that the concentration of this molecule is increased in the blood of patients with a milder form of the disease, which is characterized by better sensitivity to ICS. Moreover, miRNA-155, in addition to predicting the effectiveness of the steroids used, can be used as a future therapeutic target in asthma (chapter: miRNA and new directions in asthma treatment). The above paragraph provides examples of miRNA molecules (miRNA-155-5p and miRNA 532-5p) that may be biomarkers for assessing the response to ICS in pediatric asthma.

Another molecule tested in the pediatric population was miRNA-21. It has been shown that miRNA-21 can be successfully used as a non-invasive diagnostic marker of asthma, showing a positive correlation with blood and sputum eosinophilia and IL-4, and a negative correlation with the FEV1 and IL-12 concentration [24]. Among the pediatric group aged 9.0 ± 2.7 years, its expression was increased in children not receiving ICS and in steroid-resistant cases of the disease (SR). In the above study, serum miRNA-21 has a high therapeutic predictive value in differentiating patients with ICS-sensitive and ICS-resistant asthma. It additionally reduces the risk of side effects and spares patients the disappointment of treatment failure. NF-κB and activator protein 1 (AP-1) show affinity for the promoter of the miRNA-21 gene, enhancing its synthesis. Using inhaled steroids reduces the expression of the above transcription factors, explaining the decrease in the serum miRNA-21 concentration [25]. In another project conducted on *mice*, miRNA-21 has been shown to play an essential role in developing severe, steroid-resistant asthma. It is well-known that infections are one of the leading causes of exacerbation in chronic inflammatory diseases of the respiratory tract. An infection leads to an increase in miRNA-21, the molecular goal of which is to inhibit *PTEN* (phosphatase and tensin homologue), belonging to the group of suppressor genes, which results in the activation of one of the isoforms of phosphatidylinositol 3-kinase (PI3K), which leads to the phosphorylation and reduction of HDCA2 (histone deacetylate 2) levels. The miRNA-21/*PTEN*/PI3K/HDAC2 pathway is responsible for developing severe, treatment-resistant disease. Experimental use of the intranasal form of miRNA-21, anti-miRNA-21, and PI3K inhibitor leads to a similar therapeutic effect as the use of steroids in asthma without signs of infection [26]. In 160 children aged 5–12 years using ICS for a year and selected from the CAMP (Childhood Asthma Management Program) group, miRNA-146b, miRNA-206, and miRNA-720 together with the patient's clinical characteristics helped to identify episodes of exacerbations, and in assessing the effectiveness and the need to escalate therapy. This will allow for the selection of people who are more likely to respond to steroids, minimizing the risk of disease complications, and reducing the cost of medical services. The relationship between the above molecules and the pathways responsible for the development of asthma has been proven [27].

The paragraphs mentioned above highlight studies where children were the focus. The following sections focus on adult patients and an induced experimental model of asthma in *mice*.

Another molecule associated with steroid treatment is miRNA-144-3p. In adult patients with severe disease requiring high daily doses of inhaled ICS, oral steroids, or leukotriene receptor antagonist drugs, higher levels of this molecule in the lung tissue have been shown. Its level correlates with worse lung function, eosinophilia, and type 2 inflammatory cytokines. miRNA-144-3p is responsible for the increased expression of genes whose protein products lead to the development of asthma. A single biomarker in conjunction with the clinical characteristics may be helpful in the selection of patients who are more likely to require therapy based on biological drugs [28,29]. miRNA-146a is an important modulator of the inflammatory process; it inhibits the activation of NF-κB via the TLR (toll-like receptor) and reduces the deposition of mucoproteins that contribute to airway remodeling in asthma [30]. Increased levels of this molecule have been reported in the airway epithelial cells and serum of adult patients, necessitating higher doses of inhaled steroids. ICS caused a decrease in the expression of NF-κB and miRNA-146a, while the exogenous introduction of this molecule enhanced the anti-inflammatory effect of the steroids. Increased inflammation induces the expression of miRNA-146a, the primary task of which is its reduction (feedback mechanism). Monitoring the level of this biomarker can help predict the response to glucocorticosteroids, and its synthetic equivalent can be used therapeutically to overcome steroid resistance in patients with severe asthma [31]. In another study by Swedish scientists among adults diagnosed with mild and moderate asthma, the increased expression of miRNA-155 and -146 was found in the blood serum, which additionally increased after the introduction of ICS therapy. Since these medications mainly suppress the type 2 immune response, they may increase the synthesis of post-transcriptional regulators such as miRNA-155 and -146. These biomolecules alleviate asthma symptoms by inhibiting the cells involved in the type 2 response. In addition, common molecular target genes responsible for leukocyte regulation and the response to glucocorticosteroids have been demonstrated for these molecules [32].

In addition, miRNAs can affect the translation of the target protein of the steroid receptor (GR), making steroid treatment ineffective and leading to the development of a resistant form of the disease [33]. One of these molecules is miRNA-9, which is increased upon exposure to IFN (interferon) and LPS (lipopolysaccharide) in macrophages isolated from *mice* lung tissue. The molecular mechanism of steroid resistance is complex. There is a decrease in PP2A (protein phosphatase 2A) activity, which leads to the activation of JNK1 (c-Jun N-terminal kinase), which is responsible for increased steroid receptor phosphorylation. The consequence of these events is impaired translocation to the cell nucleus, leading to ineffectiveness of the steroid therapy. This process is reversed using antagomir (anti-miRNA-9), which restores the PP2A function [34]. A similar relationship to that of the miRNA-9 molecule was shown for miRNA-433 [35] and miRNA-18a [36]. In studies conducted on *mice*, the increase in these biomolecules inhibited GR translocation into the cell nucleus.

Inhaled glucocorticosteroids continue to be the basic drugs controlling the course of asthma. Patients with a severe form of the disease are forced to use them in high doses, often with or without additional treatment, to reduce their activity. Oral corticosteroids (OCS) are drugs reserved for such cases, e.g., uncontrolled asthma and its exacerbations. The need for their long-term use is associated with the occurrence of side effects (e.g., obesity/overweight, gastrointestinal bleeding, hirsutism, steroid acne, muscle mass reduction, psychiatric disorders) and the development of other diseases (such as diabetes, cataracts, osteoporosis, hypertension, iatrogenic Cushing's syndrome) [37,38]. A study conducted on a group of adult patients with severe asthma who used OCS or did not take the drug showed a difference in the expression of 11 miRNA molecules between the mentioned groups (increased their concentration in the serum of subjects receiving oral steroids). The analysis was performed using Next Generation Sequencing (NGS). Then, after validation by RT-qPCR, it was confirmed that five biomolecules (miRNA-148b-3p, -221-5p, -618, -941, and -769-5p) in blood serum and three in lung tissue samples (miRNA-144-3p, -144-5p, and -451a) showed an increased concentration among OCS recipients. In the multivariate logistic regression model, the combination of miRNA-221-5p and -miRNA-769-5p in the serum was the best for differentiating severe asthma patients treated and untreated with OCS. The participation of these molecules in the pathways responsible for the development of asthma has been confirmed. They were responsible for the changed expression of FOXO3 (forkhead box O3a), PTEN, and MAPK3 (mitogen-activated protein kinase 3), causing, among others, an increased concentration of IgE and excessive synthesis of pro-inflammatory cytokines [39,40]. This is one of the first publications showing the relationship between miRNAs and oral steroids. The obtained results of the above study help differentiate people receiving and not taking OCS. This may be practical in assessing the compliance in treated patients who do not respond to the prescribed oral steroid therapy [41].

As a brief summary of the results of the above publications combining miRNAs with steroid treatment, their potential, huge role in predicting effectiveness, assessing the risk of exacerbations, and observing patient compliance with medical recommendations has been confirmed. They have a broad impact not only on the biological processes responsible for the development of asthma, but also on the mechanism of action of glucocorticosteroids (as described earlier, the effect on the expression of the steroid receptor protein); thus, they are of great importance in the final therapeutic effect.

β2-agonists (SABAs, short acting beta agonists) are another group of drugs that are used in the temporary relief of severe asthma symptoms [42]. Regardless of the severity of the disease and often despite treatment, patients can experience episodes of acute exacerbations and loss of control of the disease, which is where these preparations are required. Their mechanism of action is based on inhibiting the contraction and then promoting relaxation of the smooth muscles of the respiratory tract through the activation of beta-2 adrenergic receptors (ADRB2), and the increased production of cyclic adenosine monophosphoric acid is responsible for this [43]. A study in adult asthmatics who were sensitive and unresponsive to salmeterol, an example of a β2-agonist, showed higher serum miRNA-16 expression levels in drug-resistant patients. This molecule leads to a decrease in ADRB2 expression and, consequently, results in lower FEV1 values, which confirms the positive correlation between the level of the β2-agonist receptor and the lung function index. The results of the above publication show miRNA-16 as a potential biomarker for assessing the effectiveness of therapy [44]. Another molecule proven to be interestingly related to the given group of drugs is miRNA-let-7f. Airway smooth muscle cells were collected from healthy individuals and asthmatics. They were then cultured in a laboratory setting and simultaneously exposed to the β-agonist for 18 h. An increase in the concentration of miRNA-let-7f was observed only for cells derived from asthmatics, which resulted in the inhibition of ADRB2 translation (decrease in the protein product by nearly 90%). CREB (cAMP receptor element-binding protein) is involved in this process, where its activation leads to binding to the promoter of the miRNA-let-7f gene, which in turn results in the increased production of this molecule. The reversal of the process described above was obtained by inhibiting CREB and the miRNA. Interestingly, this phenomenon has been described only in the case of cells taken from sick people and illustrates one of the possible mechanisms of tachyphylaxis among β2-agonists. This confirms its key importance in the pathogenesis of the disease and in the ineffectiveness of treatment [45].

The cited results of studies show an interesting relationship between miRNAs and ad hoc preparations used in asthma exacerbations. Due to the small number of publications on this subject, this is a further interesting research direction.

In recent years, biological drugs available on the market have been mainly used in severe and uncontrolled asthma among the pediatric population. They alleviate the symptoms of the disease, improve lung function, reduce the need for oral glucocorticosteroids, and optimize patients' quality of life [46] The mechanism of their action is based on the inhibition of the inflammatory response with the involvement of Th2 (T helper 2 cells), IgE immunoglobulins, and IL-4, 5, and 13 [47]. After showing many interesting relationships between miRNAs and drugs used in asthma, such an attempt was made for modern biological preparations. In a group of adult patients with severe eosinophilic asthma, mepolizumab and reslizumab were included in the treatment. These are humanized immunoglobulin G (IgG) monoclonal antibodies with a high affinity for IL-5, neutralizing this cytokine by binding to epitopes on the IL-5-Rα binding domain. At 8 weeks from the start of the project, an increase in the expression of miRNA-338-3p was noted in the subjects. In this study, the relationship of this molecule with the regulation of the expression of genes involved in the development of asthma, including *MAPK* and *TGF-β* (transforming growth factor beta 1), was confirmed by bioinformatics analysis. They are related to the remodeling of the respiratory tract. Thus, miRNA-338-3p may become a new, useful biomarker of the early response to mepolizumab and reslizumab [48], like L-selectin and KL-6 molecules (Krebs von den lungen-6). Their potential role as new markers of the early response only to mepolizumab was confirmed [49]. Benralizumab is a monoclonal antibody that binds to the alpha subunit of the IL-5 receptor and significantly reduces the number of eosinophils in the peripheral blood. In addition, after 8 weeks of therapy, a change in the concentration of three molecules—miRNA-1246, miRNA-5100, and miRNA-338-3p—was observed in adults diagnosed with severe eosinophilic asthma. The detailed analysis showed that they are regulators of the *MAPK* signaling pathway, which controls crucial genes involved in the pathogenesis of asthma. By monitoring the levels of these miRNAs after the first dose of benralizumab, we can easily assess whether the mechanisms responsible for controlling asthma in these patients have been restored and, subsequently, if their health has improved [50]. An interesting relationship was demonstrated in the group of adults receiving benralizumab at the beginning and in experimental week 4, 8, 16, and 24. This drug has been found to affect the miRNA-21/*PTEN*/PI3K/HDAC2 pathway, which is involved in the development of steroid-resistant asthma, as mentioned in earlier sections of the paper. There was a significant decrease in miRNA-21-5p, which subsequently inactivated PI3K/AKT, ultimately leading to an increase in the expression of genes related to the steroid response, including the glucocorticosteroid receptor. The study provides interesting data where using a biological drug increases the chances of restoring the effectiveness of steroid therapy in patients with severe asthma [51]. Dupilumab is a monoclonal antibody that blocks the receptor for IL-4 and IL-13 and is also approved for treating severe asthma with a predominance for type 2 inflammation. This medicine was used in four subjects with a confirmed diagnosis of chronic sinusitis with nasal polyps (without a diagnosis of asthma). At the end of the treatment, a decrease in the expression of miRNAs-25-3p and -185-5p was shown in the blood. As was proven, they are involved in the inflammatory process, T lymphocyte regulation, and drug resistance, thanks to which we note a significant clinical improvement among the subjects after using the drug [52]. Currently, no studies show the relationship between Dupilumab and miRNAs in asthmatics. It is a promising future research direction.

3. miRNAs and Pathways Responsible for the Development of Asthma: The Future Direction of Therapeutic Action

The use of synthetic miRNAs and antagomirs (anti-miRNAs) to inhibit the pathways responsible for the pathogenesis of asthma is a promising target for future therapeutic activities. Anti-miRNA molecules are small interfering oligonucleotides that bind to the target miRNA and prevent them from being processed by the RNA-induced silencing complex (RISC). miRNA mimetics are made of two strands and restore the function of their endogenous counterpart [53]. Below, the results of some of the latest publications revealing yet unknown pathways involved in the development of asthma are presented. This chapter focuses on potential translational research in *mice*, as previous sections have focused on experiments and discoveries in humans. Figure 2 below shows the processes in the airways of people with asthma, a potential target for therapeutic actions.

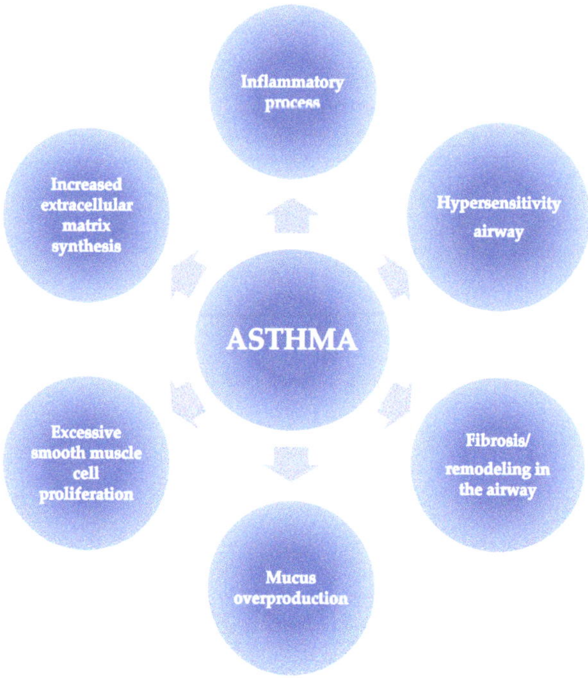

Figure 2. Target direction of therapeutic actions in asthma.

In addition to the prognostic role of miRNA-155-5p in assessing the response to treatment, this molecule may become an object of therapeutic action in the future. In an experimental study conducted on female *mice*, an increase in the concentration of the miRNA-155 molecule in the lung tissue of animals was noted after inducing asthma by the supply of ovalbumin. It was responsible for the increased synthesis of interleukins 4, 5, and 13 in the bronchoalveolar lavage fluid and the hyperreactivity of the airways. The inflammatory process was reduced when the anti-miRNA-155 counterpart was delivered to the airways with a lentiviral vector. In this way, the miRNA-155 molecule may become a new target for treating allergic inflammatory diseases [54]. In laboratory conditions, asthma was induced in *mice* by exposure to ovalbumin, and the resulting miRNA profile was then analyzed. It has been shown that miRNA-3162-3p is associated with the increased production of active beta-catenin, intensifying the inflammatory process and hyperreactivity of the airways. The use of anti-miRNA-3162-3p was compared in the use of dexamethasone and resulted in the inhibition of the Wnt/beta-catenin pathway [55,56]. Similar results between the target Wnt/beta-catenin pathway and miRNA-3162-3p were observed in *mice* lung cells for miRNA-943-3p, and the antagomir application of this molecule [57]. Among pediatric patients with an episode of asthma exacerbation, a decrease in miRNA-29c was noted in the blood, and its target point of action was the molecule B7-H3 (B7 protein homolog 3), the level of which was increased. It regulates Th2 and Th17 differentiation, cytokine synthesis, increased eosinophilic infiltration, and excessive mucus production. Infection of human monocytes obtained from laboratory cultures with a molecule that was a synthetic equivalent of miRNA-29c led to reduced expression of B7-H3 in an immunofluorescence assay in these cells [58,59]. Another complex research project was carried out on *mouse* lung cells and human bronchial epithelial cells (BEAS-2B). A complex pathway that could be a future therapeutic target for asthma is the miRNA-106b-5p/E2F1/SIX1. Increased levels of TGF-β1 were shown in asthmatics, which is responsible for the remodeling of the airways, transforming epithelial cells into myo/fibroblasts, subsequently leading to a decrease in the sensitivity to steroid therapy. Moreover, it intensifies the expression of SIX1 (sine oculis homeobox homolog 1), which mediates the change in the architecture of the bronchial tree. This study showed a decrease in the expression of miRNA-106b-5p, which directly results in an increase in SIX1 and E2F1 (E2F transcription factor 1), and TGF-β1 induces this process. Introducing the mimetic miRNA-106b-5p led to the inhibition of remodeling within the airways [60]. In airway smooth muscle cells obtained from asthmatics during bronchoscopy, miRNA-204-5p was found to play a similar role to the molecule described above. It inhibits TGF-β1-induced proliferation and extracellular matrix production within these cells by downregulating SIX1 [61]. A group of Chinese scientists exposed the epithelial cells of the *mouse* respiratory tract to house dust mites, causing an allergic inflammatory process. Subsequently, increased expression of miRNA-145-5p was observed, and the direct molecular target of this molecule was kinesin family member 3A (KIF3A), which was noted to be reduced in epithelial cells. This led to the increased production of IL-4, IL-5, and IL-13 derived from Th2 lymphocytes, which correlated with increased eosinophilic infiltration, mucus hypersecretion, and bronchial hyperresponsiveness. After using anti-miRNA-145-5p intranasally, all the above processes were inhibited [62]. Recently, it has been emphasized that exosomes may be involved in the pathogenesis of asthma [63,64]. This group of extracellular vesicles enables intercellular communication by carrying various molecules, including the same miRNA. The role of macrophage-derived exosomes was investigated using a *mouse* model of ovalbumin-induced asthma. It has been shown that after exposure of lung cells to their action, the fibrosis and inflammatory process are inhibited. The molecule responsible for this process is miRNA-370, which inactivates the MAPK/STAT (signal transducer and activator of transcription) pathway, which has key importance in the development of asthma [65]. In ASMC (airway smooth muscle cells) taken from adult subjects suffering from asthma during bronchoscopy, the increased production of the lncRNA (long non-coding RNA) NEAT1 (nuclear paraspeckle assembly transcript 1) was shown, which reduces the expression of miRNA-139 and is a signal for the phosphorylation of JAK3/STAT5 (Janus kinase 3/signal transducer and activator of transcription 5), leading to the synthesis of TNF-a, IL-6, IL-8, and IL-1b. The use of the miRNA-139 mimetic leads to the inhibition of the production of the above-mentioned pro-inflammatory substances and the development of the disease [66].

Th 17 (T helper 17 cells) are lymphocytes associated with severe asthma due to neutrophil infiltration and resistance to inhaled corticosteroids. It has been shown that the MBD2 protein (methyl-CpG-binding domain2) plays an important role in the differentiation of these cells, which allows recognition of methylated DNA and then regulation of its expression, making it a critical mediator of many epigenetic processes. Among the subjects diagnosed with severe asthma, it was shown that the level of miRNA-146a-3p was lower compared to the control group. The molecular target of this molecule was MBD2, the concentration of which was increased in asthmatics. The above relationships were simultaneously confirmed in *mouse* lung tissue. Inhalation of the mimic equivalent of miRNA-146a-3p results in a decrease in MBD2 and the number of Th17 lymphocytes involved in the inflammatory process, inhibition of mucus production, and airway hyperreactivity [67]. In the group of 68 pediatric patients with asthma, the reduction of miRNA-135b and the relationship of this molecule with the increase in CXCL12 (C-X-C motif chemokine ligand 12) were confirmed. It is responsible for increased inflammatory infiltration with the involvement of Th 17 lymphocytes and bronchial hyperreactivity. Anti-miRNA-135b and mimic miRNA-135b were applied by inhalation in *mice* in which ovalbumin supply induced the disease. In collected lung tissue samples from these *mice*, miRNA-135b was shown to lead to the inhibition of CXCL12 production and the infiltration of inflammatory cells [68]. RhoA (Ras Homolog Family Member A) is an essential protein associated with smooth muscle contraction in the airways, contributing to developing airway hyperresponsiveness. In human bronchial smooth muscle cells passaged under laboratory conditions, miRNA-140-3p was shown to be responsible for the expression of this protein. After using a mimetic of this molecule, there was a decrease in RhoA, followed by an inhibition of contraction [69]. Similar results were observed between the RhoA mRNA target gene and miRNA-140-3p in human bronchial smooth muscle cells [70]. miRNA-133a is also involved in regulating the RhoA pathway, and its increase inhibits airway constriction by limiting protein translation [71]. Another therapeutic target for asthma may be miRNA-375. In human airway epithelial cells, dexamethasone decreased this molecule and increased the synthesis of DUSP6 (dual specificity phosphatase 6), the target of miRNA-375 action. As shown, steroids, by affecting the miRNAs mentioned above (lowering the concentration), activate or inhibit several protein groups, contributing to the promotion of apoptosis within the cells of the respiratory tract, which is an important process that allows for maintaining normal function of the epithelial tissue [72]. Another critical pathway involved in the development of the disease is miRNA200a/miRNA200b/ORMDL3/ERK/MMP-9. Human respiratory epithelial cells were passaged in laboratory conditions and then mimetics and anti-miRNAs were introduced. It was observed that in samples where mimic miRNA-200a or miRNA-200b was used, there was a decrease in the expression of ORMDL3 (orosomucoid 1-like 3) and inhibition of ERK (extracellular signal-regulated kinase)/MMP-9 (matrix metalloproteinase-9) activation, and finally, a decrease in the amount of pro-inflammatory cytokines, including TNF-a, IL-4, IL-5, IL-13, and IL-1b. This is associated with stopping the inflammatory process and limiting the irreversible remodeling within the bronchial tree [73]. In vivo, after the application of dexamethasone to respiratory epithelial cells, the increased synthesis of GILZ (glucocorticoid-induced leucine zipper) was demonstrated, and decreased synthesis of miRNA-222-3p. GILZ is a transcription factor that inhibits the activation of the *MAPK* pathway, which has been shown to lead to the increased proliferation, migration, and differentiation of cells around the damaged site, playing a pivotal role in repairing the airway epithelium. The introduction of miRNA-222-3p led to a different therapeutic effect than that of steroids; a decrease in GILZ, promotion of the phosphorylation of signaling factors, and the activation of *MAPK*. The effect of these actions may be to promote repair of the damaged airway epithelium after using miRNA-222-3p [74]. Table 1 below summarizes the results of the studies presented above.

Table 1. New potential directions for the therapeutic activities of miRNAs in asthma.

Study, Year	Experiment/Population	miRNA	Molecular Effect	The Biological Effect Obtained After Using Antagomir or Mirna
C. Huilong et al. [1], 2017	Induction of asthma by ovalbumin and analysis of cells in the bronchoalveolar lavage fluid and lung tissue/*MICE*	miRNA-155	Increased production of IL-4, 5, 13, and hypersensitivity of airway	Antagomir miRNA-155; inhibited the inflammatory process
J. Liu et al. [2], 2020	Induction of asthma by ovalbumin and analysis of cells in the bronchoalveolar lavage fluid/*MICE*	miRNA-3162-3p	Wnt/beta-catenin; increased production of the active form of beta-catenin, inflammatory response, and hypersensitivity of airway	Antagomir miRNA-3162-3p; anti-inflammatory effect such as GCs
J. Shen et al. [3], 2019	Induction of asthma by ovalbumin and analysis of cells in the bronchoalveolar lavage fluid and lung tissue/*MICE*	miRNA-943-3p	Wnt/SFRP-4; increased production of the active form of beta-catenin, inflammatory response, and fibrosis in the airway	Antagomir miRNA-943-3p; anti-inflammatory effect such as GCs

Table 1. *Cont.*

Study, Year	Experiment/Population	miRNA	Molecular Effect	The Biological Effect Obtained After Using Antagomir or Mirna
X. Zhang et al. [4], 2018	Analysis of blood sample/*CHILDREN* WITH ASTHMA EXCERBATIONS	miRNA-29c	B_7-H_3; increased eosinophilic infiltration, cytokine synthesis, mucus production, regulation of Th differentiation	Mimic miRNA-29c-inhibited the inflammatory process
S. Liu et al. [5], 2021	Induction of asthma by ovalbumin in *mice* and analysis of cells in the lung; human bronchial epithelial cells/*MICE; HUMAN*	miRNA-106b-5p	TGF-B1/SIX1/E2F1; initiated remodeling, fibrosis of the airways	Mimic miRNA-106b-5p; inhibited remodeling of the respiratory tract and increased sensitivity to the GCs
Z. Yang et al. [6], 2020	Smooth muscle cells collected during bronchoscopy/*HUMAN*	miRNA-204-5p	TGF-B1/SIX1; excessive smooth muscle cell proliferation and ECM production	Mimic miRNA-204-5p; suppressed the proliferation and ECM production of airway smooth muscle cells by directly regulating SIX1
T. Xiong et al. [7], 2019	Induction of allergic airways diseases by aeroallergen and analysis bronchial epithelial cells/*MICE*	miRNA-145-5p	KIF3-increased IL-4, 5, and 13 production, eosinophil infiltration, airway hyperresponsiveness	Antagomir miRNA-145; inhibited the inflammatory process
C. Li et al. [8], 2021	Induction of asthma by ovalbumin and isolated exosomes from macrophage/*MICE*	miRNA-370	MAPK/STAT; increased fibrosis and inflammatory process	Mimic miRNA-370; prevented airway remodeling and reduced inflammation
M. Zhu et al. [9], 2021	Smooth muscle cells collected during bronchoscopy/*HUMAN*	miRNA-139	lncRNA/JAK3/STAT5; increased production of TNF-a and IL-6, 8, and 1b	Mimic miRNA-139; reduced the production of cytokines
W. Duan et al. [10], 2023	Analysis of blood sample in severe asthma; *mouse* model of Th17 predominant neutrophilic, severe asthma/ *HUMAN; MICE*	miRNA-146a-3p	MBD2; initialized severe asthma, increased Th17 lymphocytes, production of mucus, airway hyperresponsiveness	Mimic miRNA-146a-3p; suppressed production of Th17 lymphocytes and inflammatory process, development of severe asthma
Y. Liu et al. [11], 2022	Analysis of blood sample; induction of asthma by ovalbumin in *mice* and analysis bronchial epithelial cells/*CHILDREN; MICE*	miRNA-135b	CXCL-12; increased inflammatory process with Th17, airway hyperresponsiveness	Mimic miRNA-135b; decreased CXCL-12 production and inflammatory infiltration
Y. Chiba et al. [12], 2021	Bronchial smooth muscle cell/*HUMAN*	miRNA-140-3p	RhoA; increased the strength contraction of smooth muscle	Mimic miRNA-140-3p; inhibited contraction of smooth muscle
X. Duan et al. [13], 2020	Bronchial epithelial cells/*HUMAN*	miRNA-200a/200b	ORMDL3/ERK/MMP-9; increased production of TNF-a, IL-4, IL-5, IL-13, IL-1b, and the inflammatory process	Mimic miRNA-200a/200b; inhibited the inflammatory process

Table 1. *Cont.*

Study, Year	Experiment/Population	miRNA	Molecular Effect	The Biological Effect Obtained After Using Antagomir or Mirna
N. He et al. [14], 2020	Bronchial epithelial cells treated by dexamethasone/*HUMAN*	miRNA-222-3p	GILZ; inhibited proliferation, migration, and cell differentiation around the damaged epithelium	Mimic 222 3p; promoted the repair of damaged epithelium

NOTE: Fourth column—increase in the concentration of all molecules except for SFRP4 and KIF-3. Abbreviations: IL—Interleukins; GCs—glucocorticosteroids; SFRP—4-secreted frizzled-related proteins 4; s-H_3-B7 homolog 3 protein; TGF-β1—transforming growth factor β1, SIX1—sine oculis homeobox homolog 1; KIF3A—kinesin family member 3A; E2F1—E2F transcription factor 1; ECM—extracellular matrix; lncRNA—nuclear paraspeckle assembly transcript 1; JAK3/STAT5—janus kinase 3/signal transducer and activator of transcription 5; MBD-2—methyl-CpG-binding domain2; CXCL-12—C-X-C motif chemokine ligand 12; RhoA—Ras Homolog Family Member A; ORMDL3—orosomucoid 1-like 3; ERK—extracellular signal regulated kinase; MMP-9—matrix metalloproteinase-9; GILZ—glucocorticoid-induced leucine zipper.

4. miRNAs and the Course of Viral Infections as a Factor of Asthma Exacerbation

An important aspect, especially among the pediatric population, are common acute viral infections, which are the leading cause of episodes of exacerbation of chronic inflammatory diseases in the respiratory tract, including asthma [75]. These exacerbations are associated with high morbidity and even mortality worldwide. It has been confirmed that miRNAs are induced during infection, which may play a key role in modulating the antiviral response. Therefore, it is very likely that miRNA changes originate from the infected epithelium and immune cells, which could lead to further dysregulation of airway inflammation [76,77]. In this chapter, we will focus on the potential role of miRNAs as novel molecules with therapeutic potential in acute viral and bacterial infections.

Rhinovirus (RV) is one of the most common pathogens and is the etiological factor of colds, bronchiolitis, and asthma exacerbations among children. The presence of RV in the respiratory tract secretions of infected *mice* led to the increased production of miRNA-122, which ultimately affected SOCS1 (suppressor of cytokine signaling 1), leading to a decrease in its expression and ultimately resulting in an increase in neutrophil infiltration in the airways. Experimental in vivo introduction of anti-miRNA-122 in animals reduced the inflammatory infiltration of the bronchoalveolar space, hypersensitivity of the airways, and triggered innate immune mechanisms, including IFN (interferon) production. A very interesting aspect is the fact that a higher level of miRNA-122 was found in the aspirate taken from the upper respiratory tract in infants hospitalized due to bronchiolitis, and the higher the expression of this molecule, the longer the need for oxygen therapy and the higher the rate of treatment failure [78]. In addition, early-life bronchiolitis is a strong risk factor for the development of asthma, and targeted antiviral therapy may prevent the future onset of the disease [79]. After infection of respiratory epithelial cells with RSV (respiratory syncytial virus) and the influenza virus in vitro, in addition to mimetic miRNA-24, -124a, and -744, a decrease in pathogen replication in the samples was shown by approximately 75%. The molecular target of these molecules is MK2 (MAPK-activated protein kinase 2) and its limited expression, which results in inhibition of the MAPK pathway activation and the spread of infection [80]. In another study, the potential therapeutic nature of anti-miRNA-146a was demonstrated in an infection caused by the influenza virus. The experiment was conducted in vivo (infecting *mice*) and in vitro on human respiratory epithelial cells. After contact with the pathogen, an increased concentration of miRNA-146a was unanimously noted in the tissues, inhibiting the production of type I IFN, which resulted in increased viral replication. After using an antagomir, there was a reduction in infection and damage to the *mouse* lung tissue [81]. In 2020, SARS-CoV-2 (severe acute respiratory syndrome coronavirus 2) became the cause of a worldwide pandemic and led to the death of many people. COVID-19 (coronavirus disease 2019) causes dysregulation of many inflammatory pathways (such as, for example, NF-kB and JAK/STAT), which are crucial in the occurrence of post-illness complications. It has been shown that the miRNA-9, miRNA-98, miRNA-223, and miRNA-214 molecules are responsible, at various stages, for inhibiting NF-kB and JAK/STAT, which reduces the inflammatory process and prevents the development of the disease [82]. Metapneumovirus (MPV) infection is another etiological factor of lower respiratory tract infections in infants and children. In epithelial cells, after their exposure to MPV, excessive synthesis of let-7f was noted, which led to inhibition of pathogen replication. A different effect was observed after infection of cells with RSV, which showed an increase in let-7f that induced viral replication. The opposite regulatory effect of this miRNA on hMPV and RSV replication suggests that it plays a different role depending on the etiological factor. Moreover, a different let-7f binding site has been demonstrated within the MPV and RSV genomes. Accordingly, using an antagomir or a mimetic of this molecule stops the pathogen replication [83]. In addition to viral infections, bacterial infections are another risk factor for asthma exacerbations. Widespread increasing resistance to antibiotics and impaired innate immunity in response to the use of steroids in the treatment of asthma is a common clinical problem. After infection with Haemophilus influenzae, the *mice* showed a decrease in the concentration of miRNA-328 in their lung macrophages. Using an antagomir of this molecule resulted in an increase in bactericidal activity by enhancing phagocytosis and the formation of reactive oxygen species. In addition, this study showed an interesting relationship where treatment with dexamethasone inhibited phagocytosis along with other inflammatory antibacterial pathways, while anti-miRNA-328 overcomes this effect. These results suggest that an antagomir could be combined with dexamethasone in treating chronic lung diseases to better control bacterial infections. This is particularly interesting for antibiotic-resistant bacterial strains where conventional therapies tend to fail and for patients receiving steroid therapy whose infections are difficult to control [84].

Currently, drugs based on the RNA structure are undergoing intensive preclinical and clinical trials. They can be divided into three classes: gene expression inhibitory (e.g., miRNA, siRNA (small interfering RNA), lncRNAs (long noncoding RNAs)), protein coding (e.g., mRNA), and protein targeting molecules (e.g., RNA aptamers) [85]. Compared to current drugs in use, RNA-based therapy offers additional advantages such as high selectivity and potency, the ability to provide personalized treatment, and rapid bioinformatics-based design [86]. The advantage of such preparations is their direct delivery to the target site of action by inhalation, minimizing potential systemic side effects. The main RNA therapeutics being investigated in obstructive airway diseases, including asthma, are siRNAs and, recently, miRNAs.

An example of their practical use was inhaling a specific inhibitor, miRNA-141-3p, which reduced mucus production and airway hyperresponsiveness in *mice* [87]. Despite the many benefits of the potential introduction of RNA-based inhalation drugs, there are also some limitations to their widespread use in medicine. The abundant presence of the nuclease in the body leads to the rapid breakdown of the drug, there is silencing of many genes off-target, and stimulation of the immune system can also occur. The hope for eliminating these negative phenomena is the chemical modification of RNA preparations [88].

The therapy for most cases of viral infections is based on symptomatic treatment, and in the case of available and specific antiviral drugs, it is focused on conservative proteins. However, this approach puts selective pressure on the pathogen and increases the likelihood of resistance to the drug. An alternative therapeutic strategy is to target host-encoded factors essential for the spread of infection, thereby minimizing the possibility of viral mutations that elude drug action. miRNAs can become these factor-influencing molecules [89].

5. Conclusions

The above review demonstrates the relationship of miRNAs with individual groups of drugs used in asthma, their involvement in pathways related to the disease pathogenesis, and their crucial role in developing viral infections. It is a rapidly evolving medical research field, as evidenced by the increasing number of publications. Determining the concentrations of miRNA molecules in the blood before, during, and after the end of therapy, as well as during episodes of asthma exacerbations, and their change in expression will make it possible to easily predict the expected therapeutic effect, assess the response to the treatment, as well as enabling a faster diagnosis before the symptoms of the disease process intensify. In addition, by combining reproducible methods of miRNA determination and their changing levels in the blood, the clinical characteristics of the patient, and available additional tests (e.g., spirometry), monitoring the effectiveness of therapy will be characterized by incredible accuracy. miRNAs can be used to classify and divide asthma cases into individual endotypes and/or phenotypes, which will allow optimal treatment selection, becoming the basis for introducing personalized medicine and a unique approach to each patient. Experimental translational studies conducted in the laboratory have confirmed the importance of miRNAs as potential molecules for future use in general clinical practice, although further refinement is still necessary. This unique opportunity presents itself for patients with severe asthma who do not respond to commonly used treatments, undoubtedly leading to an improved quality of life and significant economic implications by reducing the financial burden and costs associated with managing severe unresponsive asthma.

Author Contributions: N.K.-G. and B.S. conceived of the review and drafted the manuscript. The review was performed under the supervision of B.S. All authors have read and agreed to the published version of the manuscript.

Funding: This research received no external funding.

Institutional Review Board Statement: Not applicable.

Informed Consent Statement: Not applicable.

Data Availability Statement: No new data were created or analyzed in this study. Data sharing is not applicable to this article.

Conflicts of Interest: The authors declare no conflict of interest.

References

1. Dharmage, S.C.; Perret, J.L.; Custovic, A. Epidemiology of Asthma in Children and Adults. *Front. Pediatr.* **2019**, *7*, 246. [CrossRef]
2. Lambrecht, B.N.; Hammad, H. The Immunology of Asthma. *Nat. Immunol.* **2015**, *16*, 45–56. [CrossRef]
3. Lommatzsch, M.; Virchow, J.C. Severe Asthma. *Dtsch. Arztebl. Int.* **2014**, *111*, 847–855. [CrossRef]
4. Foster, J.M.; McDonald, V.M.; Guo, M.; Reddel, H.K. "I Have Lost in Every Facet of My Life": The Hidden Burden of Severe Asthma. *Eur. Respir. J.* **2017**, *50*, 1700765. [CrossRef] [PubMed]
5. Ahmed, H.; Turner, S. Severe Asthma in Children—A Review of Definitions, Epidemiology, and Treatment Options in 2019. *Pediatr. Pulmonol.* **2019**, *54*, 778–787. [CrossRef]
6. Guilbert, T.W.; Bacharier, L.B.; Fitzpatrick, A.M. Severe Asthma in Children. *J. Allergy Clin. Immunol. Pract.* **2014**, *2*, 489–500. [CrossRef]
7. Gibson, P.G.; Simpson, J.L.; Hankin, R.; Powell, H.; Henry, R.L. Relationship between Induced Sputum Eosinophils and the Clinical Pattern of Childhood Asthma. *Thorax* **2003**, *58*, 116–121. [CrossRef]
8. Gibson, P.G.; Henry, R.L.; Thomas, P. Noninvasive Assessment of Airway Inflammation in Children: Induced Sputum, Exhaled Nitric Oxide, and Breath Condensate. *Eur. Respir. J.* **2000**, *16*, 1008–1015.
9. Zhang, X.; Xu, Z.; Wen, X.; Huang, G.; Nian, S.; Li, L.; Guo, X.; Ye, Y.; Yuan, Q. The Onset, Development and Pathogenesis of Severe Neutrophilic Asthma. *Immunol. Cell Biol.* **2022**, *100*, 144–159. [CrossRef] [PubMed]
10. Yamasaki, A.; Okazaki, R.; Harada, T. Neutrophils and Asthma. *Diagnostics* **2022**, *12*, 1175. [CrossRef] [PubMed]
11. Nabe, T. Steroid-Resistant Asthma and Neutrophils. *Biol. Pharm. Bull.* **2020**, *43*, 31–35. [CrossRef] [PubMed]
12. Sharma, R.; Tiwari, A.; McGeachie, M.J. Recent MiRNA Research in Asthma. *Curr. Allergy Asthma. Rep.* **2022**, *22*, 231–258. [CrossRef] [PubMed]
13. Gil-Martínez, M.; Lorente-Sorolla, C.; Naharro, S.; Rodrigo-Muñoz, J.M.; del Pozo, V. Advances and Highlights of MiRNAs in Asthma: Biomarkers for Diagnosis and Treatment. *Int. J. Mol. Sci.* **2023**, *24*, 1628. [CrossRef] [PubMed]
14. Wadhwa, R.; Dua, K.; Adcock, I.M.; Horvat, J.C.; Kim, R.Y.; Hansbro, P.M. Cellular Mechanisms Underlying Steroid-Resistant Asthma. *Eur. Respir. Rev.* **2019**, *28*, 190096. [CrossRef] [PubMed]

15. Heffler, E.; Allegra, A.; Pioggia, G.; Picardi, G.; Musolino, C.; Gangemi, S. MicroRNA Profiling in Asthma: Potential Biomarkers and Therapeutic Targets. *Am. J. Respir. Cell Mol. Biol.* **2017**, *57*, 642–650. [CrossRef]
16. Kierbiedź-Guzik, N.; Sozańska, B. The Potential Role of Serum and Exhaled Breath Condensate MiRNAs in Diagnosis and Predicting Exacerbations in Pediatric Asthma. *Biomedicines* **2023**, *11*, 763. [CrossRef]
17. Cay, P.; Singer, C.A.; Ba, M.A. Gene Network Analysis for Identification of MicroRNA Biomarkers for Asthma. *Respir. Res.* **2022**, *23*, 378. [CrossRef]
18. Wang, A.L.; Panganiban, R.; Qiu, W.; Kho, A.T.; Chupp, G.; Meyers, D.A.; Bleecker, E.R.; Weiss, S.T.; Lu, Q.; Tantisira, K.G. Drug Repurposing to Treat Glucocorticoid Resistance in Asthma. *J. Pers. Med.* **2021**, *11*, 175. [CrossRef]
19. Grzanka, A.; Jarząb, J. Niegenomowy Mechanizm Działania Glikokortykosteroidów. *Adv. Respir. Med.* **2009**, *77*, 387–393. [CrossRef]
20. Liu, T.; Zhang, L.; Joo, D.; Sun, S.-C. NF-κB Signaling in Inflammation. *Signal Transduct. Target Ther.* **2017**, *2*, 17023. [CrossRef]
21. Li, J.; Panganiban, R.; Kho, A.T.; McGeachie, M.J.; Farnam, L.; Chase, R.P.; Weiss, S.T.; Lu, Q.; Tantisira, K.G. Circulating MicroRNAs and Treatment Response in Childhood Asthma. *Am. J. Respir. Crit. Care Med.* **2020**, *202*, 65–72. [CrossRef] [PubMed]
22. Panganiban, R.P.; Wang, Y.; Howrylak, J.; Chinchilli, V.M.; Craig, T.J.; August, A.; Ishmael, F.T. Circulating MicroRNAs as Biomarkers in Patients with Allergic Rhinitis and Asthma. *J. Allergy Clin. Immunol.* **2016**, *137*, 1423–1432. [CrossRef]
23. Jardim, M.J.; Dailey, L.; Silbajoris, R.; Diaz-Sanchez, D. Distinct MicroRNA Expression in Human Airway Cells of Asthmatic Donors Identifies a Novel Asthma-Associated Gene. *Am. J. Respir. Cell Mol. Biol.* **2012**, *47*, 536–542. [CrossRef]
24. ElKashef, S.M.M.A.E.; Ahmad, S.E.-A.; Soliman, Y.M.A.; Mostafa, M.S. Role of MicroRNA-21 and MicroRNA-155 as Biomarkers for Bronchial Asthma. *Innate Immun.* **2021**, *27*, 61–69. [CrossRef]
25. Elbehidy, R.M.; Youssef, D.M.; El-Shal, A.S.; Shalaby, S.M.; Sherbiny, H.S.; Sherief, L.M.; Akeel, N.E. MicroRNA–21 as a Novel Biomarker in Diagnosis and Response to Therapy in Asthmatic Children. *Mol. Immunol.* **2016**, *71*, 107–114. [CrossRef] [PubMed]
26. Kim, R.Y.; Horvat, J.C.; Pinkerton, J.W.; Starkey, M.R.; Essilfie, A.T.; Mayall, J.R.; Nair, P.M.; Hansbro, N.G.; Jones, B.; Haw, T.J.; et al. MicroRNA-21 Drives Severe, Steroid-Insensitive Experimental Asthma by Amplifying Phosphoinositide 3-Kinase–Mediated Suppression of Histone Deacetylase 2. *J. Allergy Clin. Immunol.* **2017**, *139*, 519–532. [CrossRef]
27. Kho, A.T.; McGeachie, M.J.; Moore, K.G.; Sylvia, J.M.; Weiss, S.T.; Tantisira, K.G. Circulating MicroRNAs and Prediction of Asthma Exacerbation in Childhood Asthma. *Respir. Res.* **2018**, *19*, 128. [CrossRef]
28. Rodrigo-Muñoz, J.M.; Gil-Martínez, M.; Lorente-Sorolla, C.; García-Latorre, R.; Valverde-Monge, M.; Quirce, S.; Sastre, J.; del Pozo, V. MiR-144-3p Is a Biomarker Related to Severe Corticosteroid-Dependent Asthma. *Front. Immunol.* **2022**, *13*, 858722. [CrossRef] [PubMed]
29. Tan, R.; Liew, M.F.; Lim, H.F.; Leung, B.P.; Wong, W.S.F. Promises and Challenges of Biologics for Severe Asthma. *Biochem. Pharmacol.* **2020**, *179*, 114012. [CrossRef] [PubMed]
30. Nahid, M.A.; Satoh, M.; Chan, E.K.L. Mechanistic Role of MicroRNA-146a in Endotoxin-Induced Differential Cross-Regulation of TLR Signaling. *J. Immunol.* **2011**, *186*, 1723–1734. [CrossRef] [PubMed]
31. Lambert, K.A.; Roff, A.N.; Panganiban, R.P.; Douglas, S.; Ishmael, F.T. MicroRNA-146a Is Induced by Inflammatory Stimuli in Airway Epithelial Cells and Augments the Anti-Inflammatory Effects of Glucocorticoids. *PLoS ONE* **2018**, *13*, e0205434. [CrossRef] [PubMed]
32. Weidner, J.; Ekerljung, L.; Malmhäll, C.; Miron, N.; Rådinger, M. Circulating MicroRNAs Correlate to Clinical Parameters in Individuals with Allergic and Non-Allergic Asthma. *Respir. Res.* **2020**, *21*, 107. [CrossRef] [PubMed]
33. Clayton, S.A.; Jones, S.W.; Kurowska-Stolarska, M.; Clark, A.R. The Role of MicroRNAs in Glucocorticoid Action. *J. Biol. Chem.* **2018**, *293*, 1865–1874. [CrossRef]
34. Li, J.J.; Tay, H.L.; Maltby, S.; Xiang, Y.; Eyers, F.; Hatchwell, L.; Zhou, H.; Toop, H.D.; Morris, J.C.; Nair, P.; et al. MicroRNA-9 Regulates Steroid-Resistant Airway Hyperresponsiveness by Reducing Protein Phosphatase 2A Activity. *J. Allergy Clin. Immunol.* **2015**, *136*, 462–473. [CrossRef]
35. Smith, S.S.; Dole, N.S.; Franceschetti, T.; Hrdlicka, H.C.; Delany, A.M. MicroRNA-433 Dampens Glucocorticoid Receptor Signaling, Impacting Circadian Rhythm and Osteoblastic Gene Expression. *J. Biol. Chem.* **2016**, *291*, 21717–21728. [CrossRef] [PubMed]
36. Uchida, S.; Nishida, A.; Hara, K.; Kamemoto, T.; Suetsugi, M.; Fujimoto, M.; Watanuki, T.; Wakabayashi, Y.; Otsuki, K.; McEwen, B.S.; et al. Characterization of the Vulnerability to Repeated Stress in Fischer 344 Rats: Possible Involvement of MicroRNA-Mediated down-Regulation of the Glucocorticoid Receptor. *Eur. J. Neurosci.* **2008**, *27*, 2250–2261. [CrossRef]
37. Cataldo, D.; Louis, R.; Michils, A.; Peché, R.; Pilette, C.; Schleich, F.; Ninane, V.; Hanon, S. Severe Asthma: Oral Corticosteroid Alternatives and the Need for Optimal Referral Pathways. *J. Asthma* **2021**, *58*, 448–458. [CrossRef]
38. Volmer, T.; Effenberger, T.; Trautner, C.; Buhl, R. Consequences of Long-Term Oral Corticosteroid Therapy and Its Side-Effects in Severe Asthma in Adults: A Focused Review of the Impact Data in the Literature. *Eur. Respir. J.* **2018**, *52*, 1800703. [CrossRef]
39. Imraish, A.; Abu-Thiab, T.; Zihlif, M. IL-13 and FOXO3 Genes Polymorphisms Regulate IgE Levels in Asthmatic Patients. *Biomed Rep.* **2021**, *14*, 55. [CrossRef]
40. Alam, R.; Gorska, M.M. Mitogen-Activated Protein Kinase Signalling and ERK1/2 Bistability in Asthma. *Clin. Exp. Allergy* **2011**, *41*, 149–159. [CrossRef]
41. Gil-Martínez, M.; Lorente-Sorolla, C.; Rodrigo-Muñoz, J.M.; Lendínez, M.Á.; Núñez-Moreno, G.; de la Fuente, L.; Mínguez, P.; Mahíllo-Fernández, I.; Sastre, J.; Valverde-Monge, M.; et al. Analysis of Differentially Expressed MicroRNAs in Serum and Lung Tissues from Individuals with Severe Asthma Treated with Oral Glucocorticoids. *Int. J. Mol. Sci.* **2023**, *24*, 1611. [CrossRef]

42. Adams, B.S.; Nguyen, H. Salmeterol. In *StatPearls [Internet]*; StatPearls Publishing: Treasure Island, FL, USA, 2023. [PubMed]
43. Billington, C.K.; Penn, R.B.; Hall, I.P. B2 Agonists. *Handb. Exp. Pharmacol.* **2017**, *237*, 23–40. [CrossRef]
44. Yu, B.; Yao, L.; Liu, C.; Tang, L.; Xing, T. Upregulation of MicroRNA-16 Alters the Response to Inhaled Β-agonists in Patients with Asthma Though Modulating Expression of ADRB2. *Mol. Med. Rep.* **2019**, *19*, 4027–4034. [CrossRef]
45. Kim, D.; Cho, S.; Woo, J.A.; Liggett, S.B. A CREB-mediated Increase in MiRNA Let-7f during Prolonged Β-agonist Exposure: A Novel Mechanism of β 2 -adrenergic Receptor Down-regulation in Airway Smooth Muscle. *FASEB J.* **2018**, *32*, 3680–3688. [CrossRef] [PubMed]
46. Papi, A.; Brightling, C.; Pedersen, S.E.; Reddel, H.K. Asthma. *Lancet* **2018**, *391*, 783–800. [CrossRef] [PubMed]
47. Dragonieri, S.; Carpagnano, G.E. Biological Therapy for Severe Asthma. *Asthma. Res. Pract.* **2021**, *7*, 12. [CrossRef] [PubMed]
48. Rial, M.J.; Cañas, J.A.; Rodrigo-Muñoz, J.M.; Valverde-Monge, M.; Sastre, B.; Sastre, J.; del Pozo, V. Changes in Serum MicroRNAs after Anti-IL-5 Biological Treatment of Severe Asthma. *Int. J. Mol. Sci.* **2021**, *22*, 3558. [CrossRef] [PubMed]
49. Bergantini, L.; d'Alessandro, M.; Cameli, P.; Bianchi, F.; Sestini, P.; Bargagli, E.; Refini, R.M. Personalized Approach of Severe Eosinophilic Asthma Patients Treated with Mepolizumab and Benralizumab. *Int. Arch. Allergy Immunol.* **2020**, *181*, 746–753. [CrossRef]
50. Cañas, J.A.; Valverde-Monge, M.; Rodrigo-Muñoz, J.M.; Sastre, B.; Gil-Martínez, M.; García-Latorre, R.; Rial, M.J.; Gómez-Cardeñosa, A.; Fernández-Nieto, M.; Pinillos-Robles, E.J.; et al. Serum MicroRNAs as Tool to Predict Early Response to Benralizumab in Severe Eosinophilic Asthma. *J. Pers. Med.* **2021**, *11*, 76. [CrossRef] [PubMed]
51. Hirai, K.; Uehara, S.; Shirai, T.; Rachi, Y.; Kimura, T.; Akamatsu, T.; Itoh, K. Benralizumab Restores Gene and MicroRNA Expression Involved in Steroid Sensitivity in Severe Asthma. *Allergy* **2021**, *76*, 2589–2592. [CrossRef]
52. Mimmi, S.; Lombardo, N.; Maisano, D.; Piazzetta, G.; Pelaia, C.; Pelaia, G.; Greco, M.; Foti, D.; Dattilo, V.; Iaccino, E. Spotlight on a Short-Time Treatment with the IL-4/IL-13 Receptor Blocker in Patients with CRSwNP: MicroRNAs Modulations and Preliminary Clinical Evidence. *Genes* **2022**, *13*, 2366. [CrossRef] [PubMed]
53. Cherneva, R.V.; Kostadinov, D. Epigenetic Targets for Therapeutic Approaches in COPD and Asthma. Nutrigenomics &Ndash; Possible or Illusive. *Folia. Med.* **2019**, *61*, 358–369. [CrossRef] [PubMed]
54. Chen, H.; Xu, X.; Cheng, S.; Xu, Y.; Xuefei, Q.; Cao, Y.; Xie, J.; Wang, C.-Y.; Xu, Y.; Xiong, W. Small Interfering RNA Directed against MicroRNA-155 Delivered by a Lentiviral Vector Attenuates Asthmatic Features in a Mouse Model of Allergic Asthma. *Exp. Ther. Med.* **2017**, *14*, 4391–4396. [CrossRef] [PubMed]
55. Liu, J.; Chen, Y.; Zhang, F.; Peng, X.; Mao, X.; Lu, W.; Wu, R.; Huang, B.; Bao, Y.; Ma, L.; et al. Divergent Roles of MiR-3162-3p in Pulmonary Inflammation in Normal and Asthmatic Mice as Well as Antagonism of MiR-3162-3p in Asthma Treatment. *Int. Arch. Allergy Immunol.* **2020**, *181*, 594–605. [CrossRef]
56. Fang, C.; Lu, W.; Li, C.; Peng, X.; Wang, Y.; Huang, X.; Yao, Z.; Cai, N.; Huang, Y.; Zhang, X.; et al. MiR-3162-3p Is a Novel MicroRNA That Exacerbates Asthma by Regulating β-Catenin. *PLoS ONE* **2016**, *11*, e0149257. [CrossRef] [PubMed]
57. Shen, J.; Zhao, J.; Ye, Q.; Gu, X. Interference of MiR-943-3p with Secreted Frizzled-Related Proteins4 (SFRP4) in an Asthma Mouse Model. *Cell Tissue Res.* **2019**, *378*, 67–80. [CrossRef]
58. Zhang, X.; Zhao, X.; Sun, H.; Yan, Y.; Huang, L.; Gu, W.; Jiang, W.; Wang, Y.; Zhu, C.; Ji, W.; et al. The Role of MiR-29c/B7-H3 Axis in Children with Allergic Asthma. *J. Transl. Med.* **2018**, *16*, 218. [CrossRef]
59. Gu, W.; Zhang, X.; Yan, Y.; Wang, Y.; Huang, L.; Wang, M.; Shao, X.; Chen, Z.; Ji, W. B7-H3 Participates in the Development of Asthma by Augmentation of the Inflammatory Response Independent of TLR2 Pathway. *Sci. Rep.* **2017**, *7*, 40398. [CrossRef]
60. Liu, S.; Chen, X.; Zhang, S.; Wang, X.; Du, X.; Chen, J.; Zhou, G. MiR-106b-5p Targeting SIX1 Inhibits TGF-β1-induced Pulmonary Fibrosis and Epithelial-mesenchymal Transition in Asthma through Regulation of E2F1. *Int. J. Mol. Med.* **2021**, *47*, 4857. [CrossRef]
61. Yang, Z.; Qu, Z.; Yi, M.; Lv, Z.; Wang, Y.; Shan, Y.; Ran, N.; Liu, X. MiR-204-5p Inhibits Transforming Growth Factor-Β1-Induced Proliferation and Extracellular Matrix Production of Airway Smooth Muscle Cells by Regulating Six1 in Asthma. *Int. Arch. Allergy Immunol.* **2020**, *181*, 239–248. [CrossRef]
62. Xiong, T.; Du, Y.; Fu, Z.; Geng, G. MicroRNA-145-5p Promotes Asthma Pathogenesis by Inhibiting Kinesin Family Member 3A Expression in Mouse Airway Epithelial Cells. *J. Int. Med. Res.* **2019**, *47*, 3307–3319. [CrossRef] [PubMed]
63. Mortaz, E.; Alipoor, S.D.; Varahram, M.; Jamaati, H.; Garssen, J.; Mumby, S.E.; Adcock, I.M. Exosomes in Severe Asthma: Update in Their Roles and Potential in Therapy. *Biomed Res. Int.* **2018**, *2018*, 2862187. [CrossRef] [PubMed]
64. Cañas, J.A.; Sastre, B.; Rodrigo-Muñoz, J.M.; del Pozo, V. Exosomes: A New Approach to Asthma Pathology. *Clin. Chim. Acta* **2019**, *495*, 139–147. [CrossRef]
65. Li, C.; Deng, C.; Zhou, T.; Hu, J.; Dai, B.; Yi, F.; Tian, N.; Jiang, L.; Dong, X.; Zhu, Q.; et al. MicroRNA-370 Carried by M2 Macrophage-Derived Exosomes Alleviates Asthma Progression through Inhibiting the FGF1/MAPK/STAT1 Axis. *Int. J. Biol. Sci.* **2021**, *17*, 1795–1807. [CrossRef]
66. Zhu, M.-X.; Huang, L.-H.; Zhu, Y.-K.; Cai, X.-J. LncRNA NEAT1 Promotes Airway Smooth Muscle Cell Inflammation by Activating the JAK3/STAT5 Pathway through Targeting of MiR-139. *Exp. Lung Res.* **2021**, *47*, 161–172. [CrossRef] [PubMed]
67. Duan, W.; Huang, J.; Wasti, B.; Chen, Z.; Yuan, Y.; He, Y.; Li, D.; Jia, J.; Liu, S.; Liu, Y.; et al. MiR-146a-3p as a Potential Novel Therapeutic by Targeting MBD2 to Mediate Th17 Differentiation in Th17 Predominant Neutrophilic Severe Asthma. *Clin. Exp. Med.* **2023**, 1–16. [CrossRef]
68. Liu, Y.; Huo, S.-G.; Xu, L.; Che, Y.-Y.; Jiang, S.-Y.; Zhu, L.; Zhao, M.; Teng, Y.-C. MiR-135b Alleviates Airway Inflammation in Asthmatic Children and Experimental Mice with Asthma via Regulating CXCL12. *Immunol. Invest.* **2022**, *51*, 496–510. [CrossRef]

69. Chiba, Y.; Ando, Y.; Fujii, S.; Miyakawa, Y.; Suto, W.; Kamei, J.; Sakai, H.; Hanazaki, M. Downregulation of MiR-140-3p Is a Cause of Upregulation of RhoA Protein in Bronchial Smooth Muscle of Murine Experimental Asthma. *Am. J. Respir. Cell Mol. Biol.* **2021**, *64*, 138–140. [CrossRef]
70. Chiba, Y.; Ando, Y.; Kato, Y.; Hanazaki, M.; Sakai, H. Down-Regulation of MiR-140-3p Is a Cause of the Interlukin-13-Induced up-Regulation of RhoA Protein in Bronchial Smooth Muscle Cells. *Small GTPases* **2022**, *13*, 1–6. [CrossRef]
71. Chiba, Y.; Misawa, M. MicroRNAs and Their Therapeutic Potential for Human Diseases: MiR-133a and Bronchial Smooth Muscle Hyperresponsiveness in Asthma. *J. Pharmacol. Sci.* **2010**, *114*, 264–268. [CrossRef]
72. Zheng, X.; Li, C.; Gao, X. Overexpression of MiR-375 Reverses the Effects of Dexamethasone on the Viability, Migration, Invasion and Apoptosis of Human Airway Epithelial Cells by Targeting DUSP6. *Int. J. Mol. Med.* **2022**, *49*, 26. [CrossRef]
73. Duan, X.-J.; Zhang, X.; Li, L.-R.; Zhang, J.-Y.; Chen, Y.-P. MiR-200a and MiR-200b Restrain Inflammation by Targeting ORMDL3 to Regulate the ERK/MMP-9 Pathway in Asthma. *Exp. Lung Res.* **2020**, *46*, 321–331. [CrossRef]
74. He, N.; Liu, L.; Ding, J.; Sun, Y.; Xing, H.; Wang, S. MiR-222-3p Ameliorates Glucocorticoid-Induced Inhibition of Airway Epithelial Cell Repair through down-Regulating GILZ Expression. *J. Recept. Signal Transduct.* **2020**, *40*, 301–312. [CrossRef]
75. Jartti, T.; Bønnelykke, K.; Elenius, V.; Feleszko, W. Role of Viruses in Asthma. *Semin. Immunopathol.* **2020**, *42*, 61–74. [CrossRef]
76. Leon-Icaza, S.A.; Zeng, M.; Rosas-Taraco, A.G. MicroRNAs in Viral Acute Respiratory Infections: Immune Regulation, Biomarkers, Therapy, and Vaccines. *ExRNA* **2019**, *1*, 1. [CrossRef]
77. Tan, K.S.; Lim, R.L.; Liu, J.; Ong, H.H.; Tan, V.J.; Lim, H.F.; Chung, K.F.; Adcock, I.M.; Chow, V.T.; Wang, D.Y. Respiratory Viral Infections in Exacerbation of Chronic Airway Inflammatory Diseases: Novel Mechanisms and Insights From the Upper Airway Epithelium. *Front. Cell Dev. Biol.* **2020**, *8*, 99. [CrossRef]
78. Collison, A.M.; Sokulsky, L.A.; Kepreotes, E.; Pereira de Siqueira, A.; Morten, M.; Edwards, M.R.; Walton, R.P.; Bartlett, N.W.; Yang, M.; Nguyen, T.H.; et al. MiR-122 Promotes Virus-Induced Lung Disease by Targeting SOCS1. *JCI Insight* **2021**, *6*, 7. [CrossRef] [PubMed]
79. Çalışkan, M.; Bochkov, Y.A.; Kreiner-Møller, E.; Bønnelykke, K.; Stein, M.M.; Du, G.; Bisgaard, H.; Jackson, D.J.; Gern, J.E.; Lemanske, R.F.; et al. Rhinovirus Wheezing Illness and Genetic Risk of Childhood-Onset Asthma. *N. Engl. J. Med.* **2013**, *368*, 1398–1407. [CrossRef] [PubMed]
80. McCaskill, J.L.; Ressel, S.; Alber, A.; Redford, J.; Power, U.F.; Schwarze, J.; Dutia, B.M.; Buck, A.H. Broad-Spectrum Inhibition of Respiratory Virus Infection by MicroRNA Mimics Targeting P38 MAPK Signaling. *Mol. Ther. Nucleic Acids* **2017**, *7*, 256–266. [CrossRef] [PubMed]
81. Zhang, F.; Sun, X.; Zhu, Y.; Qin, W. Downregulation of MiR-146a Inhibits Influenza A Virus Replication by Enhancing the Type I Interferon Response In Vitro and In Vivo. *Biomed. Pharmacother.* **2019**, *111*, 740–750. [CrossRef]
82. Amini-Farsani, Z.; Yadollahi-Farsani, M.; Arab, S.; Forouzanfar, F.; Yadollahi, M.; Asgharzade, S. Prediction and Analysis of MicroRNAs Involved in COVID-19 Inflammatory Processes Associated with the NF-κB and JAK/STAT Signaling Pathways. *Int. Immunopharmacol.* **2021**, *100*, 108071. [CrossRef] [PubMed]
83. Deng, J.; Ptashkin, R.N.; Wang, Q.; Liu, G.; Zhang, G.; Lee, I.; Lee, Y.S.; Bao, X. Human Metapneumovirus Infection Induces Significant Changes in Small Noncoding RNA Expression in Airway Epithelial Cells. *Mol. Ther. Nucleic Acids* **2014**, *3*, e163. [CrossRef] [PubMed]
84. Tay, H.L.; Kaiko, G.E.; Plank, M.; Li, J.; Maltby, S.; Essilfie, A.-T.; Jarnicki, A.; Yang, M.; Mattes, J.; Hansbro, P.M.; et al. Antagonism of MiR-328 Increases the Antimicrobial Function of Macrophages and Neutrophils and Rapid Clearance of Non-Typeable Haemophilus Influenzae (NTHi) from Infected Lung. *PLoS Pathog.* **2015**, *11*, e1004549. [CrossRef]
85. Payne, D. RNA Therapies. *Nature* **2019**, *574*, S1. [CrossRef]
86. Shaffer, C. Mist Begins to Clear for Lung Delivery of RNA. *Nat. Biotechnol.* **2020**, *38*, 1110–1112. [CrossRef]
87. Siddiqui, S.; Johansson, K.; Joo, A.; Bonser, L.R.; Koh, K.D.; Le Tonqueze, O.; Bolourchi, S.; Bautista, R.A.; Zlock, L.; Roth, T.L.; et al. Epithelial MiR-141 Regulates IL-13–Induced Airway Mucus Production. *JCI Insight* **2021**, *6*, 5. [CrossRef] [PubMed]
88. Roberts, T.C.; Langer, R.; Wood, M.J.A. Advances in Oligonucleotide Drug Delivery. *Nat. Rev. Drug Discov.* **2020**, *19*, 673–694. [CrossRef]
89. Sasso, J.M.; Ambrose, B.J.B.; Tenchov, R.; Datta, R.S.; Basel, M.T.; DeLong, R.K.; Zhou, Q.A. The Progress and Promise of RNA Medicine–An Arsenal of Targeted Treatments. *J. Med. Chem.* **2022**, *65*, 6975–7015. [CrossRef]

International Journal of *Molecular Sciences*

MDPI

Review

Alterations in Circulating miRNA Levels after Infection with SARS-CoV-2 Could Contribute to the Development of Cardiovascular Diseases: What We Know So Far

Myrtani Pieri [1,2], Panayiotis Vayianos [1], Vicky Nicolaidou [1,2], Kyriacos Felekkis [1,2,*] and Christos Papaneophytou [1,2,*]

1 Department of Life Sciences, School of Life and Health Sciences, University of Nicosia, 2417 Nicosia, Cyprus
2 Non-Coding RNA Research Laboratory, School of Life and Health Sciences, University of Nicosia, 2417 Nicosia, Cyprus
* Correspondence: felekkis.k@unic.ac.cy (K.F.); papaneophytou.c@unic.ac.cy (C.P.)

Abstract: The novel coronavirus disease 2019 (COVID-19) is caused by the severe acute respiratory syndrome coronavirus 2 (SARS-CoV-2) and poses significant complications for cardiovascular disease (CVD) patients. MicroRNAs (miRNAs) are small non-coding RNAs that regulate gene expression and influence several physiological and pathological processes, including CVD. This critical review aims to expand upon the current literature concerning miRNA deregulation during the SARS-CoV-2 infection, focusing on cardio-specific miRNAs and their association with various CVDs, including cardiac remodeling, arrhythmias, and atherosclerosis after SARS-CoV-2 infection. Despite the scarcity of research in this area, our findings suggest that changes in the expression levels of particular COVID-19-related miRNAs, including miR-146a, miR-27/miR-27a-5p, miR-451, miR-486-5p, miR-21, miR-155, and miR-133a, may be linked to CVDs. While our analysis did not conclusively determine the impact of SARS-CoV-2 infection on the profile and/or expression levels of cardiac-specific miRNAs, we proposed a potential mechanism by which the miRNAs mentioned above may contribute to the development of these two pathologies. Further research on the relationship between SARS-CoV-2, CVDs, and microRNAs will significantly enhance our understanding of this connection and may lead to the use of these miRNAs as biomarkers or therapeutic targets for both pathologies.

Keywords: cardiovascular diseases; circulating miRNAs; SARS-CoV-2; COVID-19; cytokine storm; inflammation

1. Introduction

Coronavirus disease-19 (COVID-19) is an emerging infectious disease caused by the widespread transmission of the severe acute respiratory syndrome coronavirus 2 (SARS-CoV-2), which threatens human health and public safety [1]. The outbreak of SARS-CoV-2 infection worldwide was declared a public health emergency by the World Health Organization (WHO) as of February 2020 [2]. SARS-CoV-2 is transmitted via contaminated aerosol droplets during close unprotected contact between infected and uninfected people. The virus infects humans through the mucosal membranes of the nose, mouth, and eyes [3].

Infection of the host cells is facilitated by the interaction of the viral spike protein, a surface protein, with the angiotensin-converting enzyme 2 (ACE2) integral protein that is expressed on the surface of the host cells (reviewed in [4]). The virus enters the host's cells through receptor-mediated endocytosis (see [5] and references cited therein). The ACE2 protein is expressed on the surface of epithelial cells, including the alveolar lung cells [6]. Additionally, ACE2 is present in the endothelial cells of veins and arteries, the surface of enterocytes of the small intestine, and the arterial smooth muscle cells [7]. It has been proposed that this expression pattern increases the susceptibility of specific tissues to

SARS-CoV-2 and, therefore, aids in understanding how multiple tissues/organs could be affected by the virus [8].

Certain underlying health conditions have been identified as high-risk factors for COVID-19, leading to higher mortality rates among COVID-19 patients with these conditions compared to those without [9]. Clinical studies have demonstrated an association between COVID-19 and cardiovascular disease (CVD). Furthermore, pre-existing CVD has been associated with worse outcomes and increased risk of death in patients with COVID-19 (reviewed in [10]). Specifically, early clinical results suggest that both the susceptibility to and the outcomes of COVID-19 are strongly associated with CVD [11]. Moreover, a high prevalence of pre-existing CVD has been observed among patients with COVID-19, and these comorbidities are linked to increased mortality [12]. Importantly, SARS-CoV-2 infections have been implicated in the development of CVDs, such as arrhythmias, acute coronary syndrome (ACS), heart failure (HF), myocardial injury, and venous thromboembolism [13]. It has also been demonstrated that children with COVID-19 are susceptible to developing hyperinflammatory shock syndrome with features akin to Kawasaki disease, such as coronary vessel abnormalities and cardiac dysfunction [14]. Overall, these data indicate a bidirectional interaction between CVDs and SARS-CoV-2 infection.

MicroRNAs (miRNAs) are short (∼22–25 nucleotides) endogenous non-coding RNA molecules that regulate gene expression at the post-transcriptional level [15]. The miRNA biogenesis takes place in a two-step cleavage process catalyzed by RNAse III endonucleases, namely Drosha and Dicer. Drosha catalyzes the nuclear processing of the primary miRNAs into stem-loop precursors (pre-miRNAs) of ~60 to 70 nucleotides in size [16]. Exportin-5 facilitates the export of pre-miRNAs from the nucleus to the cytoplasm via a Ran-GTP–dependent pathway [17]. In the cytoplasm, the pre-miRNA is cleaved by Dicer, resulting in the formation of the 22–25 nucleotides-long mature miRNA [18]. The miRNA expression has a great significance during various immunological phenomena because these RNA molecules regulate the expression of several immune-related genes [15]. In addition to the immune system, specific miRNAs play an essential role in the cardiovascular system [19]. Several heart-specific miRNAs that play a vital role in maintaining cardiac balance have been associated with the development of CVDs. Moreover, some miRNAs expressed in the cardiovascular system are increased in response to acute cardiac stress and also during long-term responses to chronic injury in the cardiovascular system [20]. Interestingly, SARS-CoV-2 infections have been found to affect the expression of cardiac-specific miRNAs (see [21] and references cited therein). Additionally, recent data suggest that the alteration of miRNAs can influence the development of various CVDs, such as myocarditis, myocardial injury, thrombosis, and arrhythmias [20]. Inflammation- and cardiac myocyte-specific miRNAs are upregulated in patients with severe COVID-19. Notably, cardiac-specific miRNA profiles were able to distinguish COVID-19 patients from influenza-acute respiratory distress syndrome patients indicating a rather specific response and cardiac involvement of COVID-19 [22]. However, the correlation between SARS-CoV-2 infections and alterations of cardiac-specific miRNAs is not well understood.

It is, therefore, essential to elucidate the role of SARS-CoV-2 infection on alterations of miRNA levels related to the development of CVDs. This critical review investigates the interplay between SARS-CoV-2 infection and cardiac-specific miRNAs that could be detected in the circulation and provides a window into the possible mechanisms behind these correlations as well as areas that require further investigation. We searched the Pubmed, Scopus, and Science Direct databases from February 2020 to May 2022. "COVID-19", "SARS-CoV-2", "cardiovascular diseases", "micro-RNAs", "circulating miRNAs", "inflammation", and their respective Medical Subject Headings (MeSH) terms were used as keywords. We also selected studies on miRNAs linked to CVDs. Only studies (both basic and clinical) that were published in the English language were included in this study.

2. SARS-CoV-2 Infection and Cardiovascular Diseases

The clinical manifestations of COVID-19 may range from asymptomatic or mild respiratory symptoms to severe life-threatening respiratory and cardiovascular complications [23]. As previously mentioned, pre-existing CVD increases the morbidity and mortality of COVID-19, and it has been correlated with poor disease outcomes, while SARS-CoV-2 infection can cause de novo acute and chronic CVD [24]. Acute cardiac complications include HF, myocardial injury, arrhythmias, and myocarditis, which are significantly associated with higher in-hospital mortality [24]. It should be noted that acute cardiac injury is a common extrapulmonary manifestation of COVID-19 with potential chronic consequences [25]. The potential mechanism by which SARS-CoV-2 infection causes acute CVDs involves direct damage caused by viral invasion of cardiomyocytes and indirect damage through systemic inflammation [26].

SARS-CoV-2 contributes to cardiovascular damage through systemic inflammation and hypoxia, causing the following complications [27]: (1) cytokine storm (discussed further below), (2) myocardial injury, (3) coronary artery ischemia in the setting of underlying coronary artery disease and (4) elevated vascular thrombosis of small and large coronary arteries that could occur in the absence of coronary artery disease. Notably, a cardiac injury could also occur following global ischemia related to multiorgan failure, respiratory distress, and associated metabolic and hemodynamic abnormalities (see [28] and references cited therein). Overactivation of the inflammatory response and the subsequent cytokine storm, a phenomenon of hyper-inflammation that involves the overexpression of several cytokines, is crucial for the pathogenic mechanism of SARS-CoV-2 infection. For example, in patients with HF, chronic activation of the innate immune system has been detected with T-cell and macrophage myocardial infiltration and increased pro-inflammatory cytokine levels (i.e., TNF-α, IFN-γ, IL-1β, IL-6, IL-17, and IL-18) [29]. These pro-inflammatory cytokines are also present in viral myocarditis, and their sustained elevation is associated with HF progression. These cytokines are responsible for both compensatory cardiac hypertrophy and fibrosis in the setting of cardiac injury and induce further inflammation [30]. It has been proposed that in patients diagnosed with COVID-19, the cytokine storm contributes to increased myocardial oxygen consumption, suppressed cardiac function, and endothelial dysfunction [31]. Notably, the cytokine storm in COVID-19 patients is characterized by increased IL-6 [12]. Even though it has been previously demonstrated that inflammatory cytokines, such as TNF-α, induce cardiomyocyte death, the precise mechanism of cytokine-induced myocardial injury remains inconclusive [32].

Furthermore, inflamma-miRs, a set of miRNAs that are capable of regulating inflammation, including miR-146a, miR-126-3p, and miR-21-5, exhibit age-dependent changes in blood levels and could be used as indicators of a pro-inflammatory state [33]. These miRNAs target components of the nuclear factor κB (NF-κB) pathway, which is a central mediator of inflammation [34]. MiR-223-3p, miR-155-5p, and miR-21-5p are expressed in immune cells and are involved in the innate immune response as well as in HF and myocardial infarction (reviewed in [35]). These miRNAs regulate the production and secretion of pro-inflammatory cytokines in the heart by Toll-like (TRL) receptors and their downstream signaling pathway, which involves the transcription of NF-κB. Another set of miRNAs, namely miR-146a-5p and miR-155-5p, form a unique regulatory network for the fine-tuning of the macrophage inflammatory response via the regulation of NF-κB activity [36].

As the inflammatory response develops, the transcription of miR-146a-5p increases, inhibiting its targets (i.e., Interleukin-1 receptor-associated kinase 1/IRAK1 and TNF receptor-associated factor 6/TRAF6), and thus activation of NF-κB is reduced. Moreover, miR-21-5p targets programmed cell death 4 (PDCD4), a component of the NF-κB pathway, thus stimulating the production of pro-inflammatory cytokines and inhibiting the production of anti-inflammatory cytokines [37]. Also, miR-155-5p contributes to myocardial damage through the modulation of monocyte-macrophages cardiac infiltration and T-lymphocyte activation in both human and mouse viral myocarditis [35]. Alterations in the levels of a set of miRNAs, which are heart- and muscle-specific, also named myomiRs (miR-1-3p, miR-133a-3p, miR-133b-3p, miR-208a/b-3p, and miR-499-5p) have also been implicated in the development of CVDs (reviewed in [35]).

3. Are Circulating miRNAs a Linker between SARS-CoV-2 Infection and Cardiovascular Diseases?

To gain a deeper understanding of the relationship between SARS-CoV-2 infection and CVD, as well as the role of circulating miRNAs in the development of both diseases, it is essential to investigate the fundamental biological mechanisms involved in viral entry into host cells, the immune response, and organ injury. As previously mentioned, SARS-CoV-2 uses the ACE2 transmembrane protein that is highly expressed in the heart, lungs, kidneys, and gut to gain entry into the host cells [38]. A recent single-cell RNA sequencing study showed that more than 7.5% of myocardial cells express the ACE2 receptor protein, which could mediate SARS-CoV-2 entry into cardiomyocytes and, in this way, cause direct cardiotoxicity [39]. Therefore, SARS-CoV-2 can induce direct cardiotoxicity by infecting cardiomyocytes and developing myocarditis [28]. Viral particles have also been identified within the cardiac tissue of several patients [40].

Furthermore, a recent single-cell RNA-seq analysis of the heart revealed low levels of ACE2 in cardiac myocytes and elevated levels in pericytes, while it has been demonstrated that ACE2 was upregulated in failing hearts [41,42]. As aforementioned, ACE2 is also expressed in endothelial cells of several organs [6,43,44], and the infection of these cells by SARS-CoV-2 could be vital in vascular events that have been observed in COVID-19 patients [43]. Furthermore, due to its large size (80–100 nm), SARS-CoV-2 may need to infect endothelial cells to infect organs, including the heart or kidney. It has also been reported that SARS-CoV-2 infects engineered human blood vessel organoids derived from human induced pluripotent stem cells (iPSCs) [44]. Infection of the blood vessel organelle was inhibited by a human soluble recombinant ACE2 [44].

There is limited research on the connection between SARS-CoV-2 and miRNA deregulation, but some recent studies have explored this relationship. Some miRNAs which are encoded by RNA viruses may be involved in the regulation of viral and/or host gene expression to promote favorable conditions, which lead to viral replication [45]. Several hospitalized COVID-19 patients exhibit a systemic dysregulation of pro-inflammatory cytokines, a phenomenon known as a cytokine storm [46]. The overproduction of early response pro-inflammatory cytokines (TNF-α, IL-6, and IL-1β) leads to an increased risk of vascular hyperpermeability, multiorgan failure, and eventually death when the high cytokine concentrations are unabated over time [47]. The immune hyperinflammatory response, mediated by dysregulated macrophages and innate and adaptive immunity, has been considered a common condition in severe COVID-19 cases [48].

Investigating miRNAs that are involved in both SARS-CoV-2 and cardiovascular pathology could contribute to understanding the relationship and the positive correlation between COVID-19 and increased susceptibility to CVDs. However, due to the recent outbreak of the COVID-19 pandemic, only a limited number of studies focusing on elucidating the dual role of miRNAs in SARS-CoV-2 infection and CVDs have been published. Alteration of miRNA expression levels (either increase or decrease) is linked with the disturbance of normal physiological functions and contributes to numerous diseases, including CVDs [27]. For example, it has been reported that levels of miRNA-16 were elevated in the peripheral blood of patients with COVID-19, and higher levels of miRNA-16 were associated with more severe disease [49]. Interestingly, Wicik et al. [50] found that miRNA-16 may be involved in the interactions between the SARS-CoV-2 and the functional networks of ACE-2, which plays a vital role in the development of acute myocardial infarction (AMI). While the exact ways in which miRNAs contribute to viral infections are not fully understood, changes in the levels of specific miRNAs after SARS-CoV-2 infection have raised the possibility that they may be involved in the development of COVID-19 (discussed further below).

Overall, at the time of the publication of this work, only a limited number of miRNAs (miR-146a, miR-27/miR-27a-5p, miR-451, miR-486-5p, miR-21, miR-155, and miR-133a) that have been implicated in the development of CVDs were identified to be altered following infection with SARS-CoV-2 (Table 1). However, the exact roles of these miRNAs remain inconclusive, and further studies are required to fully understand their mechanisms of action and potential connections to CVDs.

Table 1. Circulating miRNAs associated with COVID-19 and cardiovascular diseases.

miRNA	Function/Alteration in CVDs	Ref	Function/Alteration in COVID-19	Ref
miRNA-146a	Altered in CVDs Promotes formation of atherosclerotic plaques through upregulation of IL-6 and TNF-α	[51,52]	Downregulated in COVID-19 patients leading to upregulation of IL6 Correlated with the severity of the disease	[53,54]
miR-27 miR-27a-5p	Overexpressed in patients with coronary artery disease Promotes fibrosis, activation of macrophages	[55–57]	Downregulated in COVID-19 patients leading to upregulation of IL6 ACE2 levels are positively correlated with miR-27a/b	[58] [59]
miR-451	Mediates the cycloocygenase-2 pathway and prevents myocardial cell death from ischemia-reperfusion injury	[60]	Downregulated in COVID-19 patients leading to overexpression of IL 6	[58,61]
miR-486-5p	Reduces cell proliferation Associated with arrhythmias Upregulated in patients with arrhythmias Downregulated in the heart upon cardiac ischemia/reperfusion	[62] [63] [64]	Downregulated in COVID-19 patients favoring viral replication by weakening the immune system	[27]
miR-21	Has been implicated in cardiac dysfunctions, including cardiac hypertrophy and fibrosis. Associated with the pro-inflammatory state. Regulates cardiac structure Upregulated during heart failure	[65] [66] [22]	Elevated levels in COVID-19 patients	[22]
miR-155	Correlated with cardiovascular damage, endothelial inflammation, fibroblast proliferation Upregulated in CVDs	[22]	Elevated levels in COVID-19 patients	[22]
miR-133a	Upregulated in patients with cardiopulmonary diseases	[67]	Elevated levels in COVID-19 patients. Associated with COVID-19 severity and 28 days mortality	[68]

The potential roles of these miRNAs in the development of both CVDs and COVID-19 are discussed in the following paragraphs.

3.1. MiR 146a

The inflammatory response to SARS-CoV-2 infection commonly results in marked activation of coagulation—the process of thromboinflammation—with evidence of systemic endothelial damage and a resultant loss of normal anticoagulant properties [69,70]. Furthermore, it has been demonstrated that COVID-19 patients with severe symptoms are characterized by hyper-inflammatory responses due to dysregulation of pro-inflammatory cytokines, resulting in excessive endothelial cell and lung damage [71]. MiRNAs regulate inflammatory responses, and miR-21-5p, miR-146a, and miR-126-3p act as pro-inflammatory state markers by targeting components of the NF-κB pathway [54]. MiR-146a is particularly noteworthy among the miRNAs that have been linked to inflammation. In immune cells, miR-146a and miR-155 are the first miRNAs that are induced after immune activation. These miRNAs modulate the TLR-signaling pathway, which triggers the production of a large variety of inflammatory cytokines and specifically IL6, Type I interferons (IFNs), and antiviral proteins through the NF-kB pathway [72]. Even though miR-146a-5p does not target the mRNAs encoding IL-6 or its receptor, elevated levels of serum IL-6 have been detected in miR-146a-5p$^{-}/^{-}$ mice indicating a functional link between miR-146a-5p and IL-6 levels [73]. The transcription of miR-146a-5p is also under the control of NF-κB despite IL-6 and miR-146a-5p playing opposite roles in the inflammatory process [74]. Interestingly, in-silico analyses show that miR-146 and -155 negatively regulate genes encoding SARS-CoV-2 cellular entry factors and their interactors are present in the oral cavity [75]. However, the role of miRNA-146a in the development of CVDs following infection with SARS-CoV-2 remains inconclusive.

Sabbatinelli et al. [54] evaluated the levels of circulating miR-146a, miR-21-5p, and miR126-3p in COVID-19 patients that were treated with the monoclonal antibody Tocilizumab (TCZ), an IL-6 signaling inhibitor, to evaluate the role of these pro-inflammatory-linked miRNAs. It has been previously demonstrated that the levels of IL-6 correlate with COVID-19 severity [76], while miR-146a-5p is a negative regulator of NF-κB, which is in turn a transcription factor of the gene encoding IL-6 [77]. Interestingly, Sabbatinelli et al. [54] reported an inverse relationship between miR-146a and IL-6. In detail, this study revealed increased levels of IL-6 and reduced levels of miR-146a-5p in COVID-19 patients compared to healthy age-matched control subjects, highlighting the role of the imbalance between IL-6 and miR-146a-5p in the pathogenesis of COVID-19. Furthermore, Sabbatinelli et al. [54] reported that COVID-19 patients responding to TCZ expressed higher levels of miR-146a compared to patients who did not respond to TCZ. Interestingly, COVID-19 severity was higher in non-responders with the lowest levels of miR-146a-5p. Together, this study demonstrated that in COVID-19 patients, the serum levels of miR-146a are altered, and this miRNA is a valuable indicator of the clinical course of COVID-19 response to treatment against hyper-inflammatory states. It has been suggested that the downregulation of circulating miR-146a observed in hypertension, diabetes, and obesity may explain the more severe COVID-19 cases occurring in these patients [78]. Therefore, it could be suggested that COVID-19 patients with downregulated miR-146a expression are expected to have elevated levels of pro-inflammatory cytokines, specifically IL-6, and increased susceptibility to cytokine storms. The role of IL-6 in COVID-19 progression has been also highlighted by Vasuri et al. [57], who reported that IL-6 levels were higher in COVID-19 patients compared to healthy controls.

Gao et al. [79] showed that miR-146a weakens sepsis-induced cardiac dysfunction. In another study, Oh et al. [52] reported that miR-146a was upregulated in failing cardiomyocytes and that overexpression of this miRNA suppresses the expression level of small ubiquitin-like modifier 1 (SUMO1), which reduces sarco-endoplasmic reticulum calcium ATPase-2 (SERCA-2) sumoylation leading to cardiac contractile dysfunction. On the contrary, Huang et al. [53] demonstrated that the expression levels of miR-146a were elevated in peripheral blood and correlated with IL-6 and TNF-α expression. This correlation is also proportional to plaque vulnerability and the degree of stenosis in carotid atherosclerosis. Furthermore, it has been demonstrated that miR-146a has a cardiomyocyte-protective function, and it increases cardiomyocyte viability and protects against oxidative stress [80].

Together, the above data suggest that miR-146 expression may be inversely correlated between the two conditions, i.e., CVDs and COVID-19. Even though the mechanism by which miR-146a regulates the expression of pro-inflammatory cytokines remains unclear, there is a strong link between miR-146a and hyper-inflammatory processes observed in both SARS-CoV-2 and atherosclerosis. Therefore, disruption of the miR-146a/IL6 balance might impair the modulation of pro-inflammatory cytokines and facilitate their secretion to induce a hyper-inflammatory state. It has been proposed that the downregulation of miR-146a might be one of the factors underlying severe COVID-19 due to an insufficient host antiviral response, showing more prolonged and excessive cytokine release and lack of a feedback mechanism to limit inflammatory damage to tissues [78]. Overall, it can be assumed that there are distinct alteration patterns of miR-146a in different pathological conditions and that alterations in miRNA-146a following infection with SARS-CoV-2 may contribute to the development of CVDs, including atherosclerosis.

3.2. MiR-27

It has been demonstrated that the levels of the miR-27 family (including miR-27a-5p) are altered in patients with cardiac remodeling and atherosclerosis, while miR-27 is over-expressed in patients with coronary artery disease [55,56]. In detail, miR-27 is linked with cardiac remodeling and the formation of atherosclerotic plaques through the activation of macrophages participating in tissue fibrosis and by positively regulating the synthesis of low-density lipoprotein (LDL) and cholesterol [55]. The upregulation of miR-27a is also linked with promoting angiogenesis, a vital adaptation mechanism in ischemia-reperfusion [81]. Interestingly, miR-27 is downregulated in patients with SARS-CoV-2 [57,58], while it has been suggested that this is a negative regulator of the IL-6 expression. More specifically, in the study by Vasulri et al. [57], an analysis of four atherosclerotic and normal femoral arteries from a COVID-19 patient with several cardiovascular comorbidities that developed pneumonia and bilateral interstitial lesions revealed a significant decline in miR-27a-5p [57]. The study also showed an increase in both the mRNA and the IL-6 protein expression, a hyperinflammatory status, and a loss of endothelial cell homeostasis and suggested that perivascular endothelialitis is linked with miR-27a-5p. The association of IL-6 expression with miR-27a-5p in this study emphasizes an essential contribution of this miRNA in SARS-CoV-2 pathology.

These results suggest an inverse relationship between the expression levels of miR-27 in the two pathological conditions, i.e., CVDs and COVID-19, suggesting that miR-27 does not contribute to atherosclerosis following infection with SARS-CoV-2. However, the precise role of miR-27 in the pathophysiological mechanisms underlying CVDs was not evaluated in the obtained articles, and no research study has found reduced miR-27 expression levels in CVD patients.

3.3. MiR-21 and miR- 155

In a recent study, Garg et al. [22] compared the levels of five cardiac-specific circulating miRNAs, including miR-21 and miR-126 (associated with cardiac fibroblast and endothelial cell dysfunction), miR-155 (linked to inflammation), and miR-208a and miR-499 (related to myocardial/cardiomyocyte damage) in COVID-19 patients, influenza-associated acute respiratory distress syndrome (Influenza-ARDS) admitted to the intensive care unit and healthy individuals. The serum concentration of inflammatory miR-155, heart muscle-specific miR-208a, and miR-499, as well as that of the fibromiR miR-21, were significantly increased in COVID-19 patients compared to healthy controls. Furthermore, the levels of miR-21, which is associated with fibrosis [82], were elevated in acute COVID-19 patients compared to Influenza-ARDS patients and healthy controls. Notably, the altered levels of miR-155 and miR-499 could further differentiate COVID-19 patients from Influenza-ARDS patients, despite both diseases being very similar phenotypically, indicating a rather specific response and cardiac involvement of COVID-19. MiR-21 regulates cardiac structure and function by regulating the ERK–MAP kinase signaling pathway in cardiac fibroblasts, and it is increased during heart failure, where it acts as a promoter of interstitial fibrosis and cardiac dysfunction [82]; i.e., its upregulation facilitates the survival of fibroblasts causing fibrosis and hypertrophy resulting in cardiac dysfunction [83].

HF has been observed in 23% of patients with COVID-19, and the regulatory pathways in which miRNAs participate, such as miR-155, may increase the risk of HF. MiR-155 is overexpressed in patients with COVID-19 and has been correlated with cardiovascular damage fibroblast proliferation, endothelial inflammation, promotion of apoptosis, and cardiomyocyte pyroptosis. These changes induce hypertrophy and ventricular dysfunction, which lead to HF. Furthermore, miR-15b is upregulated in patients with severe COVID-19, and this was associated with a significantly higher incidence of arrhythmias, indicating that this miRNA could be used as a possible marker of heart damage (see [27] and references cited therein).

3.4. MiR-133a

Gutmann et al. [68] examined the association of circulating miRNAs with COVID-19 severity and 28-day intensive care unit (ICU) mortality. The results of this study revealed that a total of 60 miRNAs, including cardiomyocyte- platelet-, endothelial-, and hepatocyte-derived miRNAs, were differentially expressed depending on the severity of the disease. Interestingly, the myocyte-derived (MyomiR) miR-133a [84] and cardiomyocyte-derived miR-208b [85] were detectable in patients with severe disease. Furthermore, miR-133a, which reflects inflammation-induced myocyte damage, was related to 28-day mortality and negatively correlated with neutrophil counts and proteins related to neutrophil degranulation, such as myeloperoxidase [68]. MiR-133a levels were also the highest in critically ill patients with cardiopulmonary diseases. It is worth mentioning that miR-133 participates in the proliferation, differentiation, survival, hypertrophic growth, and electrical conduction of cardiac cells, which are essential for cardiac fibrosis, cardiac hypertrophy, and arrhythmia [86]. Furthermore, myocardial injury [87] and chronic obstructive pulmonary disease (COPD) [88] are known to increase circulating miR-133a levels. Neutrophil degranulation and extravasation resulting in myocyte damage is a likely cause for the rise of circulating miR-133a [89]. Moreover, there is evidence that neutrophils may be a secondary source of miR-133a in circulation [90].

3.5. MiR-486-5p

MiR-486 is expressed mainly in the heart and muscle and targets proteins in cardiomyocyte survival (PI3K/PTEN/pAKT), myogenesis (MyoD), myotube survival/differentiation (DOCK3/PTEN/AKT), and cardiac progenitor cell proliferation networks. miR-486 regulates myocardial homeostasis by negatively regulating PIM1 kinase, an essential component of cardiac progenitor cell cyclin (see [91] and references cited therein). MiR-486-5p is linked with reduced cell proliferation and promotes cell death by accumulating superoxide anions and causing DNA damage. These mechanisms are tightly associated with heart failure, and cardiac remodeling as their characteristic features are increased inflammatory responses, cardiomyocyte apoptosis or necrosis, and tissue fibrosis. miR-486 also alters the activation and differentiation of B- and T-lymphocytes, prevents resolution at inflammatory sites, and downregulates the expression of the anti-inflammatory cytokine IL-10, highlighting the pro-inflammatory role of miR-486-5p [27]. It has been proposed that reduced miR-486-5p expression levels contribute to weakening the immune system and intensifying the severity of the disease by favoring viral replication [27]. Interestingly, miR-486–5p is one of the most downregulated miRNAs in lung tumor tissues and contributes to lung cancer progression [92]. The role of miR-486-5p is unclear, as it is downregulated in COVID-19 patients and upregulated in patients with atrial fibrillation, implying that COVID-19 cannot induce arrhythmias by altering miR-486-5p expression levels. The association of miR-486-5p with arrhythmias is further supported by Li and coworkers [63], whose study showed that miR-486-5p is overexpressed in patients with arrhythmias and is involved in depressing the sinoatrial node function. Although miR-486-5p upregulation in patients with arrhythmias can be associated with prolonged inflammatory responses, this highlights the notion that CVD patients with upregulated miR-486-5p can be susceptible to severe COVID-19 states characterized by hyper-inflammatory state or cytokine storm. It has been demonstrated that miR-486 is significantly downregulated in the heart upon cardiac ischemia/reperfusion (I/R) injury [64].

3.6. *MiR-451*

MiR-451 is expressed in multiple systems, including the urinary, respiratory, and digestive systems, and regulates various physiological and pathological processes, including hematopoiesis, epithelial cell polarity, and embryogenesis [93]. MiR-451 is reported to be upregulated in both the right and left atria of patients diagnosed with atrial fibrillation. One of the multiple functions of miR-451 is to prevent cell death from ischemia-reperfusion injury by mediating the cyclooxygenase-2 pathway, which is heavily linked with arrhythmias [60]. Although the role of this miRNA in atrial fibrillation can be regarded as an adaptation mechanism, the exact mechanism of action remains unknown. miR-451 was found to be downregulated in COVID-19 patients and is associated with severe outcomes, presumably by preventing the expression of pro-inflammatory cytokines, similar to miR-486-5p, and favoring viral replication [58]. The association of miR-451 with the protection against apoptosis and the strong link with severe COVID-19 phenotypes suggests a decline in miR-451 that COVID-19 patients can also increase the susceptibility to CVDs, presumably due to the lack of protection against apoptosis. Aberrant apoptotic mechanisms can contribute to extensive tissue damage, fibrosis, and cardiac remodeling [94]. Yang et al. [61] reported that compared to healthy donors, COVID-19 patients had significantly higher mRNA expression of IL-6R while miR-451a, a known negative regulator of IL-6R translation, was down-regulated. This may lead to increased IL-6 protein expression.

4. Conclusions

Identifying correlations between CVDs and COVID-19 is of particular interest because individuals suffering from CVDs may be at higher risk for severe illness or death from COVID-19. Recent evidence suggests that SARS-CoV-2 infection can severely damage the cardiovascular system. In a recent review, Izzo et al. [95] investigated the cardiovascular effects of miRNAs in COVID-19. They identified six cardio-specific miRNAs (miR-21-5p, miR-153, miR-155-5p, miR-214, miR-208a-3p, and miR-375) that are upregulated and nine cardio-specific miRNAs (miR-26B, miR30e-3p, miR-125b, miR-133a, miR-146a, mir-200c-3p, miR-223-3p, miR-500, and miR-590-3p) that are downregulated in COVID-19 patients. These miRNAs have been linked to various forms of cardiovascular damage, including acute ischemic injury, nonischemic injury, stress cardiomyopathy, heart failure, myocardial fibrosis, thromboembolism, vascular inflammation, and Kawasaki disease (see [95] and reference cited therein). Our critical review also indicates that alteration of the expression levels of some COVID-19-related miRNAs are correlated with CVDs and are linked with various mechanisms (Figure 1). However, how an infection with SARS-CoV-2 affects the profile and/or expression levels of cardiac-specific miRNAs, and what are the exact mechanisms by which the alteration of these miRNAs in response to SARS-CoV-2 infection leads to CVDs or worsening pre-existing CVDs remain inconclusive. Similar to Izzo et al. [95], our work highlights the potential implication of miR-21, miR-155, 133a, and miR-146a in the pathophysiology of both CVDs and COVID-19. Furthermore, in our work, miR-27/miR27a-5p, miR-451, miR-486-5p, miR-21, and miR-155 that have been implicated in the development of CVDs were identified to be altered following infection with SARS-CoV-2 (Table 1). Despite the alterations in the expression levels of these miRNAs, the identified changes and contributions to cellular processes do not correlate clearly between SARS-CoV-2 infections and cardiovascular pathologies. The results of this work also highlight the central role of IL-6 in the pathogenesis of both CVDs and COVID-19. This work emphasizes the importance of conducting further research to understand the relationship between CVDs and COVID-19 in order to develop strategies for preventing and managing COVID-19 in individuals with CVDs. This could involve studying the relationship between cardio-specific miRNAs and immune function, inflammation, or other biological pathways that may be involved in the severity of COVID-19. To effectively identify correlations between miRNAs, CVDs, and COVID-19, it will be necessary to use systematic and collaborative approaches involving basic researchers, clinical investigators, industry partners, and governmental agencies. This will help to translate biomarkers into clinical use.

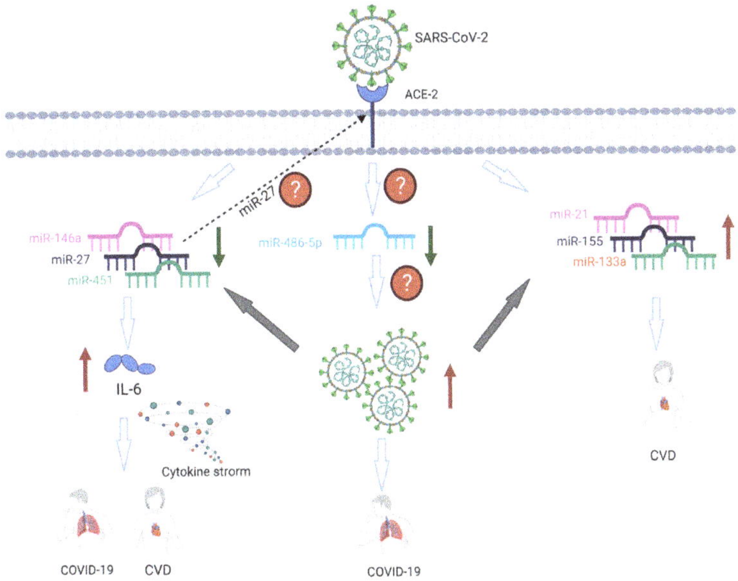

Figure 1. Potential associations between alterations of miRNA levels after infection with SARs-CoV-2 virus and the development of CVDs. Viral infection causes downregulation of miR-146a, miR-27, and miR-451, which are well-established inhibitors of IL-6 protein and IL-6 mRNAs, leading to the overproduction of several pro-inflammatory cytokines, a process known as cytokine storm. This phenomenon has been implicated in the pathogenesis of coronaviruses disease-19 (COVID-19) and cardiovascular diseases (CVDs). Expression levels of miR-27 could also correlate with the expression of the ACE-2 receptor, which provides the entry point of SARS-CoV-2. On the other hand, SARS-CoV-2 infection causes the upregulation of miR-21, miR-155, and miR-133a which are implicated in the development of CVD. A SARS-CoV-2 infection causes downregulation of miR-486-5p, facilitating the replication of the virus by weakening the immune system response. Question marks (?) represent factors that should be examined further.

Author Contributions: Conceptualization, M.P., P.V., K.F. and C.P.; methodology, M.P., P.V. and C.P.; software, V.N. and K.F.; validation, M.P., V.N. and C.P.; formal analysis, P.V., K.F. and C.P., investigation, M.P. and P.V.; resources, P.V. and V.N.; data curation, P.V. and C.P.; writing—original draft preparation, P.V., V.N., M.P. and C.P.; writing—review and editing, M.P., V.N. and C.P.; visualization, M.P. and K.F.; supervision, C.P.; project administration, C.P. and K.F.; funding acquisition, C.P. All authors have read and agreed to the published version of the manuscript.

Funding: This research received no external funding.

Institutional Review Board Statement: Not applicable.

Data Availability Statement: Not applicable.

Conflicts of Interest: The authors declare no conflict of interest.

References

1. Lai, C.-C.; Shih, T.-P.; Ko, W.-C.; Tang, H.-J.; Hsueh, P.-R. Severe acute respiratory syndrome coronavirus 2 (SARS-CoV-2) and coronavirus disease-2019 (COVID-19): The epidemic and the challenges. *Int. J. Antimicrob. Agents* **2020**, *55*, 105924. [CrossRef] [PubMed]
2. Zhu, H.; Wei, L.; Niu, P. The novel coronavirus outbreak in Wuhan, China. *Glob. Health Res. Policy* **2020**, *5*, 6. [CrossRef] [PubMed]
3. Adhikari, S.P.; Meng, S.; Wu, Y.J.; Mao, Y.P.; Ye, R.X.; Wang, Q.Z.; Sun, C.; Sylvia, S.; Rozelle, S.; Raat, H.; et al. Epidemiology, causes, clinical manifestation and diagnosis, prevention and control of coronavirus disease (COVID-19) during the early outbreak period: A scoping review. *Infect. Dis. Poverty* **2020**, *9*, 29. [CrossRef] [PubMed]
4. Zhang, Q.; Xiang, R.; Huo, S.; Zhou, Y.; Jiang, S.; Wang, Q.; Yu, F. Molecular mechanism of interaction between SARS-CoV-2 and host cells and interventional therapy. *Signal Transduct. Target Ther.* **2021**, *6*, 233. [CrossRef]
5. Jackson, C.B.; Farzan, M.; Chen, B.; Choe, H. Mechanisms of SARS-CoV-2 entry into cells. *Nat. Rev. Mol. Cell Biol.* **2022**, *23*, 3–20. [CrossRef] [PubMed]
6. Hamming, I.; Timens, W.; Bulthuis, M.L.; Lely, A.T.; Navis, G.; van Goor, H. Tissue distribution of ACE2 protein, the functional receptor for SARS coronavirus. A first step in understanding SARS pathogenesis. *J. Pathol.* **2004**, *203*, 631–637. [CrossRef]
7. Bourgonje, A.R.; Abdulle, A.E.; Timens, W.; Hillebrands, J.-L.; Navis, G.J.; Gordijn, S.J.; Bolling, M.C.; Dijkstra, G.; Voors, A.A.; Osterhaus, A.D.; et al. Angiotensin-converting enzyme 2 (ACE2), SARS-CoV-2 and the pathophysiology of coronavirus disease 2019 (COVID-19). *J. Pathol.* **2020**, *251*, 228–248. [CrossRef] [PubMed]
8. Beyerstedt, S.; Casaro, E.B.; Rangel, É.B. COVID-19: Angiotensin-converting enzyme 2 (ACE2) expression and tissue susceptibility to SARS-CoV-2 infection. *Eur. J. Clin. Microbiol. Infect. Dis.* **2021**, *40*, 905–919. [CrossRef]
9. Dessie, Z.G.; Zewotir, T. Mortality-related risk factors of COVID-19: A systematic review and meta-analysis of 42 studies and 423,117 patients. *BMC Infect. Dis.* **2021**, *21*, 855. [CrossRef]
10. Nishiga, M.; Wang, D.W.; Han, Y.; Lewis, D.B.; Wu, J.C. COVID-19 and cardiovascular disease: From basic mechanisms to clinical perspectives. *Nat. Rev. Cardiol.* **2020**, *17*, 543–558. [CrossRef]
11. Madjid, M.; Safavi-Naeini, P.; Solomon, S.D.; Vardeny, O. Potential effects of coronaviruses on the cardiovascular system: A review. *JAMA Cardiol.* **2020**, *5*, 831–840. [CrossRef] [PubMed]
12. Huang, C.; Wang, Y.; Li, X.; Ren, L.; Zhao, J.; Hu, Y.; Zhang, L.; Fan, G.; Xu, J.; Gu, X.; et al. Clinical features of patients infected with 2019 novel coronavirus in Wuhan, China. *Lancet* **2020**, *395*, 497–506. [CrossRef] [PubMed]
13. Shi, S.; Qin, M.; Cai, Y.; Liu, T.; Shen, B.; Yang, F.; Cao, S.; Liu, X.; Xiang, Y.; Zhao, Q.; et al. Characteristics and clinical significance of myocardial injury in patients with severe coronavirus disease 2019. *Eur. Heart J.* **2020**, *41*, 2070–2079. [CrossRef] [PubMed]
14. Case, S.M.; Son, M.B. COVID-19 in pediatrics. *Rheum. Dis. Clin. N. Am.* **2021**, *47*, 797–811. [CrossRef] [PubMed]
15. Felekkis, K.; Touvana, E.; Stefanou, C.; Deltas, C. microRNAs: A newly described class of encoded molecules that play a role in health and disease. *Hippokratia* **2010**, *14*, 236–240.
16. Lee, Y.; Ahn, C.; Han, J.; Choi, H.; Kim, J.; Yim, J.; Lee, J.; Provost, P.; Rådmark, O.; Kim, S.; et al. The nuclear RNase III Drosha initiates microRNA processing. *Nature* **2003**, *425*, 415–419. [CrossRef]
17. Lund, E.; Güttinger, S.; Calado, A.; Dahlberg, J.E.; Kutay, U. Nuclear export of microRNA precursors. *Science* **2004**, *303*, 95–98. [CrossRef]
18. Bartel, D.P. MicroRNAs: Genomics, biogenesis, mechanism, and function. *Cell* **2004**, *116*, 281–297. [CrossRef]
19. Romaine, S.P.R.; Tomaszewski, M.; Condorelli, G.; Samani, N.J. MicroRNAs in cardiovascular disease: An introduction for clinicians. *Heart* **2015**, *101*, 921–928. [CrossRef]
20. Peters, L.J.F.; Biessen, E.A.L.; Hohl, M.; Weber, C.; van der Vorst, E.P.C.; Santovito, D. Small things matter: Relevance of microRNAs in cardiovascular disease. *Front. Physiol.* **2020**, *11*, 793. [CrossRef]
21. Marchi, R.; Sugita, B.; Centa, A.; Fonseca, A.S.; Bortoletto, S.; Fiorentin, K.; Ferreira, S.; Cavalli, L.R. The role of microRNAs in modulating SARS-CoV-2 infection in human cells: A systematic review. *Infect. Genet. Evol.* **2021**, *91*, 104832. [CrossRef] [PubMed]
22. Garg, A.; Seeliger, B.; Derda, A.A.; Xiao, K.; Gietz, A.; Scherf, K.; Sonnenschein, K.; Pink, I.; Hoeper, M.M.; Welte, T.; et al. Circulating cardiovascular microRNAs in critically ill COVID-19 patients. *Eur. J. Heart Fail.* **2021**, *23*, 468–475. [CrossRef]
23. Dhakal, B.P.; Sweitzer, N.K.; Indik, J.H.; Acharya, D.; William, P. SARS-CoV-2 infection and cardiovascular disease: COVID-19 heart. *Heart Lung Circ.* **2020**, *29*, 973–987. [CrossRef] [PubMed]
24. Sato, K.; Sinclair, J.E.; Sadeghirad, H.; Fraser, J.F.; Short, K.R.; Kulasinghe, A. Cardiovascular disease in SARS-CoV-2 infection. *Clin. Transl. Immunol.* **2021**, *10*, e1343. [CrossRef] [PubMed]
25. Chung, M.K.; Zidar, D.A.; Bristow, M.R.; Cameron, S.J.; Chan, T.; Harding, C.V.; Kwon, D.H.; Singh, T.; Tilton, J.C.; Tsai, E.J.; et al. COVID-19 and cardiovascular disease. *Circ. Res.* **2021**, *128*, 1214–1236. [CrossRef]
26. Ning, Q.; Wu, D.; Wang, X.; Xi, D.; Chen, T.; Chen, G.; Wang, H.; Lu, H.; Wang, M.; Zhu, L.; et al. The mechanism underlying extrapulmonary complications of the coronavirus disease 2019 and its therapeutic implication. *Signal Transduct. Target Ther.* **2022**, *7*, 57. [CrossRef]
27. Bautista-Becerril, B.; Pérez-Dimas, G.; Sommerhalder-Nava, P.C.; Hanono, A.; Martínez-Cisneros, J.A.; Zarate-Maldonado, B.; Muñoz-Soria, E.; Aquino-Gálvez, A.; Castillejos-López, M.; Juárez-Cisneros, A.; et al. miRNAs, from evolutionary junk to possible prognostic markers and therapeutic targets in COVID-19. *Viruses* **2021**, *14*, 41. [CrossRef]
28. Knowlton, K.U. Pathogenesis of SARS-CoV-2 induced cardiac injury from the perspective of the virus. *J. Mol. Cell Cardiol.* **2020**, *147*, 12–17. [CrossRef]

29. Devaux, B.; Scholz, D.; Hirche, A.; Klövekorn, W.P.; Schaper, J. Upregulation of cell adhesion molecules and the presence of low grade inflammation in human chronic heart failure. *Eur. Heart J.* **1997**, *18*, 470–479. [CrossRef]
30. Shirazi, L.F.; Bissett, J.; Romeo, F.; Mehta, J.L. Role of inflammation in heart failure. *Curr. Atheroscler. Rep.* **2017**, *19*, 27. [CrossRef]
31. Moore, J.B.; June, C.H. Cytokine release syndrome in severe COVID-19. *Science* **2020**, *368*, 473–474. [CrossRef] [PubMed]
32. Krown, K.A.; Page, M.T.; Nguyen, C.; Zechner, D.; Gutierrez, V.; Comstock, K.L.; Glembotski, C.C.; Quintana, P.J.; Sabbadini, R.A. Tumor necrosis factor alpha-induced apoptosis in cardiac myocytes. Involvement of the sphingolipid signaling cascade in cardiac cell death. *J. Clin. Investig.* **1996**, *98*, 2854–2865. [CrossRef] [PubMed]
33. Mensà, E.; Giuliani, A.; Matacchione, G.; Gurău, F.; Bonfigli, A.R.; Romagnoli, F.; De Luca, M.; Sabbatinelli, J.; Olivieri, F. Circulating miR-146a in healthy aging and type 2 diabetes: Age- and gender-specific trajectories. *Mech. Ageing Dev.* **2019**, *180*, 1–10. [CrossRef] [PubMed]
34. Mussbacher, M.; Salzmann, M.; Brostjan, C.; Hoesel, B.; Schoergenhofer, C.; Datler, H.; Hohensinner, P.; Basílio, J.; Petzelbauer, P.; Assinger, A.; et al. Cell type-specific roles of NF-κB linking inflammation and thrombosis. *Front. Immunol.* **2019**, *10*, 85. [CrossRef]
35. Greco, S.; Madè, A.; Gaetano, C.; Devaux, Y.; Emanueli, C.; Martelli, F. Noncoding RNAs implication in cardiovascular diseases in the COVID-19 era. *J. Transl. Med.* **2020**, *18*, 408. [CrossRef]
36. Mann, M.; Mehta, A.; Zhao, J.L.; Lee, K.; Marinov, G.K.; Garcia-Flores, Y.; Lu, L.F.; Rudensky, A.Y.; Baltimore, D. Author Correction: An NF-κB-microRNA regulatory network tunes macrophage inflammatory responses. *Nat. Commun.* **2018**, *9*, 3338. [CrossRef] [PubMed]
37. Das, A.; Ganesh, K.; Khanna, S.; Sen, C.K.; Roy, S. Engulfment of apoptotic cells by macrophages: A role of microRNA-21 in the resolution of wound inflammation. *J. Immunol.* **2014**, *192*, 1120–1129. [CrossRef] [PubMed]
38. Samavati, L.; Uhal, B.D. ACE2, much more than just a receptor for SARS-COV-2. *Front. Cell Infect. Microbiol.* **2020**, *10*, 317. [CrossRef]
39. Nicin, L.; Abplanalp, W.T.; Mellentin, H.; Kattih, B.; Tombor, L.; John, D.; Schmitto, J.D.; Heineke, J.; Emrich, F.; Arsalan, M.; et al. Cell type-specific expression of the putative SARS-CoV-2 receptor ACE2 in human hearts. *Eur. Heart J.* **2020**, *41*, 1804–1806. [CrossRef]
40. Yao, X.H.; Li, T.Y.; He, Z.C.; Ping, Y.F.; Liu, H.W.; Yu, S.C.; Mou, H.M.; Wang, L.H.; Zhang, H.R.; Fu, W.J.; et al. A pathological report of three COVID-19 cases by minimal invasive autopsies. *Zhonghua Bing Li Xue Za Zhi* **2020**, *49*, 411–417. [CrossRef]
41. Goulter, A.B.; Goddard, M.J.; Allen, J.C.; Clark, K.L. ACE2 gene expression is up-regulated in the human failing heart. *BMC Med.* **2004**, *2*, 19. [CrossRef] [PubMed]
42. Chen, L.; Li, X.; Chen, M.; Feng, Y.; Xiong, C. The ACE2 expression in human heart indicates new potential mechanism of heart injury among patients infected with SARS-CoV-2. *Cardiovasc. Res.* **2020**, *116*, 1097–1100. [CrossRef] [PubMed]
43. Varga, Z.; Flammer, A.J.; Steiger, P.; Haberecker, M.; Andermatt, R.; Zinkernagel, A.S.; Mehra, M.R.; Schuepbach, R.A.; Ruschitzka, F.; Moch, H. Endothelial cell infection and endotheliitis in COVID-19. *Lancet* **2020**, *395*, 1417–1418. [CrossRef] [PubMed]
44. Monteil, V.; Kwon, H.; Prado, P.; Hagelkrüys, A.; Wimmer, R.A.; Stahl, M.; Leopoldi, A.; Garreta, E.; Hurtado Del Pozo, C.; Prosper, F.; et al. Inhibition of SARS-CoV-2 infections in engineered human tissues using clinical-grade soluble human ACE2. *Cell* **2020**, *181*, 905–913. [CrossRef] [PubMed]
45. Skalsky, R.L.; Cullen, B.R. Viruses, microRNAs, and host interactions. *Annu. Rev. Microbiol.* **2010**, *64*, 123–141. [CrossRef]
46. Zhou, F.; Yu, T.; Du, R.; Fan, G.; Liu, Y.; Liu, Z.; Xiang, J.; Wang, Y.; Song, B.; Gu, X.; et al. Clinical course and risk factors for mortality of adult inpatients with COVID-19 in Wuhan, China: A retrospective cohort study. *Lancet* **2020**, *395*, 1054–1062. [CrossRef]
47. Jose, R.J.; Manuel, A. COVID-19 cytokine storm: The interplay between inflammation and coagulation. *Lancet Respir. Med.* **2020**, *8*, e46–e47. [CrossRef]
48. Schultze, J.L.; Aschenbrenner, A.C. COVID-19 and the human innate immune system. *Cell* **2021**, *184*, 1671–1692. [CrossRef]
49. Li, C.; Hu, X.; Li, L.; Li, J.H. Differential microRNA expression in the peripheral blood from human patients with COVID-19. *J. Clin. Lab. Anal.* **2020**, *34*, e23590. [CrossRef]
50. Wicik, Z.; Eyileten, C.; Jakubik, D.; Simões, S.N.; Martins, D.C., Jr.; Pavão, R.; Siller-Matula, J.M.; Postula, M. ACE2 interaction networks in COVID-19: A physiological framework for prediction of outcome in patients with cardiovascular risk factors. *J. Clin. Med.* **2020**, *9*, 3743. [CrossRef]
51. Shen, L.; Li, C.; Zhang, H.; Qiu, S.; Fu, T.; Xu, Y. Downregulation of miR-146a contributes to cardiac dysfunction induced by the tyrosine kinase inhibitor sunitinib. *Front. Pharmacol.* **2019**, *10*, 914. [CrossRef] [PubMed]
52. Oh, J.G.; Watanabe, S.; Lee, A.; Gorski, P.A.; Lee, P.; Jeong, D.; Liang, L.; Liang, Y.; Baccarini, A.; Sahoo, S.; et al. miR-146a suppresses SUMO1 expression and induces cardiac dysfunction in maladaptive hypertrophy. *Circ. Res.* **2018**, *123*, 673–685. [CrossRef] [PubMed]
53. Huang, P.; He, X.Y.; Xu, M. The role of miRNA-146a and proinflammatory cytokines in carotid atherosclerosis. *Biomed. Res. Int.* **2020**, *2020*, 6657734. [CrossRef]
54. Sabbatinelli, J.; Giuliani, A.; Matacchione, G.; Latini, S.; Laprovitera, N.; Pomponio, G.; Ferrarini, A.; Svegliati Baroni, S.; Pavani, M.; Moretti, M.; et al. Decreased serum levels of the inflammaging marker miR-146a are associated with clinical non-response to tocilizumab in COVID-19 patients. *Mech. Ageing Dev.* **2021**, *193*, 111413. [CrossRef] [PubMed]

55. Polyakova, E.A.; Zaraiskii, M.I.; Mikhaylov, E.N.; Baranova, E.I.; Galagudza, M.M.; Shlyakhto, E.V. Association of myocardial and serum miRNA expression patterns with the presence and extent of coronary artery disease: A cross-sectional study. *Int. J. Cardiol.* **2021**, *322*, 9–15. [CrossRef]
56. Barsanti, C.; Trivella, M.G.; D'Aurizio, R.; El Baroudi, M.; Baumgart, M.; Groth, M.; Caruso, R.; Verde, A.; Botta, L.; Cozzi, L.; et al. Differential regulation of microRNAs in end-stage failing hearts Is associated with left ventricular assist device unloading. *BioMed Res. Int.* **2015**, *2015*, 592512. [CrossRef] [PubMed]
57. Vasuri, F.; Ciavarella, C.; Collura, S.; Mascoli, C.; Valente, S.; Degiovanni, A.; Gargiulo, M.; Capri, M.; Pasquinelli, G. Adventitial microcirculation is a major target of SARS-CoV-2-mediated vascular inflammation. *Biomolecules* **2021**, *11*, 1063. [CrossRef]
58. De Gonzalo-Calvo, D.; BenÍTez, I.D.; Pinilla, L.; CarratalÁ, A.; MoncusÍ-Moix, A.; Gort-Paniello, C.; Molinero, M.; GonzÁLez, J.; Torres, G.; Bernal, M.; et al. Circulating microRNA profiles predict the severity of COVID-19 in hospitalized patients. *Transl. Res.* **2021**, *236*, 147–159. [CrossRef]
59. Mukhopadhyay, D.; Mussa, B.M. Identification of novel hypothalamic microRNAs as promising therapeutics for SARS-CoV-2 by regulating ACE2 and TMPRSS2 expression: An in silico analysis. *Brain Sci.* **2020**, *10*, 666. [CrossRef]
60. Slagsvold, K.H.; Johnsen, A.B.; Rognmo, O.; Høydal, M.; Wisløff, U.; Wahba, A. Comparison of left versus right atrial myocardium in patients with sinus rhythm or atrial fibrillation—An assessment of mitochondrial function and microRNA expression. *Physiol. Rep.* **2014**, *2*, e12124. [CrossRef]
61. Yang, P.; Zhao, Y.; Li, J.; Liu, C.; Zhu, L.; Zhang, J.; Yu, Y.; Wang, W.-J.; Lei, G.; Yan, J.; et al. Downregulated miR-451a as a feature of the plasma cfRNA landscape reveals regulatory networks of IL-6/IL-6R-associated cytokine storms in COVID-19 patients. *Cell. Mol. Immunol.* **2021**, *18*, 1064–1066. [CrossRef] [PubMed]
62. Slagsvold, K.H.; Johnsen, A.B.; Rognmo, O.; Hoydal, M.A.; Wisloff, U.; Wahba, A. Mitochondrial respiration and microRNA expression in right and left atrium of patients with atrial fibrillation. *Physiol. Genomics* **2014**, *46*, 505–511. [CrossRef] [PubMed]
63. Li, N.; Artiga, E.; Kalyanasundaram, A.; Hansen, B.J.; Webb, A.; Pietrzak, M.; Biesiadecki, B.; Whitson, B.; Mokadam, N.A.; Janssen, P.M.L.; et al. Altered microRNA and mRNA profiles during heart failure in the human sinoatrial node. *Sci. Rep.* **2021**, *11*, 19328. [CrossRef] [PubMed]
64. Bei, Y.; Lu, D.; Bär, C.; Chatterjee, S.; Costa, A.; Riedel, I.; Mooren, F.C.; Zhu, Y.; Huang, Z.; Wei, M.; et al. miR-486 attenuates cardiac ischemia/reperfusion injury and mediates the beneficial effect of exercise for myocardial protection. *Mol. Ther.* **2022**, *30*, 1675–1691. [CrossRef] [PubMed]
65. Surina, S.; Fontanella, R.A.; Scisciola, L.; Marfella, R.; Paolisso, G.; Barbieri, M. miR-21 in human cardiomyopathies. *Front. Cardiovasc. Med.* **2021**, *8*, 767064. [CrossRef]
66. Kura, B.; Kalocayova, B.; Devaux, Y.; Bartekova, M. Potential clinical implications of miR-1 and miR-21 in heart disease and cardioprotection. *Int. J. Mol. Sci.* **2020**, *21*, 700. [CrossRef]
67. Slagsvold, K.H.; Rognmo, O.; Høydal, M.; Wisløff, U.; Wahba, A. Remote ischemic preconditioning preserves mitochondrial function and influences myocardial microRNA expression in atrial myocardium during coronary bypass surgery. *Circ. Res.* **2014**, *114*, 851–859. [CrossRef]
68. Gutmann, C.; Khamina, K.; Theofilatos, K.; Diendorfer, A.B.; Burnap, S.A.; Nabeebaccus, A.; Fish, M.; McPhail, M.J.W.; O'Gallagher, K.; Schmidt, L.E.; et al. Association of cardiometabolic microRNAs with COVID-19 severity and mortality. *Cardiovasc. Res.* **2022**, *118*, 461–474. [CrossRef]
69. Ackermann, M.; Verleden, S.E.; Kuehnel, M.; Haverich, A.; Welte, T.; Laenger, F.; Vanstapel, A.; Werlein, C.; Stark, H.; Tzankov, A.; et al. Pulmonary vascular endothelialitis, thrombosis, and angiogenesis in Covid-19. *N. Engl. J. Med.* **2020**, *383*, 120–128. [CrossRef]
70. Connors, J.M.; Levy, J.H. Thromboinflammation and the hypercoagulability of COVID-19. *J. Thromb. Haemost.* **2020**, *18*, 1559–1561. [CrossRef]
71. Rabaan, A.A.; Al-Ahmed, S.H.; Muhammad, J.; Khan, A.; Sule, A.A.; Tirupathi, R.; Mutair, A.A.; Alhumaid, S.; Al-Omari, A.; Dhawan, M.; et al. Role of inflammatory cytokines in COVID-19 patients: A review on molecular mechanisms, immune functions, immunopathology and immunomodulatory drugs to counter cytokine storm. *Vaccines* **2021**, *9*, 436. [CrossRef] [PubMed]
72. Tsitsiou, E.; Lindsay, M.A. microRNAs and the immune response. *Curr. Opin. Pharmacol.* **2009**, *9*, 514–520. [CrossRef] [PubMed]
73. Boldin, M.P.; Taganov, K.D.; Rao, D.S.; Yang, L.; Zhao, J.L.; Kalwani, M.; Garcia-Flores, Y.; Luong, M.; Devrekanli, A.; Xu, J.; et al. miR-146a is a significant brake on autoimmunity, myeloproliferation, and cancer in mice. *J. Exp. Med.* **2011**, *208*, 1189–1201. [CrossRef]
74. Luly, F.R.; Lévêque, M.; Licursi, V.; Cimino, G.; Martin-Chouly, C.; Théret, N.; Negri, R.; Cavinato, L.; Ascenzioni, F.; Del Porto, P. MiR-146a is over-expressed and controls IL-6 production in cystic fibrosis macrophages. *Sci. Rep.* **2019**, *9*, 16259. [CrossRef] [PubMed]
75. Roganović, J.R. microRNA-146a and -155, upregulated by periodontitis and type 2 diabetes in oral fluids, are predicted to regulate SARS-CoV-2 oral receptor genes. *J. Periodontol.* **2021**, *92*, e35–e43. [CrossRef]
76. Aziz, M.; Fatima, R.; Assaly, R. Elevated interleukin-6 and severe COVID-19: A meta-analysis. *J. Med. Virol.* **2020**, *92*, 2283–2285. [CrossRef]
77. Su, Y.-L.; Wang, X.; Mann, M.; Adamus, T.P.; Wang, D.; Moreira, D.F.; Zhang, Z.; Ouyang, C.; He, X.; Zhang, B.; et al. Myeloid cell-targeted miR-146a mimic inhibits NF-κB-driven inflammation and leukemia progression in vivo. *Blood* **2020**, *135*, 167–180. [CrossRef]

78. Roganović, J. Downregulation of microRNA-146a in diabetes, obesity and hypertension may contribute to severe COVID-19. *Med. Hypotheses* **2021**, *146*, 110448. [CrossRef]
79. Gao, M.; Wang, X.; Zhang, X.; Ha, T.; Ma, H.; Liu, L.; Kalbfleisch, J.H.; Gao, X.; Kao, R.L.; Williams, D.L.; et al. Attenuation of cardiac dysfunction in polymicrobial sepsis by microRNA-146a is mediated via targeting of IRAK1 and TRAF6 expression. *J. Immunol.* **2015**, *195*, 672–682. [CrossRef]
80. Palomer, X.; Capdevila-Busquets, E.; Botteri, G.; Davidson, M.M.; Rodríguez, C.; Martínez-González, J.; Vidal, F.; Barroso, E.; Chan, T.O.; Feldman, A.M.; et al. miR-146a targets Fos expression in human cardiac cells. *Dis. Model Mech.* **2015**, *8*, 1081–1091. [CrossRef]
81. Xu, Y.; Zhou, M.; Wang, J.; Zhao, Y.; Li, S.; Zhou, B.; Su, Z.; Xu, C.; Xia, Y.; Qian, H.; et al. Role of microRNA-27a in down-regulation of angiogenic factor AGGF1 under hypoxia associated with high-grade bladder urothelial carcinoma. *Biochim. Biophys. Acta* **2014**, *1842*, 712–725. [CrossRef]
82. Thum, T.; Gross, C.; Fiedler, J.; Fischer, T.; Kissler, S.; Bussen, M.; Galuppo, P.; Just, S.; Rottbauer, W.; Frantz, S.; et al. MicroRNA-21 contributes to myocardial disease by stimulating MAP kinase signalling in fibroblasts. *Nature* **2008**, *456*, 980–984. [CrossRef]
83. Zhou, S.S.; Jin, J.P.; Wang, J.Q.; Zhang, Z.G.; Freedman, J.H.; Zheng, Y.; Cai, L. miRNAS in cardiovascular diseases: Potential biomarkers, therapeutic targets and challenges. *Acta Pharmacol. Sin.* **2018**, *39*, 1073–1084. [CrossRef]
84. Horak, M.; Novak, J.; Bienertova-Vasku, J. Muscle-specific microRNAs in skeletal muscle development. *Dev. Biol.* **2016**, *410*, 1–13. [CrossRef] [PubMed]
85. Callis, T.E.; Pandya, K.; Seok, H.Y.; Tang, R.-H.; Tatsuguchi, M.; Huang, Z.-P.; Chen, J.-F.; Deng, Z.; Gunn, B.; Shumate, J.; et al. MicroRNA-208a is a regulator of cardiac hypertrophy and conduction in mice. *J. Clin. Investig.* **2009**, *119*, 2772–2786. [CrossRef]
86. Li, N.; Zhou, H.; Tang, Q. miR-133: A suppressor of cardiac remodeling? *Front Pharmacol.* **2018**, *9*, 903. [CrossRef] [PubMed]
87. Schulte, C.; Barwari, T.; Joshi, A.; Theofilatos, K.; Zampetaki, A.; Barallobre-Barreiro, J.; Singh, B.; Sörensen, N.A.; Neumann, J.T.; Zeller, T.; et al. Comparative analysis of circulating noncoding RNAs versus protein biomarkers in the detection of myocardial injury. *Circ. Res.* **2019**, *125*, 328–340. [CrossRef]
88. Donaldson, A.; Natanek, S.A.; Lewis, A.; Man, W.D.; Hopkinson, N.S.; Polkey, M.I.; Kemp, P.R. Increased skeletal muscle-specific microRNA in the blood of patients with COPD. *Thorax* **2013**, *68*, 1140–1149. [CrossRef]
89. Klinke, A.; Nussbaum, C.; Kubala, L.; Friedrichs, K.; Rudolph, T.K.; Rudolph, V.; Paust, H.J.; Schröder, C.; Benten, D.; Lau, D.; et al. Myeloperoxidase attracts neutrophils by physical forces. *Blood* **2011**, *117*, 1350–1358. [CrossRef]
90. Slezak, S.; Jin, P.; Caruccio, L.; Ren, J.; Bennett, M.; Zia, N.; Adams, S.; Wang, E.; Ascensao, J.; Schechter, G.; et al. Gene and microRNA analysis of neutrophils from patients with polycythemia vera and essential thrombocytosis: Down-regulation of micro RNA-1 and -133a. *J. Transl. Med.* **2009**, *7*, 39. [CrossRef] [PubMed]
91. Beg, F.; Wang, R.; Saeed, Z.; Devaraj, S.; Masoor, K.; Nakshatri, H. Inflammation-associated microRNA changes in circulating exosomes of heart failure patients. *BMC Res. Notes* **2017**, *10*, 751. [CrossRef] [PubMed]
92. Wang, J.; Tian, X.; Han, R.; Zhang, X.; Wang, X.; Shen, H.; Xue, L.; Liu, Y.; Yan, X.; Shen, J.; et al. Downregulation of miR-486-5p contributes to tumor progression and metastasis by targeting protumorigenic ARHGAP5 in lung cancer. *Oncogene* **2014**, *33*, 1181–1190. [CrossRef]
93. Bai, H.; Wu, S. miR-451: A novel biomarker and potential therapeutic target for cancer. *OncoTargets Ther.* **2019**, *12*, 11069–11082. [CrossRef] [PubMed]
94. Van der Weg, K.; Prinzen, F.W.; Gorgels, A.P.M. Reperfusion cardiac arrhythmias and their relation to reperfusion-induced cell death. *Eur. Heart J. Acute Cardiovasc. Care* **2019**, *8*, 142–152. [CrossRef] [PubMed]
95. Izzo, C.; Visco, V.; Gambardella, J.; Ferruzzi, G.J.; Rispoli, A.; Rusciano, M.R.; Toni, A.L.; Virtuoso, N.; Carrizzo, A.; Di Pietro, P.; et al. Cardiovascular implications of microRNAs in coronavirus disease 2019. *J. Pharmacol. Exp. Ther.* **2023**, *384*, 102–108. [CrossRef] [PubMed]

MDPI AG
Grosspeteranlage 5
4052 Basel
Switzerland
Tel.: +41 61 683 77 34

International Journal of Molecular Sciences Editorial Office
E-mail: ijms@mdpi.com
www.mdpi.com/journal/ijms

www.ingramcontent.com/pod-product-compliance
Lightning Source LLC
Chambersburg PA
CBHW040836250926
51850CB00007BA/1326
9783725856671